NeuroPhytomedicine

Neurological disorders categorized as neurodegenerative, neuropsychiatric, and neurotraumatic impose a substantial health burden and cause a frail impact on life attributes. The use of synthetic drugs can have undesirable side effects, making neurotherapeutics challenging. Research on neuroprotection aiming for the utilization of safer natural compounds; phytochemicals specifically – is a cutting-edge approach. *NeuroPhytomedicine* intends to present readers with a wide-ranging and state-of-the-art appearance at the beneficial properties of phytochemicals on various neurological ailments.

It additionally contains sections explaining:

- Applications of phyto-nanotechnology in neurological ailments,
- Phytoconstituents related deep learning and machine learning-based solutions for neurological disorders,
- Epigenetic relevance pertaining to modulation by phytochemicals.

The goal of *NeuroPhytomedicine* is to give readers a thorough and up-to-date look at how phytochemicals affect the brain and neurological illnesses in a way that is helpful for research scholars, academicians, scientists, neuroscientists, physicians, and researchers from the pharmaceutical industry.

NeuroPhytomedicine

Edited by
Mithun Singh Rajput, Tewarit Sarachana,
Purnima Dey Sarkar, and Manan Raval

CRC Press
Taylor & Francis Group
Boca Raton London New York

CRC Press is an imprint of the
Taylor & Francis Group, an **informa** business

First edition published 2024
by CRC Press
2385 NW Executive Center Drive, Suite 320, Boca Raton FL 33431

and by CRC Press
4 Park Square, Milton Park, Abingdon, Oxon, OX14 4RN

CRC Press is an imprint of Taylor & Francis Group, LLC

Library of Congress Cataloging-in-Publication Data

Names: Singh Rajput, Mithun, editor. | Sarachana, Tewarit, editor. | Dey Sarkar, Purnima, editor. | Raval, Manan, editor.
Title: Neurophytomedicine / edited by Mithun Singh Rajput, Tewarit Sarachana, Purnima Dey Sarkar and Manan Raval.
Description: First edition. | Boca Raton, FL : CRC Press, 2024. | Includes bibliographical references and index.
Identifiers: LCCN 2023044509 (print) | LCCN 2023044510 (ebook) | ISBN 9781032485935 (hardback) | ISBN 9781032485959 (paperback) | ISBN 9781003389781 (ebook)
Subjects: LCSH: Neuropharmacology--Research. | Phytochemicals--Therapeutic use--Research. | Nervous system--Degeneration--Chemoprevention--Research. | Materia medica, Vegetable--Research.
Classification: LCC RM315 .N4815 2024 (print) | LCC RM315 (ebook) | DDC 615.7/8--dc23/eng/20231107
LC record available at https://lccn.loc.gov/2023044509
LC ebook record available at https://lccn.loc.gov/2023044510

ISBN: 978-1-032-48593-5 (hbk)
ISBN: 978-1-032-48595-9 (pbk)
ISBN: 978-1-003-38978-1 (ebk)

DOI: 10.1201/9781003389781

Typeset in Times
by KnowledgeWorks Global Ltd.

Contents

List of Figures and Tables

Preface

Neurological disorders categorized as neurodegenerative, neuropsychiatric, and neurotraumatic impose a substantial health burden and cause a frail impact on the life attributes. Neuronal cell damage occurring from oxidative stress, inflammation, and other molecular etiologies contributes to various chronic and acute neurological ailments. The tactics involved in neurotherapeutics are complex and challenging and bear a shortcoming of side effects of synthetic chemicals. Research on neuroprotection aiming utilization of safer natural compounds, phytochemicals specifically, is a cutting-edge approach. Phytochemicals are known to exhibit various biological activities in humans and are naturally found in plants, herbs, vegetables, and edible fruits. Alkaloids, flavonoids, isoflavonoids, phytosterols, saponins, terpenes, polyphenols, etc., are the phytochemicals known to have neuroprotective efficacy. The molecular mechanisms behind the curative effects of phytoconstituents rely largely on subsiding oxidative stress, neuro-inflammation, and signal transduction related with protein folding. However, the particular molecular mechanisms connected with advantageous effects of phytochemicals still continue as a matter of quest.*"NeuroPhytomedicine"* intends to present readers with a wide-ranging and state-of-the-art appearance at the beneficial properties of phytochemicals on the neurodegenerative (Parkinson's disease, Alzheimer's disease, Huntington's disease, amyotrophic lateral sclerosis, etc.), neuropsychiatric (anxiety, seizures, cognitive deficit, palsies, addictions, depression, etc.), and neurotraumatic (traumatic brain injury, cerebral ischemia, etc.) ailments. A recent advancement in the discovery of phytochemicals as neuro-protectives, their mechanism of action, study on semisynthetic derivatives, and their conventional and nanotechnology-based formulations has been presented. This comprehension will be helpful to neuroscientists, physicians, ethnopharmacologists, and researchers in understanding the neuroprotection by natural plant-derived compounds.

Editor Biography

Dr. Mithun Singh Rajput is working with the Institute of Pharmacy, Nirma University, Ahmedabad, Gujarat, India as a capacity of Assistant Professor. He has expertise in neuropharmacology and received a doctoral degree from the Faculty of Medicine, M.G.M. Medical College, Indore, India. He is a recipient of two years' National Postdoctoral Fellowship from SERB, Department of Science and Technology, India and associated with the research project as principal investigator. He has an overall research experience of 12 years. His area of interest includes diagnosis and mitigating strategies for various neurodegenerative disorders like Alzheimer's disease and cerebral ischemia, diabetes-associated cognitive impairment, etc. Mithun has one patent and three book chapters published and edited four books. Mithun has been credited with 30 publications in peer-reviewed indexed journals and is currently bagged with an h-index of eighteen. He has been serving many government funding agencies and journals and as a reviewer and editorial board member.

Dr. Tewarit Sarachana is the Head of Chulalongkorn SYstems Neuroscience of Autism and PSychiatric disorders (SYNAPS) Research Unit. He is teaching at the Department of Clinical Chemistry, Medical Technology Program, Faculty of Allied Health Sciences, Chulalongkorn University. He pursued a doctoral degree in Neuroscience at School of Medicine and Health Sciences, The George Washington University, Washington, DC, USA and received postdoctoral research training at The Office of Blood Research and Review, Center for Biologics Evaluation and Research, U.S. Food and Drug Administration (FDA), Bethesda, Maryland, USA. He also serves as a technical assessor for the medical laboratory standards ISO 15189:2012 and ISO 15190:2003, Bureau of Laboratory Quality Standards, Department of Medical Sciences, Ministry of Public Health, Thailand. He has handled more than 20 government-funded research grants and supervised ample postdoctoral, doctoral, and master's candidates for their research. He has published more than 30 research articles, review articles, and book chapters and is a reviewer for more than 15 international journals, including Molecular Autism, Current Medicinal Chemistry, and Frontiers in Neuroscience.

Prof. Purnima Dey Sarkar is serving as Head of the Department of Biochemistry, M.G.M. Medical College, Indore, India and has vast experience of teaching and research of 40 years in various medical colleges of Madhya Pradesh, India. She

received a doctoral degree from Devi Ahilya Vishwavidyalaya, Indore, India in year 1991. Her area of interest in research includes biochemistry and pharmacology of various ailments, viz. neurological and cardiovascular disorders, cancer and diabetes. She has supervised 30 Ph.D. candidates for their dissertation work. Also, she has been supervisor for 14 M.D. candidates. She has more than 75 publications in journals of international and national repute and serving as a reviewer for many funding agencies and journals. Prof. Sarkar has acquired various administrative responsibilities and served government universities as chairperson and member of board of studies, chairperson of various scientific committees, subject expert for faculty recruitment, etc.

Prof. Manan Raval is presently designated as Principal at Ramanbhai Patel College of Pharmacy, Charotar University of Science and Technology, Changa, Anand, India. He received master's and doctoral credentials from Maharaja Sayajirao University, Baroda, Gujarat. His area of interest includes chemistry of bioactive constituents isolated from natural products. Precisely, his research interest is on screening natural compounds for their efficacy in improving male infertility and exploring their molecular mechanisms. He exhibits vast academic, administrative, and industrial experience. His work is regularly being presented and appreciated before a number of experts of national and international repute. Dr. Raval has received many project grants from various government agencies and supervised many doctoral and master's candidates for their research work. He has published and presented ample research articles and serving the national and international scientific community in many capacities.

Contributors

Imomotimi Ajoko
Department of Chemical Sciences, Niger Delta University
Wilberforce Island, Nigeria

Likhit Akotkar
Department of Pharmacology, Poona College of Pharmacy, Bharati Vidyapeeth (Deemed to be University)
Pune, India

Saravanan Alamelu
Department of Biochemistry and Biotechnology, Faculty of Science, Annamalai University
Chidambaram, Tamil Nadu

Urmila Aswar
Department of Pharmacology, Poona College of Pharmacy, Bharati Vidyapeeth (Deemed to be University)
Pune, India

Malik Badshah
Department of Microbiology, Faculty of Biological Sciences, Quaid-i-Azam University
Islamabad, Pakistan

Sumanta Bhattacharya
Maulana Abul Kalam Azad University of Technology
Kolkata, India

Karthikeyan Subbiahanadar Chelladurai
Regeneration and Stem Cell Biology Lab, Centre for Molecular and Nanomedical Sciences, International Research Centre, Sathyabama Institute of Science & Technology (Deemed to be University)
Chennai, India

Jackson Durairaj Selvan Christyraj
Regeneration and Stem Cell Biology Lab, Centre for Molecular and Nanomedical Sciences, International Research Centre, Sathyabama Institute of Science & Technology (Deemed to be University)
Chennai, India

Shareena Gadde
Department of Pharmaceutical Chemistry, Poona College of Pharmacy, Bharati Vidyapeeth (Deemed to be University)
Pune, India

Aditya Ganeshpurkar
Shri Ram Institute of Technology-Pharmacy
Jabalpur, India

Ankit Ganeshpurkar
Department of Pharmaceutical Chemistry, Poona College of Pharmacy, Bharati Vidyapeeth (Deemed to be University)
Pune, India

Nivedha Jayaseelan
Department of Biochemistry and Biotechnology, Faculty of Science, Annamalai University
Chidambaram, Tamil Nadu

Kanimozhi Kaliyamoorthi
Department of Biochemistry and Biotechnology, Faculty of Science, Annamalai University
Chidambaram, Tamil Nadu

Sathish-Kumar Kamaraj
Instituto Politécnico Nacional (IPN) – Centro de Investigación en Ciencia Aplicada y Tecnología Avanzada (CICATA), Unidad Altamira
Altamira, Mexico

Samiullah Khan
Department of Microbiology, Faculty of Biological Sciences, Quaid-i-Azam University
Islamabad, Pakistan

Ameya Kothekar
Department of Pharmacy, Birla Institute of Technology and Science
Pilani, India

Dileep Kumar
UC Davis Comprehensive Cancer Center, Department of Entomology and Nematology, University of California Davis
Davis, California

Inbathamizh L
Department of Biotechnology, School for Bio and Chemical Engineering, Sathyabama Institute of Science & Technology (Deemed to be University)
Chennai, India

Alina Majid
Department of Microbiology, Faculty of Biological Sciences, Quaid-i-Azam University
Islamabad, Pakistan

Aarthi Narasimhan
Department of Zoology, Sri Sarada College for Women
Salem, India

Aneela Nawaz
Department of Microbiology, Faculty of Biological Sciences, Quaid-i-Azam University
Islamabad, Pakistan

Odangowei Inetiminebi Ogidi
Department of Biochemistry, Faculty of Basic Medical Sciences, Bayelsa Medical University
Yenagoa, Nigeria

Sonam Pandey
Christian Medical College Campus
Vellore, India
Centre for Stem Cell Research (a unit of inStem)
Bengaluru, India

Rashmi Patil
Department of Pharmacology, Poona College of Pharmacy, Bharati Vidyapeeth (Deemed to be University)
Pune, India

Atmaram Pawar
Department of Pharmaceutics, Poona College of Pharmacy, Bharati Vidyapeeth (Deemed to be University)
Pune, India

Rajesh Pradhan
Department of Pharmacy, Birla Institute of Technology and Science
Pilani, India

Srishti Raja
Department of Biotechnology, School for Bio and Chemical Engineering, Sathyabama Institute of Science and Technology (Deemed to be University)
Chennai, India

Kamarajan Rajagopalan
Regeneration and Stem Cell Biology Lab, Centre for Molecular and Nanomedical Sciences, International Research Centre, Sathyabama Institute of Science & Technology (Deemed to be University)
Chennai, India

Sudha S
Department of Biotechnology, School for Bio and Chemical Engineering, Sathyabama Institute of Science and Technology (Deemed to be University)
Chennai, India

Umme Habiba Saeeda
Department of Microbiology, Faculty of Biological Sciences, Quaid-i-Azam University
Islamabad, Pakistan

Bhagawati Saxena
Department of Pharmacology, Institute of Pharmacy, Nirma University
Ahmedabad, India

Chandini Sengupta
Department of Biotechnology, School for Bio and Chemical Engineering, Sathyabama Institute of Science and Technology (Deemed to be University)
Chennai, India

Muqaddas Shahzadi
Department of Microbiology, Faculty of Biological Sciences, Quaid-i-Azam University
Islamabad, Pakistan

Sri Raagavee Sivakumar
Department of Zoology, PSGR Krishnammal College for Women
Coimbatore, India

Manoj Kumar Srinivasan
Department of Biochemistry and Biotechnology, Faculty of Science, Annamalai University
Chidambaram, Tamil Nadu

Rajeev Taliyan
Department of Pharmacy, Birla Institute of Technology and Science
Pilani, India

Kamalesh Balakumar Venkatesan
Department of Biochemistry and Biotechnology, Faculty of Science, Annamalai University
Chidambaram, Tamil Nadu

1 Phytochemicals

Recent Trends Pertaining to Beneficial Effects on Neurological Disorders

Sonam Pandey

1.1 INTRODUCTION

The prevalence in chronic degenerative diseases in the past few decades has prompted a stronger emphasis on the environmental variables that contribute to their origin and progression. A neuronal cell disorder known as neurodegenerative disease (ND) is characterised by protein accumulation, oxidative stress (OS), hereditary predispositions, and viral infections. Loss of neurons can cause memory loss, motor impairments, and problems with the central and peripheral nervous systems' ability to function. Neuronal degeneration has an effect on a patient's social network, family, and quality of life. The most common NDs are Alzheimer's disease (AD) and Parkinson's disease (PD) (Paramanick et al, 2022). AD accounts for roughly 60%–70% of the approximately 50 million people worldwide who suffer from dementia, which is a major cause of neurological damage and dependency among the elderly. More than ten million people worldwide suffer from PD, and the prevalence rises with age. Other NDs include multiple sclerosis (MS), Huntington's disease (HD), amyotrophic lateral sclerosis (ALS), Creutzfeldt-Jakob disease (CJD), HIV-associated neurocognitive disorders (HAND), and stroke-induced secondary neurodegeneration (SND). Neurological diseases are the largest cause of disability-adjusted life years and the second major cause of mortality worldwide. The names of the diseases labelled in honour of distinguished physicians, subtyping according to pathological and clinical findings, overlapping histopathological classification, and numerous molecular discoveries all obscure a basic understanding of the diseases and may obstruct carefully targeted research into the pathogenesis and new drug discovery for NDs (Masaru et al, 2020).

The past decade has also seen an increase in interest in herbal medications, which include phytochemical components that may have long-term health-promoting or therapeutic effects. In addition, many medicinal plants have specific medicinal effects and can be utilised temporarily or permanently to treat a variety of health issues. Phytochemicals abundant in vegetables and fruits are considered to minimise the risk of a wide range of serious medical conditions, including cardiovascular disease, cancer, and neurological disorders (Lobo et al, 2010; Selvam, 2008). Numerous medications have been developed to date for the prevention and

DOI: 10.1201/9781003389781-1

treatment of neurodegenerative disorders (NDDs), but the quest still persists (Raj et al, 2020).

Hence, there is a significant demand for new therapies and medications for treating and preventing these disorders. Currently, scientists are focusing on using bioactive substances derived from natural sources to target the pathophysiology of disease. The major NDs are covered in detail in this chapter, along with potential therapeutic targets, therapeutic plants, and bioactive substances that can be used to treat NDs.

This chapter was accomplished by utilising an electronic search of existing literature, with PubMed, Science Direct, Google Scholar, Research Gate, and Scopus, as the major sources for relevant review papers, books and randomised clinical trials published in the recent 5 years on the use of ethnomedicinal plants to treat various neurological diseases. Reference lists from the most recent reviews were used, as well as additional primary literature sources and references provided by relevant articles.

1.2 GLOBAL BURDEN OF NEUROLOGICAL DISORDERS

In the world, neurological disorders are the primary cause of disability and the second greatest cause of mortality. The absolute number of deaths and persons with disabilities caused by neurological disorders has increased significantly over the past 30 years, notably in low- and middle-income nations, and further increases are projected globally as a result of population expansion and ageing. The burden of neurological disorders-related mortality and disability is rapidly being recognised as a global public health concern, and this burden is expected to rise over the next few decades. Despite declines in infectious neurological illnesses, the absolute number of deaths has climbed by 39% over the past 30 years, and disability-adjusted life-years (DALYs), the sum of years of life lost and years lived with impairment have increased by 15%. The burden is greater in low- and middle-income countries (LMICs). Therefore, estimates of the burden of neurological disorders and the factors influencing variations in that burden across various populations and nations are required on a regular basis by funders and policymakers. Such estimates would allow for resource allocation and cost-efficient planning for healthcare, as well as evaluations of the effects of effective policy actions. Additionally, understanding how sociodemographic shifts (for instance, at the national level) affect neurological illnesses and general population health is necessary to set priorities for research and the development of healthcare services, as well as to guide policy (Feigin et al, 2019).

Traditionally, epidemiological examinations of neurological disorders have been limited to specific disorders (e.g., stroke and dementia) or geographical areas (e.g., district, city, and region), database settings (e.g., official reports, mortality or hospitalisation statistics) or demographics (e.g., specific ethnicities or age groups). Comparisons between studies are complicated by differences in study designs, methodology, diagnostic criteria, and case ascertainment completeness. The Global Burden of Diseases, Injuries, and Risk Factors (GBD) Study 2 seeks to address these issues in the most systematic, comprehensive, and consistent manner possible. Metrics of the worldwide, regional income (e.g., LMICs or HICs), and national (and in some cases subnational) burden of neurological illnesses are available for

195 nations, organised by age, gender, and time (Roth et al, 2018). The GBD global, regional, and national estimates of the burden of neurological disorders, risk factors, and trends should be used by funding organisations and policymakers to establish priority areas for healthcare and research as well as to advance prevention, management, service, and workforce development (Dorsey et al, 2018).

According to the GBD findings, neurology services should be a top priority for healthcare systems, and epidemiological monitoring (observational and experimental) studies, the evaluation of preventive and disease-modifying interventions, and translational studies should be top priorities for research funding organisations. Future GBD estimates and other epidemiological studies will be crucial for providing regular updates and evaluation of the burden of neurological disorders at different levels, which will be crucial for determining the appropriate allocation of resources, improving existing policies, and developing new ones (Valery et al, 2019, 2020).

As stated in the December 2017 UN General Assembly report, progress in lowering the burden of NCDs, including neurological illnesses, is insufficient to reach the UN Sustainable Development Goals targets by 2030 (Feigin et al, 2019). Global, national, and regional neurological and disease-specific organisations, such as the World Federation of Neurology, the World Stroke Organization, and Alzheimer's International, must work together with the World Health Organization (WHO) to gain the momentum needed to confront this expanding problem (Carroll, 2017). At all ages and levels of human activity, brain health is one of humanity's most valuable resources.

1.3 PATHOPHYSIOLOGY OF NDs

A set of pathologies known as NDDs, including AD, PD, and HD, are characterised by various etiologies with specific pathophysiological and morphological aspects. Multifactorial conditions such as OS and production of free radicals, protein misfolding with impaired protein aggregation and degradation, exposure to pesticides and metal toxicity, altered bioenergetics, and mitochondrial dysfunction can all contribute to the development of these disorders (Küpeli et al, 2021). A common pathognomonic finding of the brain regions reflecting damage and dysfunction of the related motor, autonomic, and/or cognitive nervous systems is neurodegeneration, which is the structural and/or functional loss of neurons in the central nervous system (CNS) (Gerez and Riek, 2020). The post-mortem brain specimens' neurodegenerative lesions and the results of structural and functional imaging tests are highly correlated. Magnetic resonance imaging (MRI) and positron emission tomography (PET) may be used to identify afflicted domains and diagnose disorders using the regional patterns of brain shrinkage (Shimizu et al, 2018).

There has been huge progress in identifying the primary factors that trigger neuronal damage in several NDDs, including AD, PD, MS, ALS, and HD, over the past few years. Preventing neuronal damage and death will offer a significant clinical advantage (Solanki et al, 2015).

NDs' molecular approaches highlight intriguing elements of their pathogenesis. The basic causes of neurodegeneration appear to be aberrant protein processing,

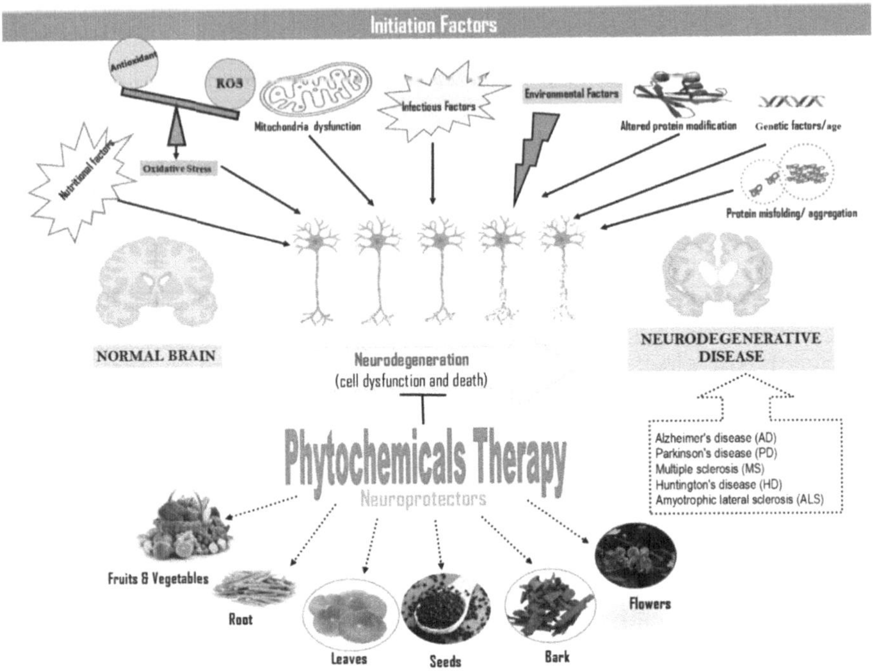

FIGURE 1.1 Mechanism involved in the development of neurodegenerative diseases and neuroprotectors.

genetic diseases, misfolding and aggregation of different proteins, activating cellular apoptosis, triggering mitochondrial failure, free radical production, and OS (Rekatsina et al, 2020). A summary of the pathophysiology of NDs and the role of neuroprotectors are shown in Figure 1.1.

NDDs are currently incurable and are rapidly increasing mortality and morbidity rates in industrialised nations, with a similar situation projected in poorer countries. In the twenty-first century, the major health issues are neuropsychiatric and neurological disorders such as schizophrenia, depression, anxiety, AD, PD, HD, ALS, cerebrovascular impairment, seizure disorders, and head injuries, as well as other systemic disorders such as cardiovascular, respiratory, renal, gastrointestinal, and many others (Hussain et al, 2018).

A brief overview of NDs and their development is helpful in comprehending the therapeutic approaches and will be discussed.

1.3.1 Alzheimer's Disease

AD is one of the most prominent NDDs, characterised by the progressive degradation of memory, learning, and other cognitive skills, as well as the production of disoriented plaques and neurofibrillary masses, which result in severe neuronal death. The most prominent component of senile plaques, amyloid-β peptide (Aβ), is primarily implicated in the development of AD (Schmitt et al, 2014). AD has been associated to neurofibrillary tangles of intracellular tau and extracellular deposits of amyloid-β (Aβ) plaques in the brain. Plaque formations have been associated to OS, which may play a significant role in the pathogenesis of AD. This disease is also considered as a short-term memory loss (Rekatsina et al, 2020).

1.3.2 Parkinson's Disease

PD is an irreversible, progressive neurological disease that affects around 1% of adults over the age of 50 years. It is commonly linked to dementia. The patient loses 50%–70% of all dopaminergic neurons in the substantial nigra as PD advances. With the loss of dopaminergic fibres in the brain, the patient has progressively increasing motor symptoms, which are a hallmark of PD. Growing research has linked PD pathogenesis to both OS and immunological alterations. As reactive oxygen species and other free radicals collect, they overwhelm dopaminergic neurons and cause them to degenerate, establishing the foundation for PD pathogenesis.

A number of potential contributors to PD have been identified in the disease process, including mitochondria, endoplasmic reticulum, α-synuclein, and dopamine. It appears that their interactions, rather than their individual actions, contribute to progressive neurodegeneration. Neuroinflammation may be necessary in the pathogenesis of PD because it damages the nigrostriatal dopaminergic pathways. This, in turn, causes the brain's glial cells (mostly microglia and astrocytes) to release a variety of soluble substances that may be pro-inflammatory and/or neurotoxic (Rekatsina et al, 2020).

1.3.3 Amyotrophic Lateral Sclerosis (ALS)

ALS is characterised by the irreversible degeneration of motor neurons in the cerebral cortex, brain stem, and spinal cord (National Institute of Neurological Disorders and Stroke, 2017). The majority of people diagnosed with this syndrome die within 3–4 years of onset of symptoms (D'Amico et al, 2013). The world-renowned physicist Stephen Hawking is an exception, having lived for more than 50 years after being diagnosed with ALS. ALS, as the name implies, is characterised by muscular atrophy (amyotrophic) and spinal cord scarring (lateral sclerosis). ALS patients experience increasing paralysis, dysphagia, and respiratory insufficiency (Malaspina et al, 2015). ALS is considered to be caused by one or more of the 150 mutations in the superoxide dismutase 1 (SOD1) antioxidant system (Hooten et al, 2015), which induce motor neuron damage and death as well as mitochondrial malfunction (Li et al, 2011). The SOD1 mutation causes protein misfolding (D'Amico et al, 2013).

1.3.4 Huntington's Disease

HD is the most frequent hereditary cause of chorea and is also known as Huntington's chorea, chorea major or simply HD. This incurable neurodegenerative condition was named after George Huntington, an American physician who precisely described it in 1872. Prevalence per country can be as high as seven individuals per 100,000 (in populations of Western European ancestry) and can be much higher in specialised areas. Physical symptoms can appear at any age; however, the average age of onset is 35–44 years old. When beginning occurs before the age of 20, the disorder is characterised as juvenile HD (also known as akinetic-rigid HD or Westphal variant HD), which advances faster and has slightly different symptoms.

The disease's mechanism is unknown, but a variety of factors have been identified. A Huntingtin gene mutation results in the creation of the mutant protein huntingtin, which causes cell and macroscopical alterations in the brain. There is no cure for HD; however, there are therapies available to alleviate some of its symptoms. Chora, or jerky, erratic, and uncontrollable movements are the most common initial physical signs. As the condition worsens, rigidity and dystonia become more noticeable, eventually becoming the major physical symptoms (Sandhya et al, 2010).

1.3.5 Multiple Sclerosis

MS is a CNS autoimmune disease that causes severe disability and healthcare costs. MS is a chronic neurological disease marked by inflammation and demyelination. The destruction of the nerves' myelin sheaths, as well as axonal damage and glial scarring, causes the disease's characteristic symptoms of muscular spasms, neuropathic pain, paralysis, and visual neuritis. An inflammatory state is promoted throughout this process, mostly by myelin antigen-specific T_H cells. Monocytes are recruited as lesions appear, producing reactive oxygen species. Demyelination and neurodegeneration may result in the presence of oxidised lipids in myelin membranes, apoptotic oligodendrocytes, and neuronal axons, according to observations of white matter and cerebral cortex lesions (Rekatsina et al, 2020).

1.4 MEDICAL TREATMENT STRATEGIES FOR NEUROLOGICAL DISORDERS

In terms of basic principles, treatment options for neurological disorders are identical to non-comorbid patients, with the exception that pharmaceutical interactions require particular vigilance (Antony, 2001). There are several therapeutic methods available, including pharmaceutical treatments, cognitive behavioural therapy, somatic interventions, and electroconvulsive therapy (Gorman, 1996). Some of the therapies indicated aim to improve both anxiety and depression at the same time, while others are unique to the individual disease. Anxiolytic drugs such as benzodiazepines are normally started for a short amount of time to ease anxiety, but they are only used as a last resort due to their unfavourable influence on cognitive disorders in AD and balance problems in PD (El-Guebaly et al, 2010). Oral antidepressant therapy may have several limitations, such as drug interactions, limited tolerance owing to side effects, and low efficacy (Santarsieri and Schwartz, 2015).

1.5 NEUROPROTECTIVE EFFECTS OF MEDICINAL PLANTS

Despite significant improvements in the causal mechanisms that trigger certain NDDs, no therapies that benefit patients with these diseases are often available. Because most NDs appear late and remain asymptomatic for the majority of their progression, medications beginning in advanced stages of the disease are of limited benefit to patients. If therapy begins early in the disease's course, it may be feasible to significantly slow or stop disease progression. By lowering or even removing the original stressor, such therapies have the potential to restore neuronal function. Phytoconstituents are key molecules in the creation of a new generation of therapeutically effective treatment medicines for NDDs. Herbal supplements have been linked to a lower risk of NDs when consumed on a regular basis (Solanki et al, 2015).

There are several prominent traditional Indian plants that can be used to treat neurological disorders like Alzheimer's and dementia. Ayurvedic Rasayana medications are abundant in anti-oxidants and immunomodulatory active ingredients. Since the majority of diseases are triggered by a disruption in the delicate balance of oxidants and antioxidants, the ability to scavenge free radicals or activate cellular oxidant defences can be considered as their primary mechanism of action. Many plants, including Ashwagandha, Brahmi, Mandukaparni, Shankapushpi, Vacha, Jatamansi, and Jyotshmati, belong to the category of Rasayana plants. These herbs are categorised as brain tonics or rejuvenators since they are specifically formulated for brain tissues (Farooqui and Farooqui, 2017).

Some ayurvedic herbs like *Guduchi, Yashtimadhuk, Padma (Nelumbo nucifera), Vacha, Convolvulus pluricaulis, Shankhpushpi, Pancha-Tikta-Ghruta Gugguli, Amalaki, Musta Arjun, Amalaki, Ashwagandha, Galo Satva, Kutaj*, and others are excellent herbs for slowing down the brain cell degeneration caused by Alzheimer's. They enhance the brain's ability to function, and therefore, provide stability when used consistently.

1.5.1 *Crocus sativus*

Crocus sativus L (*C. sativus*), widely known as saffron, is a member of the *Crocoideae* superfamily and is farmed in various countries including Iran, Afghanistan, Turkey, and Spain (Abdullaev, 1993). Saffron contains more than 150 different substances, including carbohydrates, polypeptides, lipids, water, minerals, and vitamins. Crocins, a family of red-coloured and water-soluble carotenoids that are all glycosides of crocetin, are the major physiologically active constituents of saffron. Saffron also contains four major bioactive components: Crocin, crocetin, picrocrocin, and safranal (Bathaie and Mousavi, 2010).

C. sativus is used to treat cognitive disorders in Iranian, as a traditional medicine. *C. sativus* components have recently been used to treat various neurological disorders and to relax smooth muscle (Hosseinzadeh et al, 2007; Khazdair et al, 2015; Mokhtari-Zaer et al, 2015). Saffron extract has been shown to have anticonvulsant and anti-Alzheimer effects in both human and animal models (Khazdair et al, 2015). The efficacy of *C. sativus* in the treatment of mild to moderate depression in clinical trial research, as well as its influence on brain neurotransmitter concentrations and interaction with the opioid system, was discussed (Khazdair et al, 2015). *C. sativus* and its major component, crocin, have strong antioxidant properties via lowering malondialdehyde (MDA) levels (Karimi et al, 2010; Tamaddonfard et al, 2013).

1.5.2 *Nigella sativa*

The annual herb *Nigella sativa L.*, also known as *N. sativa*, is a member of the *Ranunculaceae* family and is commonly grown in the Mediterranean region as well as in Western Asia, the Middle East, and Eastern Europe. Oil, protein, glucose, and fibre are all chemical components of *N. sativa* seeds. Linoleic acid, oleic acid, palmitic acid, arachidic acid, eicosadienoic acid, stearic acid, linoleic acid, and myristic acid are the chemical components of *N. sativa*'s fixed oil (El-Tahir and Bakeet, 2006). *N. sativa* enhanced scopolamine e-induced learning and memory deficits in rats while also decreasing acetylcholinesterase (AChE) activity and OS in the brain (Hosseini et al, 2015). In the other clinical trial, 40 healthy volunteers were randomly assigned to either *N. sativa* (500 mg) or placebo (500 mg) capsules twice a day for 9 weeks. In comparison to the placebo group, *N. sativa* improved memory, attention, and cognition (Sayeed et al, 2013). After 4 weeks, *N. sativa* (500 mg) reduced anxiety, stabilised mood, and modulated cognition in a human model (Sayeed et al, 2014). The neuroprotective effects of *N. sativa* and thymoquinone (TQ) (its main components) on a variety of nervous system illnesses such as AD, epilepsy, and neurotoxicity have been studied (Khazdair, 2015).

1.5.3 *Bacopa Monniera* (Brahmi)

Since ancient times, Brahmi (*Bacopa monniera*; family: *Scrophularaceae*) has been particularly known as a brain tonic for reviving intellect, an anti-stress remedy for anxiety, and a way to improve cognitive abilities. According to prior research, this medicinal herb can be used to treat neurological and mental diseases since it operates as a nervine and a mental tonic (Thakur et al, 2018).

1.5.4 *Withania Somnifera* (Ashwagandha)

A member of the Solanaceae family, ashwagandha is a shrub by the botanical name of *Withania somnifera*. It is classified as an adaptogen – a non-toxic stimulant that normalises physiological processes in response to prolonged stress by activating the immunological and endocrine systems. By extending its neurite outgrowth, ashwagandha may help in the repair of broken neural circuits (Thakur et al, 2018).

1.5.5 *Centella asiatica* (Mandookaparni)

Centella asiatica, a plant in the family Apiaceae, is known as gotu kola or mandookaparni (Umbelliferae). It has been used to enhance focus, improve memory, and increase alertness. It is a psychotropic medicinal herb that is employed in the treatment of stress and anxiety. It has been used to boost the nervous system, revive youth, and improve memory. It has been applied to memory improvement (Thakur et al, 2018).

1.5.6 *Santalum album* (Chandan)

Chandan (*Santalum album*) is a member of the Santalaceae family. This plant has the potential to enhance memory and cognitive function (as described in Siddha). Licorice has a considerable memory-enhancing activity and promotes learning and memory in scopolamine-induced dementia (Thakur et al, 2018).

1.5.7 *Curcuma Longa* (Haldi)

Curcuma longa (Haldi) is utilised to treat the majority of chronic disorders, including neurological, cardiovascular, pulmonary, metabolic, autoimmune, and neoplastic diseases. Curcumin is an herbal medicine with therapeutic potential in AD. The blood-brain barrier (BBB) is crossed by curcumin, which also prevents the development of A plaques, the elimination of Aβ fibrils that have already formed, and inhibits the extension of those fibrils. Curcumin therapy has a higher therapeutic impact on AD and can restore the distorted neuritic morphology that is observed close to plaques (Thakur et al, 2018).

Curcumin's therapeutic potential for NDs has piqued the curiosity of many researchers in recent years (Fu et al, 2015). According to Kulkarni, curcumin water-soluble extract can increase dopamine, norepinephrine, and 5-hydroxy tryptamine levels in the CNS (Kulkarni et al, 2012). Curcumin derived from *Curcuma longa* has been shown in cell culture and animal models to prevent PD, reactive oxygen species production, apoptosis, platelet aggregation, cytokine production, cyclooxygenase enzyme activity, brain oxidative damage, and cognitive impairments (Yang et al, 2005, 2014).

1.5.8 *Ferula Assafoetida* (Asafoetida)

Asafoetida (*F. assafoetida L.*) is a plant in the Apiaceae family that grows from the exudates of the plant's living beneath rhizome or tap roots. In Iran, *F. assafoetida*, also known as gum-resin, is known as "Anghouzeh," "Khorakoma," and "Anguzakoma." In India and Nepal, it has been used in traditional medicine and as a spice in many dishes (Iranshahy and Iranshahi, 2011).

According to scientific data, *F. asafoetida* resin could be able to inhibit the monoamine oxidase B (MAO-B) enzyme and be used to treat NDs including Parkinson's and Alzheimer's (Zarmouh et al, 2016). Meanwhile, AChE inhibition by *F. asafoetida* has been demonstrated in vitro as well as in vivo on the neurological system of the snail. According to researchers, the ability of *F. asafoetida* to improve memory may be due to its ability to suppress AChE in rat brain tissue (Kumar et al, 2009).

1.5.9 *Thymus vulgaris* (Thyme)

Thymus vulgaris (T. vulgaris) is a plant in the *Lamiaceae* family that is strongly aromatic. This plant has roughly 38 species and is found in subtropical areas. *T. vulgaris* is mostly composed of phenols, thymol (40%) and carvacrol (15%). During the winter, it has lower levels of phenol. The essential oil also contains thymol methyl ether (2%), cineol, cymen, pinene, borneol, and esters (Azaz et al, 2004). The neuroprotective and ameliorative benefits of thymol, a bioactive monoterpene derived from *T. vulgaris*, on amyloid β or scopolamine-induced cognitive impairment in rats have recently been reported (Deng et al, 2015). Researchers believe that thymol's neuroprotective benefits are due to its potential effect on gamma-aminobutyric acid (GABA)-mediated regulation of synaptic transmission (Marin et al, 2011). Meanwhile, scientists discovered that 1-Methyl-4-propan-2-ylbenzene isolated from Thymus vulgaris might influence cholinergic function by increasing synaptic acetylcholine (Ach) and nicotinic Ach receptor activity (Sammi et al, 2016).

1.5.10 *Coriandrum sativum* (Coriander)

Coriander (*Coriandrum sativum L.*) is a parsley family annual herb (Apiaceae). In Persian, this plant is known as Geshniz. *Coriandrum sativum* (*C. sativum*) is a Mediterranean native that is widely grown all across the world (Lawrence, 1993; Small, 1997). *C. sativum* was frequently used as a digestive agent in folk medicine. *C. sativum* seed extract has antibacterial and antirheumatoid properties and has been used in lotions and shampoos (Yusuf et al, 1994). The plant's seeds and leaves have been shown to have antioxidant, diuretic, cholesterol-lowering, anxiolytic, sedative-hypnotic, and anticonvulsant properties. Furthermore, as the major component of coriander, linalool has a variety of neuropharmacological actions, including anti-anxiety, sedative, anticonvulsant, and anti-disease Alzheimer's activity (Hosseini et al, 2021).

1.5.11 Shankhpushpi

Shankhpushpi is broadly categorised as a "Medhya Rasayana" in Ayurveda, which means a medication that rejuvenates, preserves, and enhances intellect and memory. Four botanicals, *Canscora decussata Schult. (CD) (Gentianaceae), Clitorea ternatea Linn. (CT) (Leguminosae), Convolvulus pluricaulis Choisy. (CP) (Convulvulaceae),* and *Evolvulus alsinoides Linn. (EA) (Convulvulaceae)*, are referred Shankhpushpi by Indian practitioners. The findings support the traditional usage of Shankhpushpi for its neuropharmacological activity and have significant implications for its application in the prevention and treatment of memory and CNS dysfunctions (Sethiya et al, 2019).

1.5.12 *Angelica Archangelica L. (Umbelliferae)*

Angelica archangelica L., also known as Dudhachoraa (Laghu Coraka), contains several compounds that show similar activity to AD medications. These compounds do not produce the same adverse effects as medicines do, such as nausea, stomach ache, insomnia, and so on. *Angelica archangelica* contains phytochemicals that can enhance blood flow to the brain. According to one study, the chloromethane subfraction of a methanol extract inhibits AChE in vitro (Singhal et al, 2012).

1.5.13 *Melissa officinalis* L. (Lamiaceae)

M. officinalis extract can improve memory in both healthy and scopolamine-induced rats. These effects of *M. officinalis* extract could be linked to the AChE-inhibitory activity identified in this investigation or to the previous research findings of nicotinic receptor activity, while other mechanisms of action could also be implicated. This extract's memory-enhancing effect via the cholinergic system, as well as its high antioxidant activity, make it a potential for AD treatment (Soodi et al, 2014).

1.5.14 *Commiphora Whighitti* (Burseraceae)

Commiphora whighitti (Guggulu), a plant resin, includes guggulsterone, the main ingredient of guggulipid. The guggulipid has been shown to be a possible cognitive enhancer for memory improvement in scopolamineinduced memory deficits (Lannert and Hoyer, 1998). *Commiphora whighitti* affects on learning and memory impairment as well as decreased choline acetylcholine esterase levels in the hippocampus. However, *C. whighitti* has the greatest impact on memory functioning and the risk of dementia (Rubio et al, 2011).

1.5.15 *Glycyrrhiza glabra* (Fabaceae)

AD is described by neuronal death and the formation of extracellular senile plaques, the main component of which is Aβ. In this study, we looked at the effects of a licorice water extract (Yashtimadhuka) on $A\beta_{25\text{-}35}$-induced apoptosis in PC12 cells. The findings show that licorice water extract protects against apoptotic neuronal cell death caused by Aβ fragments. Licorice root extract has been found to treat or even prevent brain cell loss in disorders such as Alzheimer's and its related symptoms (Bilge and Ilkay, 2005).

1.6 NANODRUG DELIVERY OF NATURAL COMPOUNDS

Over the past few decades, there has been a lot of interest in the nanotechnology approach to disease treatment. One of the most significant advantages of nanodrug delivery is that it increases bioavailability, hence increasing the therapeutic index of the drug by directly targeting specific cells or tissues. This helps to lessen the drug's overall negative effects (Shi et al, 2010).

The nanoparticles contain the small drug molecules and carry them to the desired site. Although there are many benefits to treating NDs, the therapeutic approach only provides short-term relief since it is difficult to transfer drugs to the brain. Utilising nanoparticles for NDs is one of the most recent developments in the field of nanotechnology (De Jong and Borm, 2008). The size range of the nanoparticles makes it possible for them to pass across a variety of biological barriers in the body, including the BBB, which is a particularly difficult problem to solve (De Jong and Borm, 2008; Shilo et al, 2015; Sriraman et al, 2014).

Several studies are being done to create natural compound nanoformulations, but it is yet unknown if the compound's action is the same when it is nanoencapsulated or not. For many decades, the medical benefits of curcumin, an old ayurvedic drug made from the turmeric herb, have been well documented. Due to curcumin's poor bioavailability and low water solubility, which are some of its drawbacks, curcumin nanoparticles are employed to address these problems. Wet-milling approach, in which the curcumin was sprayed into boiling water while being stirred and sonicated, is one typical way for creating curcumin nanoparticles. Additionally, research suggests that when compared to raw curcumin, nanocurcumin showed better solubility, antibacterial, and antifungal activity (Basniwal et al, 2011). Nanocurcumin particles can also be made using different techniques. The emulsion-diffusion evaporation approach used by Duan et al (2010) and Shaikh et al (2009) to manufacture nanoparticles resulted in stable, spherical nanoparticles. When the nanoparticles are taken orally, the bioavailability of the chemical increases dramatically (ninefold) (Shaikh et al, 2009). Nanoprecipitation is a different strategy that has been employed to encase curcumin in polymer (poly(lactic acid-co-glycolic acid)–poly(ethylene glycol)) (Anand et al, 2010; Yallapu et al, 2010).

Many literature reviews have been put forward to improve the bioavailability of resveratrol, much as curcumin. According to studies, resveratrol becomes more soluble and easier to transport across the plasma membrane as its size approaches the nanoscale (Ansari et al, 2011). Resveratrol has a number of drawbacks, including limited bioavailability, low solubility, and quick metabolism. These disadvantages are eliminated by the nanotechnique. Resveratrol nanoparticles are frequently prepared using a high-shear homogenisation procedure that first creates microparticles and then uses an ultrasonic method to create nanoparticles (Neves et al, 2013). When resveratrol is loaded onto lipid core nanoparticles, the tissue concentration in the brain, liver, and kidney improves. Resveratrol's anti-glioma activity has been documented when it is included in biodegradable nanoparticles (Shao et al, 2009).

According to Khan et al (2013), withaferin-A, an active component of *Withania somnifera*, tends to have a more anxiolytic effect when it is nanoencapsulated. Ginseng extract powder was processed at different speeds and energies during high-energy ball milling to create nanoscaled ginseng (Wen et al, 2009). When compared to raw Ginseng powder extract, the antioxidant capacity and capacity for cellular development were determined to be surprisingly high (Lee et al, 2013). Shinji et al (2011) investigated the effects of Ginkgo nanoparticles on brain cells. Combinatorial dry (gas phase grinding) and wet methods were used to create *Ginkgo biloba* nanoparticles (liquid phase grinding). The Ach release from the cortical synapse of the brain's cerebral hemispheres is increased by nanosized ginkgo (Shinji et al, 2011). Additionally, studies have indicated that natural herbs such as *Bacopa monnieri, Ashwagandha, Mucuna pruriens Linn*, and *Panax ginseng* root leaf extract may be used to create gold and silver nanoparticles (Arulkumar and Sabesan, 2012; Khan et al, 2013; Lee et al, 2013; Babu et al, 2013).

Polymeric nano-micelles are cutting-edge delivery colloid systems that can be used to nanoencapsulate phenolics that are amphiphilic and poorly water soluble.

They are composed of a hydrophilic shell and a hydrophobic core in a copolymer diblock structure. Two factors work together to generate micelles. The interaction between molecules causes their connection, and the repulsion between molecules prevents micelles from expanding indefinitely to reach a separate macroscopic phase. In order to generate stable polymeric micelles, amphiphilic block copolymers minimise free energy; generally speaking, change in entropy is thought to be the most crucial component (Ratheesh et al, 2017).

1.7 MEDICINAL PROPERTIES OF HERBS AND THEIR CLINICAL USES

Various medicinal plants showed antioxidant and anti-inflammatory properties that could potentially have therapeutic benefits in treating a range of nervous system disorders. The findings of the research also suggest that interactions between the cholinergic, dopaminergic, and glutamatergic systems are primarily responsible for the therapeutic benefits of plants on NDs including Alzheimer's and PD. The plant's interaction with the GABA and opioid system may have analgesic and anticonvulsant properties (Khazdair et al, 2018). Different mechanisms of medical properties of medicinal herbs are summarised in Figure 1.2.

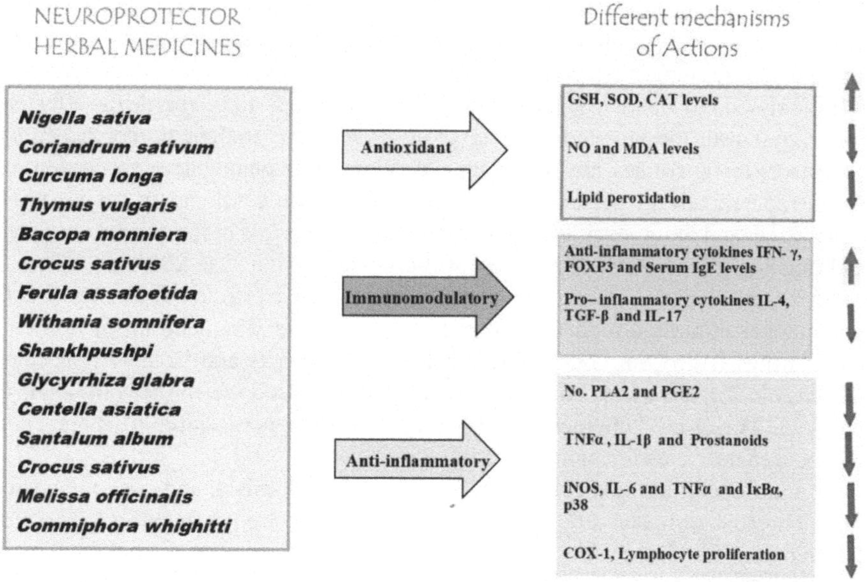

FIGURE 1.2 Medical properties of medicinal herbs with different mechanisms. GSH, glutathione; SOD, superoxide dismutase; CAT, catalase; NO, nitric oxide; MDA, malondialdehyde; PLA2: phospholipase A2; PGE2: prostaglandin-E2; IL-1β, interleukin-1β; COX-1, cyclooxygenase-1; iNOS, inducible nitric oxide synthase.

Modified from Khazdair et al (2018).

1.8 FUTURE DIRECTION

Both brain deterioration and neuropharmacological disorders involve the process of neurodegeneration. Cognitive impairment is a significant public health issue in the twenty-first century, and a number of neurodegenerative and neuropsychiatric conditions, including Parkinsonism, AD, depression, schizophrenia, cerebrovascular impairment, head injuries, dementia, and seizure disorders, can be profoundly functionally incapacitating.

Phytochemicals protect neurons by interacting with many pathogenic factors associated with NDs. Dietary habits have been shown in epidemiological research to increase health and longevity, as well as to protect cognitive and motor function, as well as prevent depressive disorders. Food-derived chemicals have been considered as a treatment method for age-related neurodegeneration. Food-derived chemical bioavailability is reduced by fast metabolism, insufficient permeability across the BBB and decreased bioavailability and stability in the brain (Pandareesh et al, 2015). Lipophilic flavonoids, alkaloids, and terpenes can cross the BBB into the brain, where they have high bioavailability and affinity for brain receptors. Among these phytochemicals and their derivatives, the development of novel neuroprotective substances is envisaged (Gomes et al, 2009). Several delivery strategies have been suggested to improve bioavailability and pharmacological formulation, including nanoparticles, liposomes, complexes with phospholipids and amphiphilic polymers, as well as conjugation with amino acids and glycosides (Gomes et al, 2009).

Diet and food-derived compounds should have diverse impacts on individuals with diverse genetic backgrounds, metabolic profiles, ages, and environmental exposure. These evidences indicate that the findings of an epidemiological study should be carefully analysed. To identify the potent bioactive compounds and estimate the effective concentration in the targeted brain areas, very sensitive analytical approaches for pharmacokinetic studies are necessary. Additionally, a quantitative assessment of the therapeutic advantages and consequences on disease development has yet to be established, and more sensitive in vivo quantitative measures of the survival or death and function of individual neurons are still necessary (Naoi et al, 2019).

Recent research has demonstrated the neuroprotective effects of phytochemicals in a number of animal models of neurological disorders. The need for phytotherapeutic drugs is growing, but before plant-derived extracts and isolated molecules responsible for the action are widely accepted and used, scientific validation is required. Therefore, "phytochemicals" might assure a new source of therapeutic neurodegenerative medications.

The most debilitating difficulty in the world is ND diseases, and due to the deterioration of lifestyle and stress, additional ND patients are predicted in the future. Based on the fact that existing ways effectively deliver drugs to these patients' parts of the brain.

1.9 CONCLUSION

Numerous NDs have mysterious origins. Neurodegeneration is a process that occurs in response of both neuropharmacological diseases and brain aging. Cognitive dysfunction is a serious public health issue in the twenty-first century, and various neurodegenerative and neuropsychiatric disorders have been linked to it. AD, depression, schizophrenia, cerebrovascular impairment, head injury, dementia, seizure disorders, and Parkinsonism are all examples of extremely functionally disabling illnesses.

Since many decades ago, the use of herbal medicines has attracted considerable interest due to their medicinal potential. Phytochemicals have recently been shown to have neuroprotective properties in several animal models of neurological disorders. While there is a growing demand for phytotherapeutic drugs, scientific validation is required before plant-derived extracts and isolated molecules responsible for the action achieve widespread recognition and application. As a conclusion, "phytochemicals" could provide a new supply of helpful neurodegenerative medications. Most of the phytochemicals studied had decreased activity when compared to traditional criteria. Due to their anti-inflammatory, antioxidative, and anticholinesterase properties, phytochemicals will likely be used in the future as a promising treatment for NDDs. The cellular and subcellular characteristics of NDs including AD, PD, Huntington's, and others are similar, as are the majority of the molecular signalling pathways that might result in apoptosis, necroptosis, and inflammation.

In general, using herbal medicine offers hopeful alternatives for the present treatments for NDs. The poor pharmacokinetic qualities of herbal medicine/natural substances, however, severely limit their potential. The herbal remedy has been included in a number of drug delivery formulations to get around these restrictions. In recent years, nanoencapsulation has become a promising new field for medication delivery. These nanoformulations can target drugs to certain cells, lowering toxicity and dose requirements. Furthermore, it has been demonstrated that using natural substances in the nanosize range as a therapeutic agent has the same effect as using them directly.

REFERENCES

Abdullaev, F.I. 1993. "Biological effects of saffron." *BioFactors (Oxford, England)* 4: 83.

Anand, P., H.B. Nair, B. Sung, and A.B. Kunnumakkara, *et al.* 2010. "Design of curcumin-loaded PLGA nanoparticles formulation with enhanced cellular uptake, and increased bioactivity in vitro and superior bioavailability in vivo." *Biochemical Pharmacology* 79: 330–338.

Ansari, K.A., P.R. Vavia, F. Trotta, and R. Cavalli. 2011. "Cyclodextrin-based nanosponges for delivery of resveratrol: In vitro characterisation, stability, cytotoxicity and permeation study." *AAPS PharmSciTech* 12(1): 279–286.

Antony, M.M. 2001. "Treatment plans and interventions for depression and anxiety disorders." Journal *of* Psychiatry *and* Neuroscience 26(5): 422–423.

Arulkumar, S., and M. Sabesan. 2012. "The behavioral performance tests of Mucuna pruriens gold nanoparticles in the 1-methyl 4-phenyl-1,2,3,6-tetrahydropyridine treated mouse model of Parkinsonism." Asian Pacific Journal *of* Tropical Disease 2(1): 499–502.

Azaz, A.D., H.A. Irtem, M. Kurkcuoğlu, and K.H. Baser. 2004. "Composition and the in vitro antimicrobial activities of the essential oils of some Thymus species." *Zeitschrift für* Naturforschung *C: A Journal of* Biosciences 59 (1–2): 75–80.

Babu, P.J., P. Sharma, S. Saranya, and U. Bora. 2013. "Synthesis of gold nanoparticles using ethanolic leaf extract of Bacopa monnieri and UV irradiation." *Materials* Letters 93: 431–434.

Basniwal, R.K., H.S. Buttar, V.K. Jain, and N. Jain. 2011. "Curcumin nanoparticles: Preparation, characterization, and antimicrobial study." Journal of Agricultural *and* Food Chemistry 59 (5): 2056–2061.

Bathaie, S.Z., and S. Mousavi. 2010. "New applications and mechanisms of action of saffron and its important ingredients." *Critical Reviews in Food Science and Nutrition* 50: 761–786.

Bilge, S., and O. Ilkay. 2005. "Discovery of drug candidates from some. Turkish plants and conservation of biodiversity." Pure *and* Applied Chemistry 77: 53–64.

Carroll, W.M. 2017. "The need for a global neurology alliance." Journal *of the* Neurological *Sciences* 379: 321–323.

D'Amico, E., P. Factor-Litvak, R.M. Santella, and H. Mitsumoto. 2013. "Clinical perspective on oxidative stress in sporadic amyotrophic lateral sclerosis." Free Radical Biology and Medicine 65: 509–527.

De Jong, W.H., and P.J. Borm. 2008. "Drug delivery and nanoparticles: Applications and hazards." International Journal *of* Nanomedicine 3(2): 133–149.

Deng, X.Y., H.Y. Li, and J.J. Chen, *et al.* 2015. "Thymol produces an antidepressant-like effect in a chronic unpredictable mild stress model of depression in mice." Behavioural Brain Research 291: 12–19.

Dorsey, E.R., A. Elbaz, and E. Nichols, *et al.* 2018. "Global, regional, and national burden of Parkinson's disease, 1990–2016: A systematic analysis for the global burden of disease study 2016." *Lancet Neurology* 17: 939–953.

Duan, J., Y. Zhang, S. Han, Y. Chen, and B. Li, *et al.* 2010. "Synthesis and in vitro/in vivo anti-cancer evaluation of curcumin-loaded chitosan/poly(butyl cyanoacrylate) nanoparticles." International Journal of Pharmaceutics 400: 211–220.

El-Guebaly, N., J. Sareen, and M.B. Stein. 2010. "Are there guidelines for the responsible prescription of benzodiazepines?" Canadian Journal of Psychiatry 55(11): 709–714.

El-Tahir, K.E.-D.H., and D.M. Bakeet. 2006. "The black seed Nigella sativa Linnaeus-A mine for multi cures: A plea for urgent clinical evaluation of its volatile oil." Journal *of* Taibah University Medical Sciences 1: 1–19.

Farooqui, T., and A.A. Farooqui. 2017. *Neuroprotective Effects of Phytochemicals in Neurological Disorders*. USA: Wiley Blackwell.

Feigin, V.L., E. Nichols, and T. Alam, *et al.* 2019. "Global, regional, and national burden of neurological disorders, 1990–2016: A systematic analysis for the global burden of disease study 2016. *Lancet Neurology* 18(5): 459–580.

Fu, W., W. Zhuang, S. Zhou, and X. Wang. 2015. "Plant-derived neuroprotective agents in Parkinson's disease." *American Journal of Translational Research* 7 (7): 1189–1202.

Gerez, J.A., and R. Riek. 2020. "Neurodegenerative diseases distinguished through protein-structure analysis." *Nature* 578 (7794): 223–224.

Gomes, N.G., M.G. Campos, J.M. Orf¨ao, and C.A. Ribeiro. 2009. "Plants with neurobiological activity as potential targets for drug discovery." *Progress in Neuro-Psychopharmacology & Biological Psychiatry* 33 (8): 1372–1389.

Gorman, J. M. 1996. "Comorbid depression and anxiety spectrum disorders." Depression *and* Anxiety 4 (4): 160–168.

Hooten, K.G., D.R. Beers, W. Zhao, and S.H. Appel. 2015. "Protective and toxic neuroinflammation in amyotrophic lateral sclerosis." *Neurotherapeutics* 12: 364–375.

Hosseini, M., M.H. Boskabady, and M.R. Khazdair. 2021. "Neuroprotective effects of Coriandrum sativum and its constituent, linalool: A review." Avicenna *Journal of* Phytomedicine 11 (5): 436–450.

Hosseini, M., T. Mohammadpour, and R. Karami, *et al.* 2015. "Effects of the hydro-alcoholic extract of Nigella sativa on scopolamine-induced spatial memory impairment in rats and its possible mechanism." Chinese Journal of *Integrative* Medicine 21(6): 438–444.

Hosseinzadeh, H., V. Motamedshariaty, and F. Hadizadeh. 2007. "Antidepressant effect of kaempferol, a constituent of saffron (*Crocus sativus*) petal, in mice and rats." *Pharmacologyonline* 2: 367–370.

Hussain, G., A. Rasul, and H. Anwar, *et al.* 2018. "Role of plant derived alkaloids and their mechanism in neurodegenerative disorders." International Journal *of* Biological Sciences 14(3): 341–357.

Iranshahy, M., and M. Iranshahi. 2011. "Traditional uses, phytochemistry and pharmacology of asafoetida (Ferula assa-foetida oleo-gum-resin) da review." *Journal of Ethnopharmacology* 134: 1–10.

Karimi, E., E. Oskoueian, R. Hendra, and H.Z. Jaafar. 2010. "Evaluation of Crocus sativus L. stigma phenolic and flavonoid compounds and its antioxidant activity." *Molecules* 15: 6244–6256.

Khan, Z.A., A.K.A. Mandal, R. Abinaya, and K. Krithika. 2013. "Nanoencapsulation of withaferin-a using poly-(lactic acid) for enhanced anxiolytic activity." Middle-East Journal *of* Scientific Research 14(4): 544–548.

Khazdair, M.R. 2015. "The protective effects of Nigella sativa and its constituents on induced neurotoxicity." *Journal of* Toxicology Article ID 841823, 1–7. doi: 10.1155/2015/841823.

Khazdair, M.R., A. Anaeigoudari, M. Hashemzehi, and R. Mohebbati. 2018. "Neuroprotective potency of some spice herbs, a literature review." *Journal of* Traditional *and* Complementary Medicine 9(2): 98–105. doi: 10.1016/j.jtcme.2018.01.002.

Khazdair, M.R., M.H. Boskabady, M. Hosseini, R. Rezaee, and A.M. Tsatsakis. 2015. "The effects of *Crocus sativus* (saffron) and its constituents on nervous system: A review." Avicenna *Journal of* Phytomedicine 5: 376–391.

Kulkarni, S., K.K. Akula, and J. Deshpande. 2012. "Evaluation of antidepressant-like activity of novel water-soluble curcumin formulations and St. John's wort in behavioural paradigms of despair." *Pharmacology* 89: 83–90.

Kumar, P., V.K. Singh, and D.K. Singh. 2009. "Kinetics of enzyme inhibition by active molluscicidal agents ferulic acid, umbelliferone, eugenol and limonene in the nervous tissue of snail Lymnaea acuminata." Phytotherapy *Research* 23: 172–177.

Küpeli, A.E., C.I. Tatlı, and K.G. Şeker, *et al.* 2021. "Natural compounds as medical strategies in the prevention and treatment of psychiatric disorders seen in neurological diseases." *Frontiers in Pharmacology* 2021 12: 638–669.

Lannert, H., and S. Hoyer. 1998. "Intracerebroventricular administration of streptozotocin causes long-term diminutions in learning and memory abilities and in cerebral energy metabolism in adult rats." Behavioral Neuroscience 112: 1199–208.

Lawrence, B.M. 1993. "A Planning Scheme to Evaluate New Aromatic Plants for the Flavor and Fragrance Industries." In *New Crops*, edited by J. Janick and J.E. Simon, 620–627. New York: Wiley.

Lee, S.B., S. Yoo, P. Ganesan, and H.S. Kwak. 2013. "Physicochemical and antioxidative properties of Korean nanopowdered white ginseng." International Journal *of* Food Science *&* Technology 48: 2159–2165.

Li, Q., N.Y. Spencer, N.J. Pantazis, and J.F. Engelhardt. 2011. "Alsin and SOD1(G93A) proteins regulate endosomal reactive oxygen species production by glial cells and proinflammatory pathways responsible for neurotoxicity." *Journal of Biological Chemistry* 286: 40151–40162.

Lobo, V., L.A. Pati, A. Phatak, and N. Chandra. 2010. "Free radicals, antioxidants and functional foods: Impact on human health." Pharmacognosy *Reviews* 4: 118–126.

Malaspina, A., F. Puentes, and S. Amor. 2015. "Disease origin and progression in amyotrophic lateral sclerosis: An immunology perspective." International Immunology 27: 117–129..

Marin, L.D., M. Sanchez-Borzone, and D.A. Garcia. 2011. "Comparative antioxidant properties of some GABAergic phenols and related compounds, determined for homogeneous and membrane systems." Medicinal Chemistry 7: 317–324.

Masaru, T., J. Toldi, and L. Vécsei. 2020. "Exploring the etiological links behind neurodegenerative diseases: Inflammatory cytokines and bioactive kynurenines." *International Journal of Molecular Sciences* 21(7): 2431.

Mokhtari-Zaer, A., M.R. Khazdair, and M.H. Boskabady. 2015. "Smooth muscle relaxant activity of Crocus sativus (saffron) and its constituents: Possible mechanisms." Avicenna *Journal of* Phytomedicine 5: 365.

Naoi, M., N. Masayo Shamoto, and M. Wakako. 2019. "Neuroprotection of multifunctional phytochemicals as novel therapeutic strategy for neurodegenerative disorders: Antiapoptotic and antiamyloidogenic activities by modulation of cellular signal pathways." *Future Neurology* 14(1): FNL9.

National Institute of Neurological Disorders and Stroke. 2017. National institute of neurological disorders and stroke. *Bethesda: Amyotrophic Lateral Sclerosis.*

Neves, A.R., M. Lucio, S. Martins, J.L. Lima, and S. Reis. 2013. "Novel resveratrol nanodelivery system based on lipid nanoparticles to enhance its oral bioavailability." International Journal *of* Nanomedicine 8: 177–187.

Pandareesh, M.D., R.B. Mythri, and M.M. Srinivas Bharath. 2015. "Bioavailability of dietary polyphenols: Factors contributing to their clinical application in CNS diseases." Neurochemistry International 89: 198–208.

Paramanick, D., V.D. Singh, and V.K. Singh. 2022. "Neuroprotective effect of phytoconstituents via nanotechnology for treatment of Alzheimer diseases." *Journal of Controlled Release* 351: 638–655.

Raj, V. Kavitha, J.R. Kumar, C. Egbuna, and C.J. Ifemeje. 2020. "Chapter 10 – Phytochemicals as therapeutic interventions in neurodegenerative diseases." *Phytochemicals as Lead Compounds for New Drug Discovery* 1(1): 161–178.

Ratheesh, G., L. Tian, and J.R. Venugopal *et al.* 2017. "Role of medicinal plants in neurodegenerative diseases." *Biomanufacturing Reviews* 2(2). doi: 10.1007/s40898-017-0004-7.

Rekatsina, M., A. Paladini, and A. Piroli, *et al.* 2020. "Pathophysiology and therapeutic perspectives of oxidative stress and neurodegenerative diseases: A narrative review." *Advances in* Therapy 37: 113–139.

Roth, G.A., D. Abate, and K.H. Abate, *et al.* 2018. "Global, regional, and national age-sex-specific mortality for 282 causes of death in 195 countries and territories, 1980–2017: A systematic analysis for the global burden of disease study 2017." *Lancet* 392: 1736–1788.

Rubio, J., W. Qiong, X. Liu, Z. Jiang, H. Dang, and S.L. Chen, *et al.* 2011. "Aqueous extract of Black Maca (Lepidium meyenii) on memory impairment induced by ovariectomy in mice." Evidence-Based Complementary *and* Alternative Medicine 2011:7, Article ID 253958. doi: 10.1093/ecam/nen063.

Sammi, S.R., S. Trivedi, and S.K. Rath, et al. 2016. "1-methyl-4-propan-2-ylbenzene from Thymus vulgaris attenuates cholinergic dysfunction." Molecular Neurobiology 54(7): 5468–5481

Sandhya, S., K.R. Vinod, and K. Sravan. 2010. "Herbs used for brain disorders." Hygeia *Journal for Drugs and Medicines* 2(1): 38–45.

Santarsieri, D., and T. Schwartz. 2015. "Antidepressant efficacy and Side-effect burden: A quick guide for clinicians." *Drugs Context* 4: 212–290.

Sayeed, M.S.B., M. Asaduzzaman, and H. Morshed, *et al.* 2013. "The effect of Nigella sativa Linn. seed on memory, attention and cognition in healthy human volunteers." *Journal of* Ethnopharmacology 148: 780–786.

Sayeed, M.S.B., T. Shams, and S.F. Hossain, *et al.* 2014. "Nigella sativa L. seeds modulate mood, anxiety and cognition in healthy adolescent males." *Journal of Ethnopharmacology* 152: 156–162.

Schmitt, F., G. Hussain, L. Dupuis, J.P. Loeffler, and A. Henriques. 2014. "A plural role for lipids in motor neuron diseases: Energy, signalling and structure." *Frontiers in* Cellular Neuroscience 25(8): 1–14.

Selvam, A.B. 2008. "Inventory of vegetable crude drug samples housed in Botanical Survey of India, Howrah." Pharmacognosy *Reviews* 2: 61–94.

Sethiya, N.K., A. Nahata, P.K. Singh, and S.H. Mishra. 2019. "Neuropharmacological evaluation on four traditional herbs used as nervine tonic and commonly available as Shankhpushpi in India." *Journal of Ayurveda and Integrative Medicine* 10(1): 25–31.

Shaikh, J., D.D. Ankola, V. Beniwal, D. Singh, and K.M.N. Ravi. 2009. "Nanoparticle encapsulation improves oral bioavailability of curcumin by at least 9-fold when compared to curcumin administered with piperine as absorption enhancer." European Journal of Pharmaceutical Sciences 37(3–4): 223–230. doi: 10.1016/j.ejps.2009.02.019.

Shao, J., X. Li, X. Lu, and C. Jiang, *et al.* 2009. "Enhanced growth inhibition effect of resveratrol incorporated into biodegradable nanoparticles against glioma cells is mediated by the induction of intracellular reactive oxygen species levels." Colloids and Surfaces B: Biointerfaces 72(1): 40–47.

Shilo, M., A. Sharon, K. Baranes, M. Motiei, J.P. Lellouche, and R. Popovtzer. 2015. "The effect of nanoparticle size on the probability to cross the blood-brain barrier: An in vitro endothelial cell model." Journal of Nanobiotechnology 13: 19.

Shimizu, S., D. Hirose, H. Hatanaka, and N. Takenoshita, *et al.* 2018. "Role of neuroimaging as a biomarker for neurodegenerative diseases." *Frontiers in Neurology* 9: 265.

Shinji, S., T. Yasukazu, W. Hatsue, and K. Kazuo, *et al.* 2011. "Analysis of brain cell activation by nanosized particles of Ginkgo biloba extract." International Journal *of* Plant Physiology *and* Biochemistry 3(3): 28–33.

Shi, J., A.R. Votruba, O.C. Farokhzad, and R. Langer. 2010. "Nanotechnology in drug delivery and tissue engineering: From discovery to applications." *Nano Letters* 10(9): 3223–3230.

Singhal, A.K., V. Naithani, and O.P. Bangar. 2012. "Medicinal plants with a potential to treat Alzheimer and associated symptoms." International Journal of Nutrition, Pharmacology, Neurological *Diseases* 2: 84–91.

Small, E. 1997. *Culinary Herbs*. Ottawa: NRC Research Press.

Solanki, I., P. Priyanka, L.M. Mohammad, and S.P. Mordhwaj. 2015. "Flavonoid-based therapies in the early management of neurodegenerative diseases." *Advances in Nutrition* 6(1): 64–72.

Soodi, M., N. Naghdi, H. Hajimehdipoor, S. Choopani, and E. Sahraei. 2014. "Memory-improving activity of Melissa officinalis extract in naïve and scopolamine-treated rats." Research *in* Pharmaceutical Sciences 9(2): 107–114.

Sriraman, S.K., B. Aryasomayajula, and V.P. Torchilin. 2014. "Barriers to drug delivery in solid tumors." *Tissue Barriers* 22(2): e29528. doi: 10.4161/tisb.29528.

Tamaddonfard, E., A.A. Farshid, E. Ahmadian, and A. Hamidhoseyni. 2013. "Crocin enhanced functional recovery after sciatic nerve crush injury in rats." Iranian Journal *of* Basic Medical Sciences 16: 83–90.

Thakur, A.K., P. Kamboj, and K. Goswami. 2018. "Pathophysiology and management of Alzheimer's disease: An overview." Journal *of* Analytical *&* Pharmaceutical *R*esearch 9(2): 226–235.

Valery, L., T.V. Feigin, and E. Nichols, *et al.* 2020. "The global burden of neurological disorders: Translating evidence into policy." *Lancet Neurology* 19: 255–265.

Valery L Feigin, Emma Nichols, Tahiya Alam, Marlena S Bannick, et al. 2019. "The data to put neurology on top of the public-health agenda." *Lancet Neurology* 18: 1.

Wen, H.W., W.C. Li, and R.J. Chung, *et al.* 2009. "Evaluation of nanofabricated ginseng extract powders." *Journal for* Nanoscience *and* Nanotechnology 9(7): 4108–4115

Yallapu, M.M., B.K. Gupta, M. Jaggi, and S.C. Chauhan. 2010. "Fabrication of curcumin encapsulated PLGA nanoparticles for improved therapeutic effects in metastatic cancer cells." Journal *of* Colloid *and* Interface Science 351(1): 19–29.

Yang, F., G.P. Lim, and A.N. Begum, *et al.* 2005. "Curcumin inhibits formation of amyloid b oligomers and fibrils, binds plaques, and reduces amyloid in vivo." *Journal of Biological Chemistry* 280(7): 5892–5901.

Yang, J., S. Song, J. Li, and T. Liang. 2014. "Neuroprotective effect of curcumin on hippocampal injury in 6-OHDA-induced Parkinson's disease rat." Pathology, Research *and* Practice 210: 357–362.

Yusuf, M., J.U. Chowdhury, M.A. Wahab, and J. Begum. 1994. Medicinal Plants of Bangladesh. Bangladesh Council of Scientific and Industrial Research. Dhaka, Bangladesh. Pp. 1–340.

Zarmouh, N.O., S.S. Messeha, F.M. Elshami, and K.F. Soliman. 2016. "Natural products screening for the identification of selective monoamine oxidase-b inhibitors." *European Journal of Medicinal Plants* 15(1): 1–16. doi: 10.9734/EJMP/2016/26453.

2 Neuroprotective Potentials of Phytochemicals

Odangowei Inetiminebi Ogidi and Imomotimi Ajoko

2.1 INTRODUCTION

An onerous problem for medical treatment, society, and the economy is the rise in people with neurodegenerative illnesses in our ageing society. The term "neuroprotection" refers to the methods and supporting mechanisms that can protect the central nervous system (CNS) from the neuronal damage brought on by a variety of neuropsychiatric and neurodegenerative disorders, including Parkinson's disease (PD), Alzheimer's disease (AD), anxiety, cerebrovascular disease, seizures, and others (Winner et al, 2011). These illnesses have similar symptomological characteristics throughout the course of the disease. Elevated oxidative/nitrosative stress, mitochondrial dysfunction, protein misfolding/aggregation, synapse loss, and reduced neuronal survival are a few of the physiological signs. Neurodegenerative disorders advance significantly as a result of a combination of genetic and environmental factors (Figure 2.1). A significant quantity of energy is required to protect immunological and neuronal cells from the accumulating oxygen and nitrogen species that stress the environment when they are exposed to hazardous proteins (Finkel, 2011).

By the year 2040, it is predicted that neurodegenerative illnesses will rank as the second leading cause of mortality among seniors (Ansari et al, 2010). Phytochemical substances are one of the methods for neuroprotection and are an important treatment for avoiding neurodegenerative disorders. Numerous types of synthetic and natural neuroprotective agents have been described; nonetheless, it is thought that synthetic neuroprotective agents may cause adverse effects such as dry mouth, fatigue, drowsiness, lethargy, anxiety or restlessness, balance issues, etc. In recent years, both nationally and internationally, research centres and enterprises have become much more conscious of herbal-based medicinal goods. Consequently, there has been a lot of attention given to possible phytochemicals that might modify neuronal activity and provide protection against neurodegeneration (Kumar et al, 2012).

Numerous bioactive phytochemicals and other organic substances may enhance the therapy of neurodegenerative illnesses, according to a variety of research (Martel et al, 2019). Numerous plants include phytochemicals such as alkaloids, steroids, terpenoids, saponins, phenolics, flavonoids, and polyphenolic substances that have important characteristics like anti-inflammatory potential, DNA repair capacity,

DOI: 10.1201/9781003389781-2

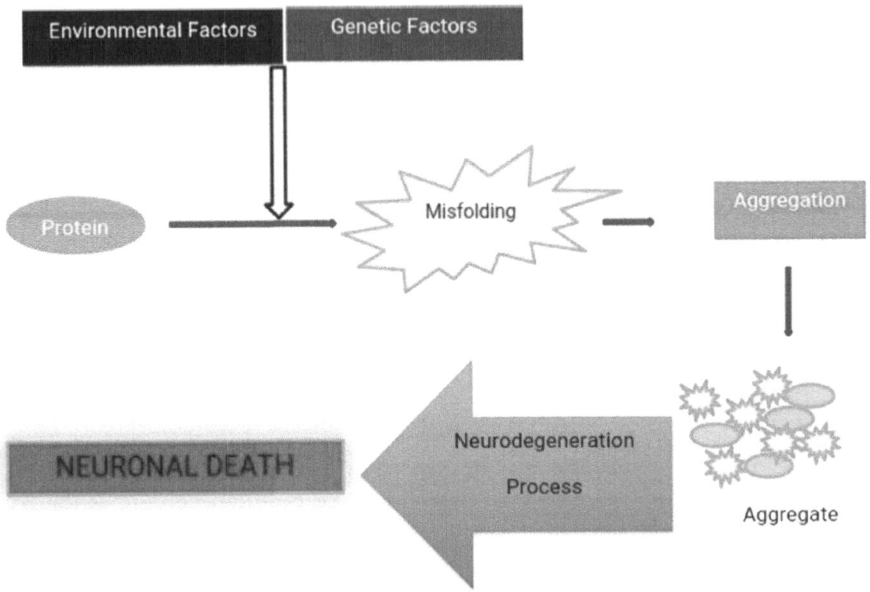

FIGURE 2.1 Factors leading to conformational changes of protein in the progression of neurodegenerative disorders.

autophagy, and antioxidant activities (Franco et al, 2019). As a result, this chapter examines neurodegenerative illnesses, an overview of phytochemicals, their potential for providing neuroprotection, as well as how they work.

2.2 NEURODEGENERATIVE DISEASES

The central and peripheral nervous systems are both affected by neurodegenerative diseases (ND), which are progressive abnormalities of the CNS. Neurodegeneration may result from the gradual and cumulative loss of brain cells. Although several variables are known to directly contribute to the beginning of neurodegeneration, the key causative element is the production of free radicals by reactive oxygen and nitrogen species (ROS and RNS). Numerous neurodegenerative illnesses are known to advance more rapidly in cases where neuroinflammatory processes are present. These illnesses' pathogenic characteristics include loss of neurons and protein misfolding and aggregation, which cause a variety of signs and symptoms (Figure 2.1). Over 600 neurological illnesses have been documented globally, according to studies from the National Institute of Neurological Disorders and Stroke (Meck et al, 1998). In this chapter, a few of them are briefly explored.

2.2.1 Anxiety

Cognitive, physical, emotional, and behavioural components are all present in the psychological and physiological state of anxiety. It may worsen to the point that it interferes with basic daily activities and makes a person fearful of what would normally happen in their life.

There are seven clinical illnesses known as anxiety disorders, including (1) generalized anxiety disorder (GAD): A person with this disorder has chronic dread, anxiety, and excessive stress over trivial concerns. (2) Panic disorder: A person has short episodes of great fear and dread that are often accompanied by disorientation, dizziness, trembling, shivering, nausea, and breathing difficulties. (3) Phobias: A condition in which a particular stimulus or circumstance causes dread and worry. Agoraphobia is the fear of being in a location or circumstance from which it would be difficult or humiliating to leave or from which aid may not be accessible. (4) Social anxiety disorder (SAD): This condition is characterized by an extreme dread of adverse public scrutiny, public humiliation, or public disgrace. (5) Obsessive-compulsive disorder (OCD): The main characteristics are recurring obsessions and compulsions. (6) Post-traumatic stress disorder (PTSD): This condition develops as a consequence of a traumatic event, such as a serious accident, child abuse, a war scenario, a natural catastrophe, a rape, hostage situations, etc. (7) Anxiety related to separation: It is the sensation of unwarranted or excessive worry brought on by being away from a person or location. The neuropeptides galanin, neuropeptide Y, arginine vasopressin, and tachykinin, as well as the monoamines dopamine, noradrenaline, and serotonin, as well as neurosteroids and cytokines, have been reported to have a modulating function in anxiety states (Pathak et al, 2011).

2.2.2 Attention Deficit/Hyperactivity Disorder

Attention deficit/hyperactivity disorder (ADHD) is a broad term for a collection of neuropsychiatric symptom clusters that first appear in infancy and often last into adulthood. People who exhibit a variety of symptoms, such as difficulty focusing, restlessness, mood swings, temperament issues, difficulties finishing work, disorganization, and an inability to handle stress, are increasingly being diagnosed with ADHD (Brue and Oakland, 2002).

2.2.3 Depression

Depression is a state of low mood and widespread sadness. One of the most prevalent mental diseases, depression may present in a variety of ways, from mild melancholy to complete anguish and despair. Both biological and environmental variables influence mood swings and the incidence of depression symptoms. The World Health Organization (WHO) estimates that 450 million people worldwide suffer from a mental or behavioural problem, and that number is expected to rise sharply (Drevets et al, 1999).

According to a WHO report, depression is among the top ten global causes of disease and death. There are a number of structural changes that occur in neurons during depressive states, including reduced frontal cortex and hippocampal volume, hypothalamic-pituitary-adrenal axis malfunction, and abnormalities in 5-hydroxy tryptamine (5-HT) and its receptors. It is generally known that the monoaminergic neurotransmitter 5-HT is dysregulated in illnesses like schizophrenia and anxiety, which are neurological disorders (Drevets et al, 1999).

2.2.4 Dementia

The most prevalent kind of human brain degenerative illness and significant motor impairment, PD affects movement. The most ubiquitous element contributing to the recognisable symptoms of PD is the loss of dopaminergic neurons in the substantia nigra (Calne, 2000). Tremor, stiffness, bradykinesia, and impairment of balance are its four primary symptoms. Lewy bodies in the substantia nigra and nerve cell loss in some of its ventral tier are the typical pathological signs. In addition to the neuropathologic symptoms, a persistent deficiency in cholinergic neurotransmission may contribute to PD neurochemically. Deficit, learning, and memory loss are the hallmarks of AD, a neurodegenerative condition that is also marked by cognitive problems such as sadness, agitation, and psychosis (Calne, 2000).

2.2.5 Epilepsy

The most prevalent neurological condition, epilepsy, affects about 50 million people globally (Scheuer and Pedley, 1990). It is distinguished by the sporadic occurrence of recurrent epileptic seizures, or the repetitive involuntary contraction of striated muscle. The grey matter of brain experiences excessive and accelerated firing of cerebral neurons during seizures.

2.2.6 Excitotoxicity

Some poisons, such as domoic acid (an amnesic shellfish poison produced by marine algae), kainic acid, etc., are intoxicated in part by excitotoxicity. Excitotoxicity is the degenerative process through which excessive glutamate stimulation kills and damages nerve cells. Excitotoxicity may also manifest in a variety of clinical situations, including multiple sclerosis, stroke, amyotrophic lateral sclerosis, spinal cord injury, and other neurological illnesses (Doble, 1999). The majority of typical brain processes, such as cognition, memory, and learning, are also affected. Excitotoxicity may result from excessive N-methyl-D-aspartate receptor activation, which causes an influx of Ca^{2+}, noise and inducible nitric oxide (NO) synthase activation, and excessive NO production (Albin and Greenamyre, 1992).

2.2.7 Schizophrenia

It is among the most significant types of mental disorders. In this situation, the patient is unaware of what is occurring right now. There are two categories of this illness' symptoms: (1) Adverse effects include aberrant behaviour, hallucinations, delusions, and thinking abnormalities. (2) Adverse effects include social isolation and a flattening of emotional reactions. In this condition, the brain's levels of neurotransmitters including dopamine, 5-HT, acetylcholine, and norepinephrine are elevated (Kumar, 2006).

2.3 OVERVIEW OF PHYTOCHEMICALS

Due to the many phytochemicals found in fruits and vegetables, there is an urgent need for regular ingestion (Somani et al, 2015). The presence of these different bioactive compounds from plant parts has been investigated in several studies (Ogidi et al, 2019a, 2019b, 2019c, 2021a, 2021b, 2021c, 2021d, 2022; Ogidi and Julius, 2021), and they have been demonstrated to protect against oxidative stress and neuroinflammation, two main characteristics of NDs (Enenebeaku et al, 2022; Kim et al, 2010). Curcumin, quercetin, diallyl trisulfide, and epigallocatechin-3-galate (EGCG) are examples of common phytochemicals that stimulate the immune system, reduce platelet aggregation, and control hormone metabolism (Farooqui, 2012; Ogidi, 2022).

Researchers have hypothesized that increasing the Mediterranean diet's intake of olive oil leads to a considerable improvement in health status, as seen by a large drop in overall mortality (13%) in PD and AD patients (Sofi et al, 2008). The condition of NDs can be improved by using phytochemicals as nutraceuticals; however, the underlying mechanism is yet unclear. Another research said that phytochemicals enhance the condition by scavenging free radicals and acting as anti-oxidants (Nokolova, 2011).

Although the cellular targets of phytochemicals are unknown, it is anticipated that these substances may activate stress response pathways that the cells would employ as a defence mechanism in addition to influencing gene expression and enzyme metabolism (Martin et al, 2004). Another theory is that phytochemicals participate in a downward signal transduction cascade as a ligand, attaching to specific receptors on cell membranes or nuclei and then exerting their antioxidant function (Si and Liu, 2007).

2.4 NEUROPROTECTIVE POTENTIAL OF PHYTOCHEMICALS

The public and scientific communities are very curious in the potential of phytoconstituents for neuroprotection and the prevention of NDs. Several phytochemicals have been found to have neuroprotective properties in animal and cell culture models of neurological disorders. The chalcone (saffron or yellow B) may protect cultured neurons from ischemic brain injury, while piceatannol may protect them against A-induced death. Phytochemicals in plants have been proven to protect the nervous system against disease, both in animal models of neurodegenerative disorders and in epidemiological studies of human populations (Liu, 2003). The fact that many of the active compounds have antioxidant activity has been the main focus of research on the health benefits of phytochemicals. Various phytochemicals' neuroprotective properties are linked to decreased oxidative stress. Reducing oxidative stress and protecting cultured hippocampal neurons from NO-mediated cell death, for instance, were achieved by resveratrol, quercetin, and catechins (Joseph et al, 2005). The following is a discussion of some of the crucial phytochemicals in neuroprotection:

2.4.1 Phenols and Polyphenols

Since the 1990s, phenolic chemicals found in food have gained a lot of attention due to mounting evidence of their positive impact on human health. Apple, grape, and citrus fruit juice polyphenols have been found to have a higher neuroprotective effect than antioxidant vitamins in *in vitro, in vivo*, and clinical investigations.

Green tea's catechins and epicatechins have been shown to be able to protect neurons from a variety of oxidative and metabolic insults, including preventing 6-hydroxy-dopamine from damaging dopaminergic neurons in a rat model of Parkinson's disease, shielding retinal neurons from ischemia-reperfusion injury, and reducing mutant huntingtin misfolding and neurotoxicity in a Huntington's disease model (Ehrnhoefer et al, 2006).

By activating protein kinase C (PKC) and transcription factors that promote the production of cell-survival genes, catechins, a kind of polyphenol, have neuroprotective properties (Mandel et al, 2005). It has been proposed that catechins may protect neurons against Alzheimer's disease processes and reduce the development of the illness. Numerous signalling pathways may be activated by catechins and their metabolites to exert cell-survival and anti-inflammatory effects. These activities include changing the expression of pro- and anti-apoptotic proteins and enhancing antioxidant defences (Sutherland et al, 2006).

EGCG, the most prevalent polyphenol in tea leaves, has captured the attention of everyone in recent years owing to its ability to slow down neuronal ageing. Tea intake is negatively linked with the prevalence of NDs (Pervin et al, 2018). In several research projects using AD models, EGCG was given. In one of these research, AD models were given the medication D-gal, and the amount of amyloid plaques was significantly decreased (Chan et al, 2016; Walker et al, 2015; Wobst et al, 2015).

Another study team hypothesized that EGCG administration reduced beta- and gamma-secretases by inhibiting ERK and NFkappaB, hence avoiding the death of neuronal cells (Liu et al, 2014a, 2014b). Another research found that EGCG administration decreased amyloid plaques by cleaving APP via -secretase (Smith et al, 2010). According to one research, persons who drank three or more cups of tea per day had a lower chance of acquiring PD. Additionally, it was shown that raising glutathione strengthens the free radical scavenging mechanism, which in turn stimulates CREB and Bcl-2 and results in beneficial effects (Choi et al, 2002). Tyrosine hydroxylase (TH)-positive cells suffer significant loss in PD, and one research found that taking EGCG and tea at the same time may stop the loss of these cells in the substantia nigra (Koh et al, 2003).

2.4.2 Flavonoids

Recent years have seen an uptick in interest in flavonoids due to the hypothesis that they may control neuronal activity and halt the progression of age-related neurodegeneration. There is promising evidence that extracts from flavonoid-rich plants or foods may improve memory and learning in a range of mammalian species, including humans, by shielding vulnerable neurons from damage, boosting existing neuronal function, or stimulating neuronal regeneration. Some flavonoids, such the citrus flavanone tangeretin, have showed promise as neuroprotective agents against the underlying pathology of Parkinson's disease. These flavonoids have been found to retain nigro-striatal integrity and functioning following lesioning with 6-hydroxydopamine (Youdim et al, 2004).

To reach the brain, flavonoids must first pass across the blood-brain barrier (BBB), which controls the entry of xenobiotics (Ehrnhoefer et al, 2006). Several dietary anthocyanins, cyanidin-3-rutinoside, and pelargonidin-3-glucoside, as well as flavanones such hesperetin and naringenin, and their in vivo metabolites, have been shown to cross the BBB in relevant in vitro and in situ models (Youdim et al, 2004). Anthocyanins could be able to pass the monolayer in in vitro BBB models (Youdim et al, 2002).

The anti-oxidant and free radical scavenging activities of flavonoids are the subject of much research, and quercetin's anti-proliferative capabilities are the subject of extensive research on quercetin (Lesjak et al, 2018). Hydrogen peroxide drastically reduced the viability of phenochromocytoma cells (PC12); however, when PC12 were re-incubated with quercetin, quercetin protected the cells against H2O2-induced toxicity even if its effects were dose-dependent (Heo and Lee, 2004). Quercetin's neuroprotective properties were seen at concentrations as low as 10 M and as high as 30 M. Quercetin is said to be able to pass the BBB and stop cytotoxicity brought on by H2O2 as well (Heo and Lee, 2004).

2.4.3 Alkaloids

A large class of naturally occurring substances called alkaloids typically comprise oxygen, nitrogen, carbon, and hydrogen. Alkaloids may have an impact on the CNS, which includes the brain and spinal cord's nerve cells, which regulate several direct bodily processes and behaviour. They may also have an impact on the autonomic nervous system, which controls respiration, heartbeat, circulation, and internal organ function.

Indole alkaloids contain the indole carbon-nitrogen ring, which is also present in the psychedelic substances lysergic acid diethylamide and the fungal alkaloids ergine and psilocybin (LSD). These alkaloids may conflict with or complement serotonin's activity in the brain (Pearson, 2001). The principal mechanism of action of ergot alkaloids was formerly believed to be their notable effects on blood flow. The spinal cord and CNS are impacted by tropane alkaloids found in Datura, including atropine, hyoscyamine, and scopolamine. Vinpocetine is a very effective vasodilator that is an alkaloid derived from Vinca minor (Halliwell, 2007).

Specifically, vinpocetine has been demonstrated to increase cerebral blood flow and metabolism, including glucose absorption, which may mitigate the negative consequences of hypoxia and ischaemia (Halliwell, 2007). The daffodil (Narcissus tazetta), snowdrop (Galanthus nivalis), and snowflake (Galanthus nivalis) all contain the tertiary alkaloid galantamine. It is a member of the phenanthrene chemical class (Leucojum aestivum). The medication may also activate nicotinic receptors to improve cognition and memory. It is a member of the pharmacological family known as cholinesterase inhibitors (Halliwell, 2007).

2.4.4 Terpenoids

Since more than a century ago, several essential oils produced from plants, including wormwood, have been recognized to have convulsant effects. Valepotriates and sesquiterpenes are the two pharmacologically active components found in the rhizome of valerian (Valerian officinalis) (valerenic acid and acetoxyvalerenic acid). The active substances have been shown to have sedating effects on mice. However, it has been shown that valerian crude extract inhibits GABA absorption and has GABA (B) receptor binding capabilities in rat synaptosomes (Ortiz et al, 1999). A sesquiterpene alkaloid known as huperzine A, which was isolated from the Chinese medicinal plant Huperzia serrata, has a variety of neuroprotective properties. Huperzine A increased spatial working memory and helped with learning and memory problems. Triterpenoid, brahminoside, and monoterpenes in Centella asiatica L. are chemical compounds that are helpful in reviving and enhancing nervous system performance (Nalini et al, 1992).

2.4.5 Fatty Acids

It has been shown that both in animals and people, consuming monounsaturated fatty acids and polyunsaturated fatty acids (PUFAs) may reduce cognitive decline. As an example, walnuts contain the monounsaturated fatty acid oleic acid (8:1) as well as the n26 and n23 PUFAs linoleic acid (LA) and a-linolenic acid (ALA), which have received the most attention in research (Crews et al, 2005).

Numerous studies have shown that eating a diet low in n23 fatty acids will have an adverse effect on cognitive performance (McCann and Ames, 2005). In order for neurons to connect with other cells, they need to have normal anchor receptors, ion channels, and the ability to release and reabsorb unmetabolized neurotransmitters. These requirements are all dependent on the structure of the cell. The neuronal membrane's fatty acid content affects these characteristics (Yehuda et al, 2002). There is evidence that the fatty acid composition of neuronal membranes declines with age, despite the fact that essential fatty acids have been found to increase membrane fluidity and PUFA content. In addition to affecting the biophysical features of membranes, PUFAs in the form of phospholipids in neuronal membranes may directly participate in signalling cascades to improve neuronal function, synaptic plasticity, and neuroprotection (Yehuda et al, 2002).

2.4.6 Berberine

According to several research using models of ND, berberine gives neuroprotection through controlling the levels of neurotrophin (Ji and Shen, 2011; Durairajan et al, 2012). The ability of berberine to halt the activity of different enzymes implicated in the development of AD has also been mentioned (Hsu et al, 2013). Neurotoxicity was generated in a number of in vitro investigations using glutamate, H_2O_2, and a cobalt chloride-driven decreased oxygen environment. It was shown in these in vitro investigations that berberine may protect neuronal cells against neurotoxicity (Cui et al, 2009; Pires et al, 2014). Additionally, it has been shown that berberine is effective in treating cobalt chloride induced hypoxia. It functions as a scavenger of ROS produced as a result of the hypoxic circumstances in this situation, which inhibits numerous agents that promote apoptosis and confers neuroprotection (Zhang et al, 2012). However, according to Kwon et al, berberine has severe side effects, including a reduction in DA neurons brought on by the cytotoxicity of 6-hydroxydopamine (Kwon et al, 2010).

By scavenging free radicals, berberine activates the PI3K/Akt/Nrf2 pathway and has neuroprotective benefits. Additionally, it has been shown that berberine has anti-apoptotic effects by upregulating Bcl-2 and downregulating caspase 1 and 3 and Bax expression (Asai et al, 2007). According to one research by Hsu et al, berberine treatment protects hydrogen peroxide-induced neurotoxicity by upregulating Bad expression while downregulating p53, caspase, and cyclin D1 expression (Hsu et al, 2012). It has been shown that berberine enhances cell survival and lessens oxidative stress when supplied at nanomolar concentrations by downregulating a number of factors such cytochrome C, Bax, and caspase. It has been discovered that administering berberine may prevent ischemic stroke by obstructing potassium currents (Maleki et al, 2018).

Another clinical investigation found that giving berberine to animals with NDs decreased their levels of the enzymes superoxide dismutase and choline acetyltransferase (Asai et al, 2007). However, several studies have also shown that berberine may cause the substantia nigra to lose dopaminergic neurons when it is administered for prolonged periods of time (Kim et al, 2014a; Shin et al, 2013). Berberine may prevent ischemic stroke by removing radical species, or in other words, lowering oxidative stress, according to some research (Kim et al, 2014b). In a rat model of AD, injection of berberin improves cholinergic enzyme activity, increases brain-derived neurotrophic factor levels, and reduces the production of inflammatory markers (Durairajan et al, 2012).

2.4.7 Limonoids

In the kingdom of plants, limonoids are rare, highly oxygenated molecules. Due to its high limonoids content, Melia toosendan, a member of the Meliaceae family, is bitter. Its extract stimulates neuronal development in a manner similar to that of NGF, which does so via activating ERK and protein kinase A (PKA) (Roy and Saraf, 2006). The seeds of Trichilia welwitschii were used to isolate the three limonoid chemicals (dregeanin DM4, rohituka 3, and trichilia lactone D5) that were the subject of this study (Tsamo et al, 2013).

In one research, it was shown that limonoids increase NGF levels in a PC12 rat cell line, promoting neuronal differentiation and proliferation (Roy and Saraf, 2006). Similar to NGF, limonoid chemicals found in Melia toosendan extract enhance neurite development (Zhang et al, 2013). The PKA and ERK pathways, which are essential in neuronal development, are hampered when the PC12 rat cell line is exposed to PKA inhibitors, according to another finding of this research (Yu et al, 2004). Limonoids stimulate ERKs and PKA, which help to enhance neuronal development (Roy and Saraf, 2006). Compared to NGF, the extract from Melia toosendan was more effective in fostering growth. However, it was discovered in that research that the plant's extract did not activate Ras or Raf1, but rather required the activation of PKA and MEK (Yu et al, 2004).

2.4.8 Sulforaphane

Cruciferous plants including broccoli, Brussels sprouts, and cauliflower are rich in the isothiocyanate sulforaphane. The neuroprotective effects of sulforaphane have been shown in animal models of both acute and chronic neurodegenerative disorders. Sulforaphane treatment in a mouse stroke model decreased brain damage and cerebral oedema while preserving the retinal pigment (Zhao et al, 2006). Cultured neurons have shown that sulforaphane protects dopaminergic neurons against mitochondrial toxins and oxidative stress (Han et al, 2007).

2.4.9 Hypericin and Pseudohypericin

Hypericin and pseudohypericin, two naphthodianthrones, are the main constituents of Hypericum perforatum. Strong evidence supports the idea that hypericin and pseudohypericin have a role in the antidepressant effect. One way several antidepressants work to raise levels of neurotransmitters like serotonin, norepinephrine, or dopamine is by inhibiting monoamine oxidase (Schulz et al, 1998). Serotonin, dopamine, and norepinephrine synaptic re-uptake seem to be blocked by this substance (Chatterjee et al, 1998). Synaptic concentration of neurotransmitters is increased by blocking their reuptake. This is an additional way that synthetic antidepressants could work (Chatterjee et al, 1998).

2.4.10 Curcumin

One of the significant components of turmeric, which has a prominent position among Indian spices, is curcumin. Curcumin is used to treat diabetes, biliary diseases, cough, and hepatic illnesses since it contains a variety of therapeutic characteristics (El-Bahr, 2015; Nam et al, 2014; Soleimani et al, 2018). Curcumin has been linked to a number of pathways, one of which being its capacity to attach to amyloid plaques by inhibiting NF-, hence lowering AD development (Nam et al, 2014).

There have been several reports of curcumin's positive effects on the neurological system. Curcumin therapy reduced behavioural impairments and protected neurons from ischemic cell death in an animal model of stroke (Xu et al, 2007). There is mounting evidence from cell culture and animal models that dietary curcumin may be useful in the prevention or treatment of age-related NDs including Alzheimer's, Parkinson's, and stroke. Curcumin has also been demonstrated to prevent chronic stress-induced impairment of hippocampal neurogenesis and increase the production of brain-derived neurotrophic factor (BDNF) in an animal model of depression (Xu et al, 2007).

Additionally, it has been shown that curcumin promotes neuronal regeneration in a PD model by enhancing BDNF levels through activating Trk/PI3K signalling pathways (Yang et al, 2014). It is believed that curcumin works by lowering TNF- and caspase levels while concurrently increasing BDNF levels (Liu et al, 2014a, 2014b; Nam et al, 2014). Curcumin nanoparticles have also been developed in recent research to treat cognitive impairment by restoring BDNF levels via Akt/GSK-3 signalling pathways (Hoppe et al, 2013). Because it works on multiple molecular elements of these disorders, curcumin presents a potential safe and affordable preventative strategy for ND.

Curcumin also decreased the development of amyloid plaque in cell cultures and in vivo, according to recent studies (Wang et al, 2018). Curcumin has been demonstrated to alter A levels and inhibit the formation of the amyloid precursor protein (APP) in mice neurons, according to a research by Zhang et al (2010). Curcumin prevents A plaques by inhibiting NF-B, which is the mechanism through which it works. Curcumin is not only effective against AD, but it may also prevent PD by causing the protein-synuclein to become unstable (Liu et al, 2014a, 2014b).

2.4.11 Resveratrol

Red grapes contain the phytophenol resveratrol, which has antioxidant properties. It is referred to as a "miracle" molecule because it has anti-inflammatory and antioxidant effects on ND (Anastacio et al, 2014). Treatment with resveratrol restored a variety of cognitive impairments generated in rat models by lowering hippocampal TNF- and IL-1 and increasing BDNF (Anastacio et al, 2014). Recent research has also revealed that resveratrol may protect neurons in the brain and spinal cord from ischemia damage after being administered peripherally. Resveratrol protected spinal cord neurons from ischemia in a rat stroke model and reduced brain ischemia damage (Huang et al, 2001). Neurons in culture may be protected against NO's role in mediating oxidative stress-related death by the antioxidant resveratrol. Dopaminergic neurons in midbrain slice cultures, a model for PD, also benefited from resveratrol's protection against oxidative and metabolic stressors (Huang et al, 2001). Protecting cells against the damage caused by mutant huntingtin, resveratrol was tested in worm and cell culture models (Parker et al, 2005). In AD models, resveratrol inhibited amyloid -peptide's ability to damage neuronal cells and promoted the clearance of amyloid-peptide from cultured cells (Marambaud et al, 2005).

2.4.12 Allium and Allicin

Allium and allicin, two organosulfur compounds found in garlic and onions, have been shown to have neuroprotective effects. The antioxidant capabilities and activation of stress-response pathways by aryl-sulfide compounds may elevate the expression of neuroprotective proteins including mitochondrial uncoupling proteins (Oi et al, 1999). Furthermore, allicin activates transient receptor potential (TRP) ion channels in the plasma membrane of neurons. There are many other phytochemicals that may trigger adaptive cellular stress responses by opening TRP channels in neurons; examples include isothiocyanates, garlic alliums, and cannabinoids (Oi et al, 1999).

2.5 MECHANISM OF ACTION OF PHYTOCHEMICALS WITH RESPECT TO NEUROPROTECTION

Phytochemicals activate cellular stress-response pathways, which in turn causes an increase in the expression of neuroprotective gene products. Emerging evidence suggests that neurotrophic factors' neuroprotective effects are mediated, at least in part, by their ability to dampen down cell death and apoptosis pathways (Haddad, 2002). There are around 50 neurotransmitters in the brain, which come from a wide variety of chemical classes. Researchers have paid a lot of attention to acetylcholine since its deficit has been related to AD and other degenerative cognitive diseases, such as senile dementia. Phytochemicals may activate the transcription factor NF-kB via elevating production of antioxidant enzymes and Bcl-2. Certain phytochemicals may activate many signalling pathways by attaching to ligands that bind to various receptors, including G-protein coupled receptors (GPCR), growth factor receptors (GFR), and insulin receptors (IR). Phosphatidylinositol-3-kinase (PI3K), PKC, and mitogen-activated protein kinases (MAPK) are some of the kinases activated by these receptors (PI3K). Flavonoids alter synaptic plasticity by stimulating the ERE-CREB and P13 kinase-mTOR signalling pathways.

Numerous phytochemicals have been demonstrated to activate one or more pathways preceding nuclear factor erythroid 2-related factor 2, and these include terpenes, diallyl sulphides, curcuminoids, and cytokines (NRF2). Ellison increases transient receptor potential (TRP) ion channels in the plasma membrane of neurons, which in turn activates neuroprotective kinase cascades via mitogen-activated protein kinases (MAPK), cAMP-response-element-binding protein (CREB), and brain-derived neurotrophic factor (BDNF). Key neurotrophic factor BDNF may initiate the MAPK/ERK and PI-3K/Akt pathways by binding to its tyrosine kinase TrkB receptor, activating the downstream molecules that might promote neurogenesis and cell survival (Cheng et al, 1999).

Allyl-sulfide sulphides may elevate mitochondrial uncoupling protein levels and promote stress-related pathways (Oi et al, 1999). Examples of flavanones that have been demonstrated to prevent oxidant-induced neuronal apoptosis include hesperetin and its metabolite, 5-nitrohesperetin. They do this by activating/phosphorylating signalling proteins crucial to the pro-survival pathways. Specific Ca^{2+} channels are activated by capsaicin (vanilloid receptors). Numerous phytochemicals, including flavonoids, terpenes, and related compounds, have been shown to have an impact on how GABA, the main inhibitory neurotransmitter in the brain, interacts with its ionotropic receptors. Fruits, vegetables, different drinks, and herbal preparations like ginseng and ginkgo biloba contain these GABA modulators (Cheng et al, 1999).

2.6 CONCLUSION

Due to a stressful lifestyle, several neurodegenerative illnesses are primarily emerging in the modern period. Serious negative effects will result from the continued use of synthetic medications in the treatment of certain illnesses. In recent years, scientists have concentrated their efforts on studying phytochemicals as a potential treatment for neurological illnesses. Alkaloids, terpenoids, phenols/polyphenols, flavonoids, and other phytochemicals protect neurons by focusing on the many pathogenic causes of NDs. Similar actions of phytochemicals have been reported in earlier studies, including (1) lowering oxidative stress-induced free radicals via an antioxidant effect, (2) enhancing immunological cells' phagocytic abilities to help clear A/senile plaques in AD and Lewy bodies in PD, (3) raising neurotransmitter concentrations close to neurons by inhibiting neurotransmitter cleaving enzymes, and (4) adjusting to the current stress conditions. Additionally, a crucial quality for a potential neuroprotective drug is its capacity to pass across the BBB and reach the desired CNS locations. Finally, while it is yet unknown if polyphenols or other phytochemicals have receptors or transporters in brain regions, substances with numerous targets seem to be a viable class of therapies for the treatment of neurodisorders.

REFERENCES

Albin, R.L., and J. Greenamyre. 1992. "Alternate excitotoxic hypothesis." *Neurology* 42 (2): 733–738.

Anastacio, J.R., C.A. Netto, and C.C. Castro, et al. 2014. Resveratrol treatment has neuroprotective effects and prevents cognitive impairment after chronic cerebral hypoperfusion. *Neurol Res* 36: 627–633.

Ansari, J., A. Siraj, and N. Inamdar. 2010. "Pharmacotherapeutic approaches of Parkinson's disease." *Int J Pharmacol* 6 (5): 584–590.

Asai, M., N. Iwata, and A. Yoshikawa, et al. 2007. Berberine alters the processing of Alzheimer's amyloid precursor protein to decrease abeta secretion. *Biochem Biophys Res Commun* 352: 498–502.

Brue, A.W., and T.D. Oakland. 2002. "Alternative treatments for attention deficit hyperactivity disorders: Does evidence support their use." *Altern Ther Med* 8 (1): 68–74.

Calne, D.B. 2000. "Parkinson's disease is not one disease." *Parkinsonism Relat Disord* 7 (1): 3–7.

Chan, S., S. Kantham, and V.M. Rao. et al. 2016. Metal chelation, radical scavenging and inhibition of abeta42 fibrillation by food constituents in relation to Alzheimer's disease. *Food Chem* 199: 185–194.

Chatterjee, S.S., S.K. Bhattacharya, M. Wonneman, A. Singer, and W.E. Muller. 1998. "Hyperforin as a possible antidepressant component of Hypericum extracts." *Life Sci* 63: 499–510.

Cheng, J., P. Kuang, W. Wu, and F. Zhang. 1999. "Effects of transient forebrain ischemia and radix Salviae miltiorrhizae (RSM) on extracellular levels of monoamine neurotransmitters and metabolites in the gerbil striatum—an in vivo microdialysis study." *J Traditional Chin Med* 19 (2): 135–140.

Choi, J.Y., C.S. Park, and D.J. Kim, et al. 2002. Prevention of nitric oxide-mediated 1-methyl-4-phenyl-1,2,3,6-tetrahydropyridine-induced Parkinson's disease in mice by tea phenolic epigallocatechin 3-gallate. *Neurotoxicology* 23: 367–374.

Crews, C., P. Hough, and J. Godward. 2005. "Study of the main constituents of some authentic walnut oils." *J Agric Food Chem* 53: 4853–4860.

Cui, H.S., K. Matsumoto, Y. Murakami, H. Hori, Q. Zhao, and R. Obi. 2009. "Berberine exerts neuroprotective actions against in vitro ischemia-induced neuronal cell damage in organotypic hippocampal slice cultures: Involvement of b-cell lymphoma 2 phosphorylation suppression." *Biol Pharm Bull* 32: 79–85.

Doble, A. 1999. "The role of excitotoxicity in neurodegenerative disease: Implications for therapy." *Pharmacol Ther* 81 (3): 163–221.

Drevets, W.C., E. Frank, and J.C. Price, et al. 1999. PET imaging of serotonin 1A receptor binding in depression. *Biol Psychiatry* 46(10): 1375–1387.

Durairajan, S.S., L.F. Liu, and J.H. Lu, et al. 2012. Berberine ameliorates beta-amyloid pathology, gliosis, and cognitive impairment in an Alzheimer's disease transgenic mouse model. *Neurobiol Aging* 33: 2903–2919.

Ehrnhoefer, D.E., M. Duennwald, P. Markovic, J.L. Wacker, and S. Engemann. 2006. "Green tea (–)-epigallocatechin-gallate modulates early events in huntingtin misfolding and reduces toxicity in Huntington's disease models." *Hum Mol Genet* 15: 2743–2751.

El-Bahr, S.M. 2015. "Effect of curcumin on hepatic antioxidant enzymes activities and gene expressions in rats ntoxicated with aflatoxin b1." *Phytother Res* 29: 134–140.

Enenebeaku, U.E., C.E. Duru, and E.N. Okotcha, et al. 2022. Phytochemical analysis and antioxidant evaluation of crude extracts from the roots, stem and leaves of Dictyandraarborescens (Welw.). *Trop J Nat Prod Res* 6 (1): 62–70.

Farooqui, A.A. 2012. Effect of Lifestyle, Aging, and Phytochemicals on the Onset of Neurological Disorders. In Phytochemicals, Signal Transduction, and Neurological Disorders, 1–29. Springer: New York, NY, USA.

Finkel, T. 2011. "Signal transduction by reactive oxygen species," *The Journal of Cell Biology* 194(1): 7–15.

Franco, R., G. Navarro, and E. Martinez-Pinilla. 2019. "Hormetic and mitochondria-related mechanisms of antioxidant action of phytochemicals." *Antioxidants* 8: 373.

Haddad, J.J. 2002. "Antioxidant and prooxidant mechanisms in the regulation of redox(y)-sensitive transcription factors." *Cell Signal* 14 (11): 879–897.

Halliwell, B. 2007. "Oxidative stress and cancer: Have we moved forward?" *Biochem J* 401: 1–11.

Han, J.M., Y.JM. Lee, S.Y. Lee, and E.M Kim. 2007. "Protective effect of sulforaphane against dopaminergic cell death." *J Pharmacol Exp Ther* 321: 249–56.

Heo, H.J., and C.Y. Lee. 2004. "Protective effects of quercetin and vitamin c against oxidative stress-induced neurodegeneration." *J Agric Food Chem* 52: 7514–7517.

Hoppe, J.B., K. Coradini, and R.L. Frozza, et al. 2013. Free and nanoencapsulated curcumin suppress beta-amyloid-induced cognitive impairments in rats: Involvement of BDNF and Akt/GSK-3 beta signaling pathway. *Neurobiol Learn Mem* 106: 134–144.

Hsu, Y.Y., C.S. Chen, S.N. Wu, Y.J. Jong, and Y.C. Lo. 2012. "Berberine activates Nrf2 nuclear translocation and protects against oxidative damage via a phosphatidylinositol 3-kinase/akt-dependent mechanism in NSC34 motor neuron-like cells." *Eur J Pharm Sci* 46: 415–425.

Hsu, Y.Y., Y.T. Tseng, and Y.C. Lo. 2013. "Berberine, a natural antidiabetes drug, attenuates glucose neurotoxicity and promotes Nrf2-related neurite outgrowth." *Toxicol Appl Pharmacol* 272: 787–796.

Huang, S.S., M.C. Tsai, C.L. Chih, L.M. Hung, and S.K Tsai. 2001. "Resveratrol reduction of infarct size in long- evans rats subjected to focal cerebral ischemia." *Life Sci* 69: 1057–1065.

Ji, H.F., and L. Shen. 2011. "Berberine: A potential multipotent natural product to combat Alzheimer's disease." *Molecules* 16: 6732–6740.

Joseph, J.A., R.T. Bartus, and D.E Clody. 2005. "Reversing the deleterious effects of aging on neuronal communication and behavior: Beneficial properties of fruit polyphenolic compounds." *Am J Clin Nutr* 81: 313S–316S.

Kim, M.K.H., M.S. Cho, and Shin, et al. 2014a. Berberine prevents nigrostriatal dopaminergic neuronal loss and suppresses hippocampal apoptosis in mice with Parkinson's disease. *Int J Mol Med* 33: 870–878.

Kim, J., H.J. Lee, and K.W Lee. 2010. "Naturally occurring phytochemicals for the prevention of Alzheimer's disease." *J Neurochem* 112: 1415–1430.

Kim, M., M. S. Shin, and J.M. Lee, et al. 2014b. Inhibitory effects of isoquinoline alkaloid berberine on ischemia-induced apoptosis via activation of phosphoinositide 3-kinase/protein kinase b signaling pathway. *Int Neurourol J* 18: 115–125.

Koh, S.H., S.H. Kim, and H. Kwon, et al. 2003. Epigallocatechin gallate protects nerve growth factor differentiated PC12 cells from oxidative-radical-stress-induced apoptosis through its effect on phosphoinositide 3-kinase/Akt and glycogen synthase kinase-3. *Brain Res Mol Brain Res* 118: 72–81.

Kumar, V. 2006. "Potential medicinal plants for CNS disorders: An overview." *Phytother Res* 20 (12): 1023–1035.

Kumar, S., R. Madaan, G. Bansal, A. Jamwal, and A. Sharma. 2012. "Plants and plant products with potential anticonvulsant activity–A review." *Pharma Comm* 2 (1): 3–84.

Kwon, I.H., H.S. Choi, and K.S. Shin, et al. 2010. Effects of berberine on 6-hydroxydopamine-induced neurotoxicity in pc12 cells and a rat model of Parkinson's disease. *Neurosci. Lett* 486: 29–33.

Lesjak, M., I. Beara, and N. Simin, N. et al. 2018. Antioxidant and anti-inflammatory activities of quercetin and its derivatives. *J. Funct Foods* 40: 68–75.

Liu, R.H. 2003. "Health benefits of fruit and vegetables are from additive and synergistic combinations of phytochemicals." *Am J Clin Nutr* 78: 517S–520S.

Liu, M., F. Chen, and L. Sha, et al. 2014a. (-)-Epigallocatechin-3-gallate ameliorates learning and memory deficits by adjusting the balance of TrkA/p75NTR signaling in APP/PS1 transgenic mice. *Mol Neurobiol* 49: 1350–1363.

Liu, D., Z. Wang, and Z Gao, et al., 2014b. Effects of curcumin on learning and memory deficits, BDNF, and ERK protein expression in rats exposed to chronic unpredictable stress. *Behav Brain Res* 271: 116–121.

Maleki, S.N., N. Aboutaleb, and F. Souri. 2018. "Berberine confers neuroprotection in coping with focal cerebral ischemia by targeting inflammatory cytokines." *J. Chem. Neuroanat* 87: 54–59.

Mandel, S.A., Y. Avramovich-Tirosh, L. Reznichenko, H. Zheng, O. Weinreb, and T. Amit. 2005. "Multifunctional activities of green tea catechins in neuroprotection. Modulation of cell survival genes, iron dependent oxidative stress and PKC signaling pathway." *Neurosignals* 14: 46–60.

Marambaud, P., H. Zhao, and P. Davies. 2005. "Resveratrol promotes clearance of Alzheimer's disease amyloid- beta peptides." *J Biol Chem* 280: 37377–37382.

Martel, J., D.M. Ojcius, and Y.F. Ko, et al. 2019. Hormetic effects of phytochemicals on health and longevity. *Trends Endocrinol Metab* 30: 335–346.

Martin, D., A.I. Rojo, and M. Salinas, et al. 2004. Regulation of heme oxygenase-1 expression through the phosphatidylinositol 3-kinase/Akt pathway and the Nrf2 transcription factor in response to the antioxidant phytochemical carnosol. *J Biol Chem* 279: 8919–8929.

McCann, J.C., and B.N. Ames. 2005. "Is docosahexaenoic acid, an n23 long chain polyunsaturated fatty acid, required for development of normal brain function? An overview of evidence from cognitive and behavioral tests in humans and animals." *Am J Clin Nutr* 82: 281–295.

Nalini, K., A. Aroor, K. Karanth, and A. Rao. 1992. "Centella asiatica fresh leaf aqueous extract on learning and memory and biogenic amine turnover in albino rats." *Fitoterapia* 63: 232–237.

Nam, S.M., J.H. Choi, and D.Y. Yoo, et al., 2014. Effects of curcumin (curcuma longa) on learning and spatial memory as well as cell proliferation and neuroblast differentiation in adult and aged mice by upregulating brain-derived neurotrophic factor and creb signaling. *J Med Food* 17: 641–649.

Ogidi, O.I. 2022. Phytochemicals of Bioactive compounds of Brassica Juncea (Brown Mustard) Seeds. In: Medicinal and Aromatic Plants of the World. Published by *Encylopedia of Life Support Systems* 1–39.

Ogidi, O.I., M.N. Ayebabogha, P.U. Eze, O. Okiemute, and C.E. Okafor. 2021c. "Determination of phytoconstituents and antimicrobial activities of aqueous and methanol extracts of neem (*Azadirachta indica*) leaves." *International Journal of Pharmacognosy and Chemistry* 2 (2): 60–67.

Ogidi, O.I., P. Chukwudi, A.I. Ibe, P.U. Eze, and T.N. Canus. 2021d. "Preliminary phytochemical profile and antimicrobial potentials of white-green African garden egg (*solanum macrocarpon*) fruits obtained from Yenagoa." *ASIO Journal of Pharmaceutical & Herbal Medicines Research* 7 (2): 01–05.

Ogidi, O.I., N.G. Esie, and O.G. Dike. 2019b. "Phytochemical, proximate and mineral compositions of *Bryophyllum pinnatum* (never die) medicinal plant." *J Pharmacogn Phytochem* 8 (1): 629–635.

Ogidi, D.G., N.G. George, and Esie. 2019a. "Ethnopharmacological properties of *Vernonia amygdalina* (bitter leave) medicinal plant." *J Med Plants Stud* 7 (2): 175–181.
Ogidi, O.I., and J.C. Julius. 2021. "Assessment of ethno-pharmacological compounds and antimicrobial efficacy of white onion (*Allium cepa*) bulb extracts against pathogenic microbial isolates." *PhytoChem & BioSub Journal* 15 (2): 167–174.
Ogidi, O.I., L.M.O. Oguoma, P.C. Adigwe, and B.B. Anthony. 2021b. "Phytochemical properties and in-vitro antimicrobial potency of wild edible mushrooms (*Pleurotus ostreatus*) obtained from Yenagoa, Nigeria." *J Phytopharma* 10 (3): 180–184.
Ogidi, O.I., C.C. Okore, U.M. Akpan, M.N. Ayebabogha, and C.J. Onukwufo. 2021a. "Evaluation of antimicrobial activity and bioactive phytochemical properties of Mango (*Mangifera indica*) stem-bark extracts." *Int J Pharmacogn* 8 (5): 189–195.
Ogidi, O.I., O. Omu, and P.A. Ezeagba. 2019c. "Ethno pharmacologically active components of *Brassica juncea* (brown mustard) seeds." *Int J Pharma Res Dev* 1 (1): 09–13.
Ogidi, O.I., P.S. Tobia, and D.N. Ijere, et al. 2022. Investigation of bioactive compounds and antimicrobial sensitivity of pawpaw (*carica papaya*) leave extracts against morbific microorganisms. *J. Appl Pharm. Res* 10 (1): 21–28.
Oi, Y.M., C. Imafuku, Y. Shishido, S. Kominato, and K. Nishimura. 1999. "Allylcontaining sulfides in garlic increase uncoupling protein content in brown adipose tissue, and noradrenaline and adrenaline secretion in rats." *J Nutr* 129 (2): 336–342.
Ortiz, J.G., J. Nieves-Natal, and P Chaves. 1999. "Effects of *Valeriana officinalis* extracts on [3H] flunitrazepam binding, synaptosomal [3H] GABA uptake, and hippocampal [3H] GABA release." *Neurochem Res* 24: 1373–1378.
Parker, J.A., M. Arango, and S. Abderrahmane, et al. 2005. Resveratrol rescues mutant polyglutamine cytotoxicity in nematode and mammalian neurons. *Nat Genet* 37: 349–350.
Pathak, N.L., B.K. Sanjay, N.M. Bhatt, and R.G. Patel. 2011. "Experimental modeling of anxiety." *Journal of Applied Pharmaceutical Science* 01 (03): 06–10.
Pearson, V.E. 2001. "Galantamine: A new alzheimer drug with a past life." *Ann Pharmacother* 35: 1406–1413.
Pervin, M., K. Unno, T. Ohishi, H. Tanabe, N. Miyoshi, and Y. Nakamura. 2018. "Beneficial effects of green tea catechins on neurodegenerative diseases." *Molecules* 23: 1297.
Pires, E.N.S., R.L. Frozza, J.B. Hoppe, B. de Melo Menezes, and C.G. Salbego. 2014. "Berberine was neuroprotective against an in vitro model of brain ischemia: Survival and apoptosis pathways involved." *Brain Res* 1557: 26–33.
Roy, A., and S. Saraf. 2006. "Limonoids: Overview of significant bioactive triterpenes distributed in plants kingdom." *Biol. Pharm. Bull* 29: 191–201.
Scheuer, M.L., and T.A. Pedley. 1990. "The evaluation and treatment of seizures." *New Engl J Med* 323 (21): 1468–1474.
Schulz, V., R. Ha¨nsel, and V Tyler. 1998. Rational Phytotherapy: A Physician's Guide to Herbal Medicine. Berlin: Springer-Verlag 1–155.
Shin, K.S., H.S. Choi, and T.T. Zhao, et al. 2013. Neurotoxic effects of berberine on long-term l-DOPA administration in 6-hydroxydopamine-lesioned rat model of Parkinson's disease. *Arch. Pharm Res* 36: 759–767.
Si, H., and D Liu. 2007. "Phytochemical genistein in the regulation of vascular function: New insights." *Curr Med Chem* 14: 2581–2589.
Smith, A., B. Giunta, P.C. Bickford, M. Fountain, J. Tan, and R. D. Shytle. 2010. "Nanolipidic particles improve the bioavailability and alpha-secretase inducing ability of epigallocatechin-3-gallate (EGCG) for the treatment of Alzheimer's disease." *Int J Pharm* 389: 207–212.
Sofi, F., F. Cesari, R. Abbate, G.F. Gensini, and A. Casini. 2008. "Adherence to mediterranean diet and health status: Meta-analysis." *BMJ* 337: a1344.
Soleimani, H., A. Amini, and S. Taheri, et al. 2018. The effect of combined photobiomodulation and curcumin on skin wound healing in type i diabetes in rats. *J Photochem Photobiol B* 181: 23–30.

Somani, S.J., K.P. Modi, A.S. Majumdar, and B.N. Sadarani. 2015. "Phytochemicals and their potential usefulness in inflammatory bowel disease." *Phytother Res* 29: 339–350.
Sutherland, B.A., and R.M. Rahman. 2006. "Appleton I. Mechanisms of action of green tea catechins, with a focus on ischemia-induced neurodegeneration." *J Nutr Biochem* 17(5): 291–306. doi: 10.1016/j.jnutbio.2005.10.005.
Tsamo, A., M.K. Langat, P. Nkounga, A.F. K. Waffo, A.E. Nkengfack, and D.A. Mulholland. 2013. "Limonoids from the West African trichilia welwitschii (Meliaceae)." *Biochem Syst Ecol* 50: 368–370.
Walker, J.M., D. Klakotskaia, and D. Ajit, et al. 2015. Beneficial effects of dietary EGCG and voluntary exercise on behavior in an Alzheimer's disease mouse model. *J Alzheimers Dis* 44: 561–572.
Wang, J.Q., T.T. Xiong, and J. Zhou, et al. 2018. Enzymatic formation of curcumin in vitro and in vivo. *Nano Res* 11:3453–3461.
Winner, B., Z. Kohl, and F. H. Gage. 2011. "Neurodegenerative disease and adult neurogenesis." *Eur J Neurosci* 33 (6): 1139–1151.
Wobst, H.J., A. Sharma, M.I. Diamond, E.E. Wanker, and J. Bieschke. 2015. "The green tea polyphenol (-)-epigallocatechin gallate prevents the aggregation of tau protein into toxic oligomers at substoichiometric ratios." *FEBS Lett* 589: 77–83.
Xu, Y., B. Ku, and L. Cui, et al. 2007. Curcumin reverses impaired hippocampal neurogenesis and increases serotonin receptor 1A mRNA and brain-derived neurotrophic factor expression in chronically stressed rats. *Brain Res* 1162: 9–18.
Yang, J.Q., S.L. Song, J. Li, and T. Liang. 2014. "Neuroprotective effect of curcumin on hippocampal injury in 6-ohda-induced Parkinson's disease rat." *Pathol Res Pract* 210: 357–362.
Yehuda, S., S. Rabinovitz, and R.L. Carasso. 2002. "The role of polyunsaturated fatty acids in restoring the aging neuronal membrane." *Neurobiol Aging* 23: 843–853.
Youdim, K.A., M.Z. Qaiser, and D.J. Begley. 2004. "Flavonoid permeability across an in situ model of the blood-brain barrier." *Free Radic Biol Med* 36: 592–604.
Youdim, K.A., J.P. Spencer, H. Schroeter, and C. Rice-Evans. 2002. "Dietary flavonoids as potential neuroprotectants." *Biol Chem* 383: 503–519.
Yu, J.C.H., Z.D. Min, and N.Y. Ip. 2004. "Melia toosendan regulates PC12 cell differentiation via the activation of protein kinase A and extracellular signal-regulated kinases." *Neurosignals* 13: 248–257.
Zhang, C., A. Browne, D. Child, and R.E. Tanzi. 2010. "Curcumin decreases amyloid-beta peptide levels by attenuating the maturation of amyloid-beta precursor protein." *J Biol Chem* 285: 28472–28480.
Zhang, Q., J.K. Li, R. Ge, J.Y. Liang, Q.S. Li, and Z.D. Min. 2013. "Novel NGF-potentiating limonoids from the fruits of melia toosendan." *Fitoterapia* 90: 192–198.
Zhang, F., Y.F. Lu, Q. Wu, J. Liu, and J.S. Shi. 2012. "Resveratrol promotes neurotrophic factor release from astroglia." *Exp Biol Med (Maywood)* 237: 943–948.
Zhao, J., N. Kobori, J. Aronowski, and P.K. Dash. 2006. "Sulforaphane reduces infarct volume following focal cerebral ischemia in rodents." *Neurosci Lett* 393: 108–112.

3 Novel Extraction and Characterization Methods for Phytochemicals Identified as Neuroprotective

Umme Habiba Saeeda, Aneela Nawaz, Alina Majid, Muqaddas Shahzadi, Malik Badshah, and Samiullah Khan

3.1 INTRODUCTION

Change in dietary habits and sedentary lifestyle that has emerged along with the dawn of modern civilization has made humans weak and vulnerable to a plethora of diseases. Recently, many scientific studies have highlighted the extraordinary potential of a high phytochemical-containing diet to prevent numerous diseases such as cancer, cardiovascular diseases, hypertension, stroke, neurodegenerative disorders, etc. This has inspired a surge of enthusiasm in researchers to work on the development of natural health products, nutraceuticals, functional foods, and the exploration of novel biomolecules for drug designing (Sharma and Kaushik, 2021). The venture begins with the collection and accurate identification of plants followed by drying, grinding, extraction, fractionation, purification, identification, and quantification of biomolecules. Plants are a desirable medicinal source due to their tremendous therapeutic potential, natural origin, relatively inexpensive, and ease of access. Herbal supplements can be an overly promising alternative in instances of drug intolerance and a high likelihood of detrimental consequences. Fundamentally, phytochemicals are byproducts of the plant's secondary metabolism and are devoid of essential nutrients. Their synthesis is aimed to preserve plants from insect and disease infestations along with extreme environmental challenges including ultraviolet (UV) exposure, drought, salinity, and other abiotic stresses. The production of both primary and secondary metabolites varies significantly since the former are byproducts of catabolic and anabolic circuits while the latter are produced by specialized cells (Yasmeen et al, 2018).

 DOI: 10.1201/9781003389781-3

Extraction from plants involves the isolation of secondary metabolites such as glycosides, alkaloids, flavonoids, saponins, terpenes, and steroids. This can be actualized via a wide array of standard extraction procedures available such as Soxhlet extraction, maceration, percolation, infusion, decoction, digestion, ultrasound-assisted extraction, and microwave-assisted extraction. Further separation and purification can be done through paper chromatography (PC), thin-layer chromatography (TLC), gas chromatography (GC), and high-performance liquid chromatography (HPLC). The selection of a suitable extraction procedure is governed by the nature of plant material, type, pH of solvent, availability of instruments, and expected use of the product (Kumar and Khanum, 2012).

Neurodegenerative diseases (NDs) have become a momentous health concern in the 20th century as they are lethal enough to cause death with no effective treatment available. NDs are most rampant in the old age population. Alzheimer's disease (AD), Parkinson's disease (PD), and dementia are some of the most widespread NDs. The progression of NDs is influenced by several environmental as well as genetic factors (Sarker and Oba, 2019). Oxidative stress, neuroinflammation, and accumulation of conglomerated proteins are some of the common pathological factors of all NDs. Neurotrophins are mandatory for the survival and maintenance of neurons in CNS and peripheral nervous system. Several researchers have figured out that a sharp decline *neurotrophin* concentration is associated with NDs. Therefore, neurotrophins have become a crucial target for phytochemicals. Henceforth, the administration of neurotrophins is a promising strategy for a cure against NDs. However, delivery across the blood-brain barrier is the major challenge in the way of CNS disorders treatment (Ma et al, 2018).

The probability of effective treatment is greatly enhanced by early diagnosis. Numerous medicines such as axona, rotigotine, levopoda, memantine, etc., are being used for the treatment of NDs. However, the current treatment strategies are incapable of curing the cause and progression of the disease; instead, they only confer symptomatic relief. Albeit drugs to manage symptoms of NDs are present, there are no preventive medicines available. This situation calls for a dire need to explore novel neuroprotectant phytochemicals and therapeutic strategies to tackle these lethal diseases. This chapter summarizes the conventional and latest techniques that have been developed for the extraction, purification, and characterization of phytochemicals (Ma et al, 2018).

3.2 PRELIMINARY PREPARATION OF PLANT MATERIAL

Pre-preparation of plant material which comprises drying, and grinding is crucial as it ensures enhanced preservation of secondary metabolites in the final extract.

3.3 EXTRACTION

Extraction involves the segregation of therapeutic components based on their solubility in various solvents and can be actualized from any part of plants such as leaves, flowers, bark, stem, roots, and fruits. Activity analysis of each extract is performed and the one with the most promising activity is subjected to further purification. The quality of extract is influenced by several parameters such as part of plant and

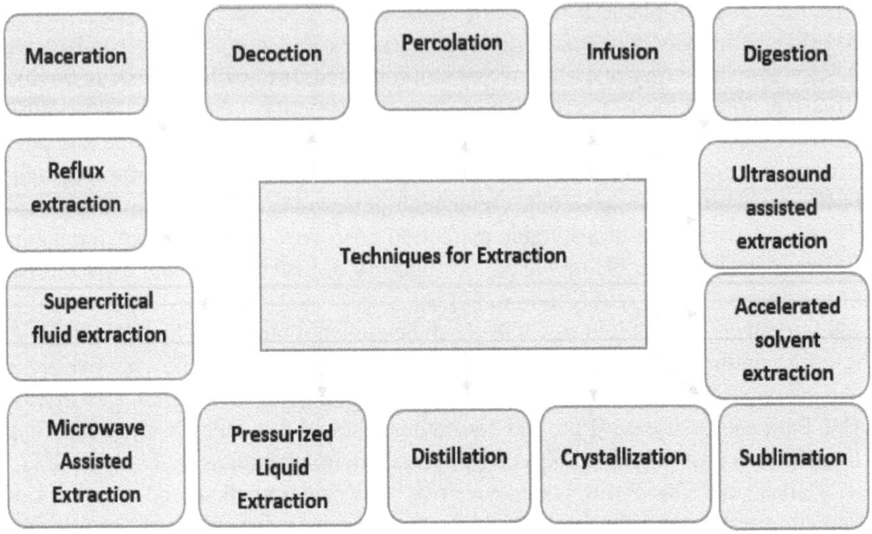

FIGURE 3.1 List of conventional and advanced methods of extraction.

type of solvent used. The interaction of solvent with plant metabolite is important for efficient interaction. This can be actualized by reducing the particle size, thus providing a large surface area for the contact. Therefore, powdered form is preferable over grinded form. Appropriate selection of solvent employed is important and must be evaluated for properties such as low boiling point, minimum toxicity, preservative action, and high solubility (Palai and Shekhawat, 2022). The major factors that govern the choice of solvent are the nature of metabolites, the nature of the solvent, potential health risks from extracts, etc. (Figure 3.1).

3.3.1 Maceration

In this method, coarsely powdered plant material, such as leaves, stems, or roots, is soaked in a suitable solvent in a stoppered container and stored at room temperature for three days with frequent stirring. Over time phytochemicals are dissolved by the solvent which can be isolated by filtration of solution or pressing. The afterward dried extract is obtained by subjecting solvent to evaporation in a water bath or oven. Mechanical procedures are not appropriate for the production of metabolites because they invariably disrupt numerous cells, and the endogenous enzymes that are released as a result harm the flavor, color, and ascorbic acid. Heat must be applied with the associated risk of damage to restrict the activity of these enzymes. This method is a preferred choice for thermo-labile substances (Altemimi et al, 2017).

3.3.2 Infusion

Infusions are prepared using the same protocol as that of maceration; however, the time taken by the process is desirable and much less. Finely grounded plant material is placed in a glass container followed by pouring hot or cold extraction solvent and is kept in this state for a few minutes. Infusions are often susceptible to bacterial and fungal colonization therefore must be administered within 12 hours of the formulation. This method is used for the extraction of high-solubility secondary metabolites. Solvent-to-sample ratio is often maintained as 16:1 or 4:1.

3.3.3 Percolation

In this method plant material moisturized with chosen extraction solvent is poured into a firmly covered container which is then allowed to stand for at least 4 hours after which it is topped off from the percolator. A thin layer is allowed to form on the mass by adding supplementary solvent, and the mixture is kept so for further 24 hours in the closed percolator. Subsequently, the outlet of the percolator is opened, permitting the gradual outward flow of filed liquid. As per requirement, the additional solvent is augmented further to the point that percolates becomes almost three-quarters volume of the total finished product. The mixture is then added to the percolator. Clarification of mixed liquid is carried out via filtration. This method is frequently used for the separation of phytochemicals in the production of fluid extracts and medicines.

3.3.4 Digestion

This process involves the utilization of heat. The powdered plant material is mixed with extraction solvent in a container. The mixture is kept in an oven or water bath at 50°C. Heat is then applied throughout the process to enhance the solubility of secondary metabolites and reduce the viscosity of the extraction solvent. This method is suitable for readily soluble plant material.

3.3.5 Decoction

The process involves the usage of a specific volume of water and continuous heat. Well-dried, grinded, and powdered material is transferred into a glass container. Subsequently, water is poured and stirred. A constant supply of heat is ensured throughout the process to accelerate extraction. The whole process takes approximately 15 minutes. The concentration of solvent and the crude drug is maintained in the ratio of 4:1 or 16:1. Preferentially, it is used for the acquisition of heat-stable and water-soluble plant material (Sheikh et al, 2017).

3.3.6 Cold Extraction

In this method, dried plant powder is measured using a weighing balance, added into a glass flask alongside selected solvent, and stored at room temperature for seven days with constant shaking every 24 hours. Subsequently, filtration of the extract is carried out under vacuum via Whatman filter paper. The obtained extract is then dried completely using a rotary evaporator. The procured dried extract is then stored in firmly capped glass vials. The weight of the final extract is observed by determining the difference between the weight of the extract before and after drying. The same procedure is applied thrice for maximum extraction (Kovačević et al, 2018).

3.3.7 Soxhlet Extraction

This technique, also named hot continuous extraction, involves the plant samples enclosed in a porous bag, also called a "thimble," composed of strong filter papers. It is then placed in the thimble chamber of the Soxhlet apparatus. The extraction solvent placed in the bottom flask is subjected to heating resulting in the emission of solvent vapors in the thimble chamber. Here they interact with plant samples and trickle down backward after condensation in the condensation chamber. The quality of this method is influenced by the speed of agitation, Temperature, and proportion of solvent sample employed. The main advantage of using this process is that it requires low concentrations of solvents. However, the emission of harmful gases and their subsequent exposure imposes serious health threats (Palai and Shekhawat, 2022).

3.3.8 Microwave-assisted Extraction

The interaction between polar molecules (of sample and solvent), microwave radiation, and the resultant transfer of heat via conduction form the basis of this method. Electromagnetic induction causes the rotation of polar molecules which results in the rupturing of the Hydrogen bond. Consequently, dissolved ions and penetration of solvents in samples are enhanced. However, the transfer of energy in non-polar solvents occurs via dielectric absorption which results in poor heating (Guntur et al, 2018; Palai and Shekhawat, 2022).

The microwave-assisted extraction process is extremely advantageous in the way that it utilizes less volume of solvent and less time. However, it is quite limited in its application as it gives the best results with small-sized molecules such as isoflavin, gallic acid, and quercetin because of their stability against thermal degradation. Other large-sized molecules such as anthocyanin and tannins are not appropriate for it because of their susceptibility to degradation.

3.3.9 Ultrasound-assisted Extraction

Also known as sonication extraction employs ultrasound waves ranging from 20 to 2000 kHz. Interaction between sample, solvent, and cell wall permeability is significantly enhanced by Acoustic activation via intense ultrasound waves. The cell wall of plant cells is disrupted as it is exposed to ultrasound waves, causing the release of bioactive metabolites from the sample and subsequent entry of solvent into the plant cell. This method is quite advantageous because of its applicability at both large and small scale, cost-effective technology, less extraction time, and minimum solvent requirement. However, ultrasound waves higher than 20 kHz lead to free radical formation, thus causing the deterioration of active phytochemical constituents (Palai and Shekhawat, 2022; Tsao and Deng, 2004).

3.3.10 Accelerated Solvent Extraction

In this method, the plant sample is packed in an inert material like sand in the extraction cell which prevents aggregation of the sample and blockage of system tubes. The diffusion of organic solvents into the sample is accelerated by providing high temperature and pressure. The extraction cell is filled swiftly under high pressure thus forcing liquid into the solid matrix. This method is preferable for phytochemical extraction because it uses a minimal amount of solvent, the temperature and pressure of each layer of the sample can be controlled, and extraction can be done within a period of less than an hour (Sarker and Oba, 2019).

3.3.11 Supercritical Fluid Extraction

Supercritical fluids are substances that manifest properties of liquid as well as solid at their critical point. Conditions of temperature and pressure at which liquid and vapor of a substance coexist are known as the critical point. In general, a supercritical fluid acts as a gas but also has the attributes of a liquid. A cylinder containing plant material is filled with supercritical fluid with the aid of a pump and the liquid-containing extract is gathered in a separation chamber. Heat-sensitive compounds are often extracted and preserved using supercritical fluids. A major advantage of this method lies in the adjustability of solvent density. This implies that the properties of supercritical fluid could be altered by a change in temperature and pressure. They can also be tailored to facilitate selective extraction by changing the solvents such as ethanol or methanol. The demand for high-pressure and expensive equipment and high energy costs render this method ineffective for long-term usage (Yasmeen et al, 2018).

3.3.12 Deep Eutectic Solvents

Two or more chemicals are often bonded together via hydrogen bonds to give rise to a solvent with a melting point lower than that of its constituent parts. These solvents known as ionic liquid equivalents are custom-made with such properties that enable them to tackle limitations imposed by conventional ionic solvents such as toxicity, high charge density, and higher melting points. The majority of these solvents commonly contain betaine derivatives or choline chloride which can associate with a versatile group of hydrogen donors such as carbohydrates, alcohols, sugars, and organic acids. These solvents have gained a lot of attention in recent years because of the myriad of benefits they offer such as biodegradability, convenient solvent preparation, cheap starting material, minimal toxicity, and adaptability. They are deemed as a greener substitute for toxic organic solvents utilized in conventional extraction methods and can also be tailored for the extraction of metabolites from plants as well as in food and pharmaceutical industries (Jakovljević et al, 2020; Ma et al, 2018).

3.3.13 Pressurized Liquid Extraction (PLE)

PLE is used for the extraction of phytochemicals using common solvents at controlled temperatures and pressure. A major advantage of PLE over conventional methods is that solvents under high pressure are capable of staying in liquid form even when the temperature exceeds their boiling point thus allowing operation at high temperatures. It has a quick extraction span and is often employed for the extraction of substances such as fatty acids and lipids from vegetable oil. It is also used for the extraction of small components which are prone to destruction at high temperatures such as phenols, sterols, anthocyanins, carotenoids, phospholipids, glycerides, and tocopherols. PLE is an advanced level technology and is beneficial in many aspects as it is quick, uses a low quantity of solvents, and provides higher yield. The rate of solvent diffusion and analyte solubility is enhanced alongside a decrease in surface tension and solvent viscosity. The automaticity of PLE makes quality control convenient. Furthermore, extraction parameters could be altered and tailored according to the target component (Kovačević et al, 2018).

3.3.14 Pulsed Electric Field (PEF) Extraction

PEF is deemed a very promising method for the disruption of wet biomass as it does not require long and energy-intensive drying with concomitant loss of thermolabile chemicals. The principal goal of this technique is to disrupt cells thus enhancing membrane permeability and facilitating the higher mass transfer of liquid. It shortens operation time, enhances extraction yield, and results in a purer extract. PEF significantly enhances the extraction of water-soluble chemicals. The extraction efficacy is enhanced by energy input, electric field intensity, and pulse polarity. The rate of diffusion of phytochemicals is significantly enhanced at room temperature by electroporation of the plasma membrane.

A decrease in heat-sensitive chemical ruin and an upsurge in extraction yield as well as a reduction in energy expenses have an environmentally friendly impact. It is often used for the extraction of proteins, vitamins, carotenoids, proteins, inulin, oats, sucrose, etc. This method is preferred over conventional methods because of its environmental friendliness and also because it enables the selective release of metabolites without debris production by allowing penetration of solvent within the cell (Kovačević et al, 2018; Poojar et al, 2017).

3.4 FRACTIONATION AND PURIFICATION TECHNIQUES

Techniques and technologies for the isolation and purification of bioactive molecules have drastically improved and evolved. The aim of a researcher in quest of a novel biomolecule is to find a simple, precise, and fast method of extraction to screen out plant material for activities such as antibacterial, antiviral, antifungal, antidiabetic, anticancerous, etc. In vitro methods are preferred over in vivo assays because experiments with animals are time taking, expensive, and vulnerable to ethical controversies. The process of separation of a mixture of phytochemicals into numerous fractions is known as fractionation. Acquired fractions are periodically segregated further into several proportions till the isolation of pure compound. The addition of solvents must be done in the order of ascending polarities. Techniques for fractionation are typically categorized into physical and chemical methods (Beulah et al, 2022; Yasmeen et al, 2018) (Figure 3.2).

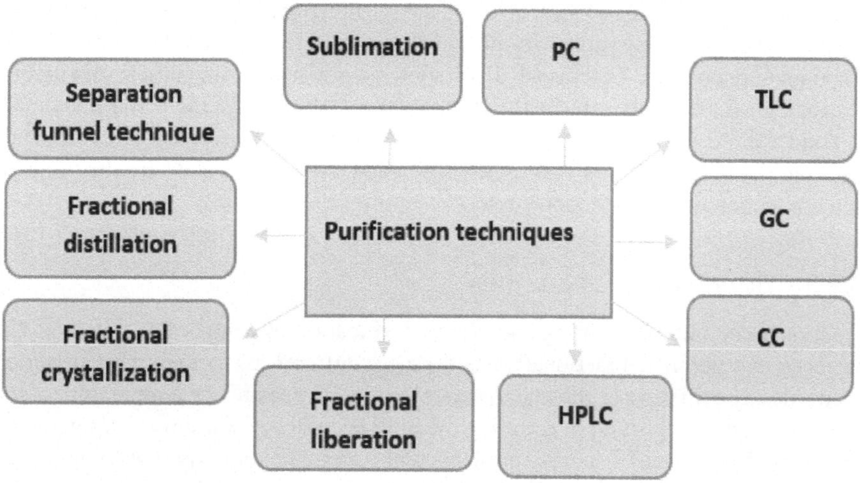

FIGURE 3.2 List of techniques used for the purification of phytochemicals.

Chemical Method:

It is based upon the kind of functional group of compounds in the mixture. Purification of compounds can be achieved by chemical reactions using suitable reagents.

Physical Methods:

Physical methods used for the segregation of compounds are chromatographic techniques, sublimation, separation funnel method, fractional distillation, fractional crystallization, and fractional liberation.

3.4.1 Separation Funnel Technique

This process begins with the selection of four different solvents such as chloroform, acetone, n-hexane, and n-butanol. Complete dissolution of the crude extract is done with 250 ml water. Subsequently, it is transferred to a separating funnel, stirred, and left to settle. Afterward, 250 ml of the least polar solvent (n-hexane) is added and stirred. The material is allowed to settle and the aqueous layer is removed by opening the bottom of the separating funnel. The rest of the material is poured into another container to acquire the n-hexane fraction. The same process is repeated periodically with n-hexane until no substantial quantity of extract appears to dissolve in it. A similar process is repeated with remaining solvents to get acetone, n-butanol, and chloroform fractions. The portion left after fractionation is termed a residual aqueous fraction (RAF) as the water was used to dissolve crude extract for the very first time (Beulah et al, 2022).

3.4.2 Fractional Distillation

Fractional distillation is mostly used for the isolation of hydrocarbons like eucalyptol, crude oil, and citral. The difference in the boiling point of components of a mixture is manipulated to carry out their purification. This involves heating the mixture of liquid. The vapors formed move upward in a glass tube known as the fractionation column. Fractional distillation is preferred over simple distillation because glass beads present in the fractionation column provide a site for vapors to condense, re-evaporate and re-condense, allowing the distillation of compounds multiple times. The vapors will eventually reach the condenser where they liquefy and are gathered in a collection vessel (Velmurugan et al, 2018).

3.4.3 Fractional Crystallization

Fractional crystallization brings about the purification of compounds based on differences in solubility. Fractionation is done via differences in their crystallization properties. If a mixture is allowed to crystallize by decreasing the temperature of the solution, the major portion of resulting precipitates will contain substances with low solubility. The value of solubility products will govern the proportion of components. For compounds with very similar solubility products, a cascade operation will be required to achieve complete separation. A large number of compounds exist in the form of crystals in plants. Their separation is actualized by the formation of crystals that occur during the process of concentration of a given extract via refrigeration or heating (Beulah et al, 2022).

3.4.4 Fractional Liberation

This method is used for the isolation of compounds that can form precipitates in the mixture. Precipitate formation usually occurs upon a change of compound in salt. For instance, when a mixture of alkaloid salts in an aqueous solution is allowed to react with an alkali, it will progressively release bases in ascending order of alkalinity. The same principle is applied to the separation of organic acids that could dissolve in water-immiscible solvents. Separation of acids from the mixture is done via the addition of mineral acids. This technique is usually employed for the separation of cinnamon alkaloids (Gil-Martín et al, 2022).

3.4.5 Sublimation

The change of a substance directly from a solid to a gaseous state without passing through the central liquid phase is known as sublimation. Material such as volatile oils and camphor can convert directly into gas upon heating. A deposition is the reverse process of sublimation that involves a transition directly from the gas phase to the solid phase (Kovačević et al, 2018).

3.4.6 Chromatographic Techniques

Chromatography involves the isolation of compounds from a mixture based on their shape, size, and charge. It involves the use of two phases. One employs the use of solvent of extraction also known as the mobile phase and the other one is the stationary phase that contains substances like Sephadex blended with calcium sulfate as a binder and silica gel. Sephadex is used for the purification of proteins and amino acids. Silica gel is utilized for the purification of sugars, fatty acids, lipids, alkaloids, amino acids, and proteins. Aluminum is applied for the separation of phenols, steroids, alkaloids, lipids, and fatty acids. Celite for the separation of steroids and organic cations and cellulose powder for phenols, alkaloids, steroids, amino acids, and food dyes. Numerous mechanisms are utilized for the separation of compounds using chromatographic techniques such as ion exchange, adsorption, partition, affinity, and size exclusion. Given below is a detailed description of each of these mechanisms (Yasmeen et al, 2018).

3.4.6.1 Adsorption Chromatography

In this technique interaction between the compounds targeted for separation is manipulated. The stationary phase pulls compounds via non-covalent Van der Waal forces and hydrophobic interactions. The weakly bonded compounds will be eluted by the mobile phase at first and so on (Lavenburg et al, 2021).

3.4.6.2 Partition Chromatography

Partition chromatography involves the separation of compounds by adding two or more immiscible solvents into the mixture. Each of the constituents of the mixture will separate by dissolving in the solvent of relatable polarity. Afterward, individual compounds are obtained by separating the immiscible solvent via a separating funnel (Guntur et al, 2018).

3.4.6.3 Affinity Chromatography (AC)

Separation of biomolecules from a mixture using AC is done based on macromolecular binding between compounds. The stationary phase comprises a ligand and is placed in a separating column. Compounds with no affinity toward the stationary phase will be washed down by the mobile phase. Likewise, compounds with high affinity for the stationary phase will be attracted toward it and thus separated (Alara et al, 2018).

3.4.6.4 Ion Exchange Chromatography

Ion exchange chromatography is preferred for the separation of polar compounds based on the kind of charge they possess. As the principle goes; "Like attracts like." Similar to other types of column-based liquid chromatography techniques, ion-exchange chromatography, which is specifically intended for the separation of differently charged or ionizable chemicals, consists of mobile and stationary phases. The combination to be resolved is placed into an aqueous buffer system in the mobile phase. The stationary phase is typically a chemically derivate of an inert organic matrix with ionizable functional groups (fixed ions) that can displace an opposing charged ion. Counter ions are ions that are in equilibrium between the stationary and mobile phases, giving birth to the two alternative formats of anion and cation exchange. Protons (H^+), hydroxide groups (OH^-), single-charged monoatomic ions (Na^+, K^+, Cl^-), double-charged monoatomic ions (Ca^{2+}, Mg^{2+}), polyatomic inorganic ions ($SO42^-$, $PO43^-$), as well as organic bases (NR2H+) and acids, are examples of exchangeable matrix counter ions (COO^-). On a column made of cation-exchange resin, cations are separated from the anion. According to variations in their net surface charge, separation is based on the binding of analytes to positively or negatively charged groups that are immobilized on a stationary phase and that are in equilibrium with free counter ions in the mobile phase.

3.4.6.5 Size Exclusion Cchromatography

The chromatographic technique known as size-exclusion chromatography (SEC), also referred to as molecular sieve chromatography (MSC), separates molecules in solution based on their size and, in some situations, molecular weight. Proteins and commercial polymers are typical examples of big molecules or macromolecular complexes to which it is typically applied. The method is typically referred to as gel-filtration chromatography when an aqueous solution is used to move the sample through the column, as opposed to the name gel permeation chromatography, which is used when an organic solvent is employed as a mobile phase. Fine, porous beads that are frequently made of dextran, agarose, or polyacrylamide polymers are crammed into the chromatography column. The separation of compounds on the base of their size is actualized by using a mesh of different diameters. Separation is done in ascending order, that is, smaller size molecules are separated at first followed by the separation of medium and larger ones (Altemimi et al, 2017). The advantages of this procedure include maintaining the biological activity of the particles to separate while providing effective separation of large molecules from small molecules with a small volume of eluate and the ability to use different solutions without interfering with the filtration process. The method is typically used in conjunction with others that further categorize molecules according to traits like charge, acidity, basicity, and affinity for particular chemicals. Since solutes don't interact with the stationary phase, there is also no sample loss.

3.4.7 Types of Chromatography

3.4.7.1 Paper Chromatography (PC)

PC operates based on the mechanism of adsorption chromatography. The apparatus constitutes a stationary phase that comprises a filter paper synthesized from cellulose and a glass chamber. The filter paper is hooked to the top and suspended in the glass chamber. The mixture is spotted at the bottom of the filter paper. This is followed by the addition of solvent in the container which will act as a mobile phase. The solvent will begin to travel upward; separation will occur during the upward movement of the solvent via capillary action. Soluble compounds will dissolve in the solvent and attach to filter paper according to their solubility. The speed of separation is governed by the type of filter paper used. Traveling of mobile phase and separation is faster with thick filter paper while it becomes quite slow with porous filter paper. Identification of the isolated compound is done by determining the retardation factor, which is calculated by taking the ratio of the distance traversed by the compound to the distance traversed by the solvent. The main benefit of this technique lies in its cost-effectiveness, simplicity, and sensitivity to a quite small quantity of substances. Because of the fragility of filter paper which makes it prone to destruction by chemicals, long operation time, and difficulty in identification and quantification, this technique is not used as frequently as before (Kumar and Khanum, 2012).

3.4.7.2 Thin-Layer Chromatography (TLC)

This purification technique also operates based on an adsorption mechanism. Separation is done on the base of the interaction between the stationary phase and compounds in the mixture. It is employed for the isolation of low-molecular-weight compounds. The stationary phase comprises a slurry made by mixing 100 g of silica gel with distilled water. However, Sephadex is also used in certain cases. The slurry is transferred to a glass plate with a dimension of 20 cm × 20 cm. It is then allowed to solidify at 105°C for 1 hour. This is followed by an injection of 10 ml of extract at the bottom of the plate. The plate is then transferred to a solvent separation chamber and kept for 30 minutes. The compounds in the mixture will travel upward at different rates depending on their solubility. Identification of each segregated compound will be done by calculation of the retardation factor. The isolated compounds are scrapped using a spatula and re-extracted using numerous solvents. The benefits of using TLC include less operation time, stability to acids as solvents, and production of clear spots (Ghotekar et al, 2020).

3.4.7.3 Column Chromatography (CC)

CC is based upon several mechanisms such as molecular sieve, ion exchange, and adsorption chromatography. The column is composed of a long glass tube with an average length of 5 cm–1 m and a diameter of 5–50 mm. The bottom is lined with a filter made of tap and glass wool. Furthermore, Sephadex, cellulose, alumina, and silica gel act as stationary phases, and the mobile phase constitutes a solvent. A total of 30 g of silica gel is packed into a translucent glass column such that no air bubbles are formed. Afterward, the extract is poured into the column from the top. Solvents were added to increase polarity, starting from n-hexane to chloroform, ethyl acetate, n-butanol, and methanol. Fractions of each of these solvents were gathered at different periods and then characterized (Tsao and Deng, 2004).

3.4.7.4 Gas Chromatography (GC)

As the name suggests, the mobile phase in GC is a carrier gas that is used to carry out the separation. Ideally, the carrier gas does not interact with the sample or harm the instrument's parts while it moves the sample molecules through the GC system. A syringe or transfer from an auto-sampler that can also extract the chemical components from solid or liquid sample matrices is used to first deliver the sample into the GC. A septum that allows for the injection of the sample mixture without losing the mobile phase is used to introduce the sample into the GC inlet. GC involves the use of two immiscible solvents: The mobile phase in gaseous form and the stationary phase comprising a liquid absorbed on the surface of an inert solvent. Compounds will dissolve in their respective phases according to their solubilities. The analytical column, a long (10–150 m), thin (0.1–0.53 mm internal diameter), fused silica or metal tube connected to the intake, holds the stationary phase coated on the interior walls. It is also connected to the inlet. The column oven in which the analytical column is kept is heated throughout the analysis to elute the less volatile components. The detector, which responds to the chemical components eluting from the column by producing a signal, is put into the column outlet. To create a chromatogram, the acquisition software on a computer records the signal. This procedure is often used for the purification of plant material contaminated with volatile pesticides and is often employed in quality control testing. Regardless of sensitivity and high resolution, the use of GC is limited because the majority of plant compounds have low volatility. Moreover, it is difficult to apply for large-scale purification (Guntur et al, 2018).

3.4.7.5 HPLC

High-pressure pumps are used in HPLC, a type of column chromatography, to move a sample mixture or analyte through a column with chromatographic packing material (stationary phase). A helium or nitrogen-moving carrier gas stream carries the sample. Compounds that are present in any sample that may dissolve in a liquid at trace amounts as low as parts per trillion can be separated and identified using HPLC. Because of its adaptability, HPLC is utilized in a wide range of scientific, industrial, and medicinal fields, including forensics, the environment, and pharmaceuticals. This technique operates on the principle of adsorption and is appropriate for the separation of organic as well as inorganic compounds. The mobile phase employs the use of a solvent whereas the stationary phase is solid. Separation is actualized via interaction between particles of solid phase and mixture. The process commences with the application of a mixture at the bottom of HPLC, followed by the addition of solvent. The tap is then opened, allowing downward movement of solvent which is pushed via a pressure pump, thus mixing it with the mixture. Finally, the mixture travels to the diode detector, waste is removed, compounds are separated, and the refined content is transferred to further processing units (Jakovljević et al, 2020). The interaction between the stationary phase, the molecules being examined, and the solvent or solvents utilized will affect sample retention time. Due to the different polarity of the analytes, the sample interacts between the two phases as it moves through the column at a different rate. Analytes that interact with the stationary phase less frequently or more frequently with the mobile phase will leave the column more quickly.

3.5 IDENTIFICATION OF PHYTOCHEMICALS

Techniques used for the identification of phytochemicals are based upon the detection of carbon and hydrogen configuration, identification of functional groups, and multiple bonds and rings present. These spectroscopic techniques provide sufficient information for quantitative as well as qualitative analysis of phytochemicals. These techniques are founded upon the principle that an organic molecule upon interaction with electromagnetic radiation absorbs waves of precise frequency according to its structure as a result of which a spectrum is formed. These spectrums are peculiar for functional groups, based on which the complete structure is laid out. Usually, spectra are formed in three or four regions, UV, infrared (IR), visible, radio frequency, and the electron beam employed. Nuclear magnetic resonance spectroscopy (NMR), IR, UV, and mass spectroscopy (MS) are some of the most commonly used spectroscopic techniques (Eswaraiah et al, 2020) (Figure 3.3).

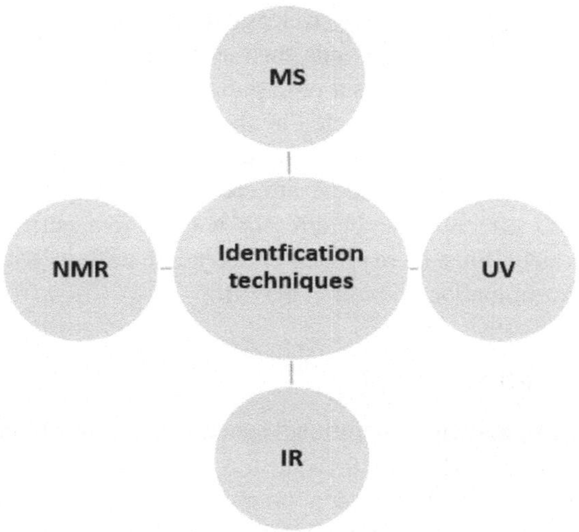

FIGURE 3.3 List of techniques used for the identification of phytochemicals.

3.5.1 UV–Visible Spectroscopy

In the UV and the adjacent, adjoining visible region of the electromagnetic spectrum, absorption spectroscopy or reflectance spectroscopy is referred to as UV spectroscopy or UV-visible spectrophotometry (UV–Vis or UV/Vis). This methodology is frequently employed in a variety of practical and theoretical applications since it is reasonably affordable and simple to execute. The sample must only be a chromophore and absorb in the UV-visible range. Fluorescence spectroscopy is enhanced by absorption spectroscopy. Aside from the measurement wavelength, variables of importance include absorbance (*A*), transmittance (%*T*), and reflectance (%*R*), as well as how they change over time. Identification of compounds both in the mixture and pure form can be done via UV-visible spectroscopy. It is preferably used for quantitative analysis as aromatic molecules are powerful chromophores in the UV range. Identification of numerous phytochemicals is carried out using this method via maximum absorption (λ_{max}) values of their respective features, for instance, phenolic acids (360 nm), anthocyanins (520 nm), flavones (320 nm), and phenolic extract (280 nm). However, this technique is less discriminatory and gives information about overall phenolic content. Observation by the mean of UV-visible spectroscopy takes very little time and is very cost-effective as well (Guntur et al, 2018). An HPLC detector is a UV/Vis spectrophotometer. When an analyte is present, a response that is thought to be proportional to concentration occurs. Similar to the use of calibration curves, the instrument's reaction to the analyte in the unknown should be compared with the response to a standard for correct results. The term "response factor" refers to the response (for example, peak height) for a specific concentration. To identify the functional groups contained within a specific molecule, it is useful to link the wavelengths of absorption peaks with the different types of bonds present in that molecule. The Woodward–Fieser rules, for instance, are a set of empirical observations used to predict λ_{max}, the wavelength of the most intense UV/Vis absorption, for conjugated organic compounds such as dienes and ketones. However, the spectrum by itself does not serve as a precise diagnostic for any particular sample. The absorption spectrum can be affected by the kind of solvent, the pH of the solution, temperature, high electrolyte concentrations, and the presence of interfering compounds. The spectrum will also be affected by experimental changes like the spectrophotometer's effective bandwidth (slit width). To identify the compounds present, these variables must be regulated or taken into account when using UV/Vis spectroscopy for examination (Guntur et al, 2018).

3.5.2 Infrared (IR) Spectroscopy

IR spectroscopy, also known as vibrational spectroscopy, aims to identify a number of bonds and the functional group present in a given compound. This helps in the elucidation of the physical and chemical properties of the compounds. The organic compound is allowed to pass through IR radiations where it is absorbed at a particular frequency. It detects the changes in vibration produced upon elongation and bending of the molecules upon interaction with the IR range of electromagnetic radiation. Various chemical bonds and functional groups have a characteristic vibrational frequency governed by the values of force constant (bond strength) and decreased mass. Therefore, analysis of the frequency absorption band on the IR spectrum of the respective functional group can be manipulated to determine the structure of a bioactive compound. Identification of solid samples is done by milling with potassium bromide and compressing it into a thin pellet whereas that of liquid samples is done using sodium chloride plates. An even high-resolution of constituent chemicals and elucidation of molecular structure can be performed using Fourier transform infrared spectroscopy (FTIR) (Guntur et al, 2018).

3.5.3 Nuclear Magnetic Resonance Spectroscopy (NMR)

Identification using NMR is done by elucidating physical properties such as the arrangement of atoms and molecules in a compound, the total count and array of carbon atoms, and the identification of isotopes of hydrogen, carbon, and protons. Magnetic properties of certain nuclei, for instance, 19F, 1H, 13C, and 31P can be determined using NMR. Interaction of magnetically active nuclei with the radio frequency range of electromagnetic radiations generates a signal with a frequency similar to the one that will be produced upon application of an external magnetic field resonance will be produced upon equivalence of oscillation frequency with intrinsic frequency. Its measurement is commonly done in terms of chemical shift, governed by the chemical environment, magnetic properties of nuclei, and power of the applied magnetic field (Tsao and Deng, 2004). NMR is a tool used by biochemists to identify complicated compounds such as proteins. NMR spectroscopy offers comprehensive details about the structure, dynamics, reaction state, and chemical environment of molecules in addition to molecular identification. Proton and carbon-13 NMR spectroscopy are the two most popular types of NMR, but they can be used with any sample that has nuclei with spin. For minuscule compounds, NMR spectra are distinctive, sharp, analytically manageable, and frequently very predictable. Different functional groups can be easily distinguished from one another and signals can still be distinguished between identical functional groups with different adjacent substituents. NMR has essentially taken the place of conventional wet chemistry tests for identification like color reagents or usual chromatography. The requirement of a relatively significant dose of a purified drug, ranging from 2 to 50 mg, notwithstanding the possibility of recovery through a workup, is a drawback. The sample should ideally be dissolved in a solvent because solids cannot be analyzed by NMR without spinning equipment, which may result in less accurately resolved spectra. NMR produces only an averaged spectrum because its timeframe is relatively long, making it unsuitable for viewing rapid processes. Although significant levels of contaminants do appear on an NMR spectrum, better methods for identifying impurities already exist because NMR is intrinsically not very sensitive, though sensitivity increases at higher frequencies (Tsao and Deng, 2004).

3.5.4 Mass Spectroscopy (MS)

MS is used for the detection of compounds based on their molecular weight and chemical structure. Substances like oligonucleotides and peptides are usually identified. It involves the conversion of organic molecules into highly energized charged species by bombarding them either with lasers or electrons. The electron ionization energy of 70 eV was first used to detect the signal. Moreover, the percentage peak of sample spectra is detected and recorded. A mass spectrum is a graph of the mass/charge ratio of ions against the relative abundance of fragmented ions. The relative molecular mass of detached ions along with the site of detachment can be determined which can be further utilized for the prediction of the molecular formula of bioactive compound. Henceforth, MS provides sufficient information required for structure determination. MS coupled with electrospray ionization (for producing charged species from macromolecules) has been proven to be a very efficient tool for the structural elucidation of phenolic compounds. MS provides plenty of information about organic molecules and is a preferred procedure for the identification of medicinal plant constituents, especially when coupled with HPLC (Beulah et al, 2022).

REFERENCES

Alara, Oluwaseun R, Nour H Abdurahman and I Chinonso. 2018. "Soxhlet extraction of phenolic compounds from Vernonia cinerea leaves and its antioxidant activity." *Journal of Applied Research on Medicinal Ukaegbu, and Aromatic Plants* 11:12–17.

Altemimi, Ammar, Naoufal Lakhssassi, Azam Baharlouei, Dennis G Watson and David A Lightfoot. 2017. "Phytochemicals: Extraction, isolation, and identification of bioactive compounds from plant extracts." *Plants* 6 (4):42.

Beulah, Gurram, Divya Deepthimahanthi, Gullipalli Simhachalam, Anjani Devi Chintagunta, Machiraju Venkata Naga Sravya and Nune Satya Sampath. 2022. "Purification and characterisation of phytochemicals extracted from Rhizophora mucronata: Their efficacy against Pseudomonas aeruginosa infection in Catla Catla." *Reviews in Analytical Chemistry* 41 (1):275–286.

Eswaraiah, G, K Abraham Peele, S Krupanidhi, M Indira, R Bharath Kumar and TC Venkateswarulu. 2020. "GC–MS analysis for compound identification in leaf extract of

Lumnitzera racemosa and evaluation of its in vitro anticancer effect against MCF7 and HeLa cell lines." *Journal of King Saud University-Science* 32 (1):780–783.

Ghotekar, Suresh, Trupti Pagar, Shreyas Pansambal and Rajeshwari Oza. 2020. "A review on green synthesis of sulfur nanoparticles via plant extract, characterization and its applications." *Advanced Journal of Chemistry-Section B* 2 (3):128–143.

Gil-Martín, Emilio, Tamara Forbes-Hernández, Alejandro Romero, Danila Cianciosi, Francesca Giampieri and Maurizio Battino. 2022. "Influence of the extraction method on the recovery of bioactive phenolic compounds from food industry by-products." *Food Chemistry* 378:131918.

Guntur, Sitaramanjaneya Reddy, NS Sampath Kumar, Manasa M Hegde and R Vijaya. 2018. "In vitro studies of the antimicrobial and free-radical scavenging potentials of silver nanoparticles biosynthesized from the extract of Desmostachya bipinnata." *Analytical Chemistry Insights* 13:1177390118782877.

Jakovljević, Martina, Jelena Vladić, Senka Vidović, Kristian Pastor, Stela Jokić, Maja Molnar and Igor Jerković. 2020. "Application of deep eutectic solvents for the extraction of rutin And rosmarinic acid from Satureja montana L. And evaluation of the extracts antiradical activity." *Plants* 9 (2):153.

Kovačević, Danijela Bursać, Marta Maras, Francisco J Barba, Daniel Granato, Shahin Roohinejad, Kumar Mallikarjunan, Domenico Montesano, Jose M Lorenzo and Predrag Putnik. 2018. "Innovative technologies for the recovery of phytochemicals from Stevia rebaudiana Bertoni leaves: A review." *Food Chemistry* 268:513–521.

Kumar, G Phani and Farhath Khanum. 2012. "Neuroprotective potential of phytochemicals." *Pharmacognosy Reviews Khanum* 6 (12):81.

Lavenburg, Valerie M, Kurt A Rosentrater and Stephanie Jung. 2021. "Extraction methods of oils and phytochemicals from seeds and their environmental and economic impacts." *Processes* 9 (10):1839.

Ma, Yaqian, Menghan Liu, Ting Tan, Aiping Yan, Lan Guo, Kun Jiang, Changheng Tan and Yiqun Wan. 2018. "Deep eutectic solvents used as extraction solvent for the determination of flavonoids from Camellia oleifera flowers by high-performance liquid chromatography." *Phytochemical Analysis* 29 (6):639–648.

Palai, Ssantiwala and Neha Shekhawat. 2022. "Novel extraction techniques for phytochemicals: A comprehensive review." *International Journal of Pharmaceutical Research* 14 (3).

Poojar, Basavaraj, B Ommurugan, S Adiga, H Thomas, RK Sori, B Poojar, N Hodlur, A Tilak, R Korde and P Gandigawad. 2017. "Methodology used in the study." *Asian Journal of Pharmaceutical and Clinical Research* 7 (10):1–5.

Sarker, Umakanta and Shinya Oba. 2019. "Nutraceuticals, antioxidant pigments, and phytochemicals in the leaves of Amaranthus spinosus and Amaranthus viridis weedy species." *Scientific Reports* 9 (1):20413.

Sharma, Meenakshi and Prashant Kaushik. 2021. "Vegetable phytochemicals: An update on extraction and analysis techniques." *Biocatalysis and Agricultural Biotechnology* 36:102149.

Sheikh, M, MY Laghari, PK Lashari, AR Khooharo, NT Narejo and Fish Aqua J, 2017. "Current status of three major carps (Labeo rohita, Cirrhinus mrigala and Catla Catla) in the downstream Indus river Sindh." *Fisheries and Aquaculture Journal* 8 (222):2.

Tsao, Rong and Zeyuan Deng. 2004. "Separation procedures for naturally occurring antioxidant phytochemicals." *Journal of Chromatography B* 812 (1–2):85–99.

Velmurugan, Bharath Kumar, Baskaran Rathinasamy, Bharathi Priya Lohanathan, Varadharajan Thiyagarajan and Ching-Feng Weng. 2018. "Neuroprotective role of phytochemicals." *Molecules* 23 (10):2485.

Yasmeen, Madiha, Shafaq Nisar, Vahid Tavallali and Talha Khalid. *International Journal of Chemical and Biochemical Sciences* 2018. "A review of phytochemicals and uses of flaxseed." 13:70–75.

4 The Neuroprotective Effect of Alkaloids and Synthetic Derivatives

Likhit Akotkar, Aditya Ganeshpurkar, and Ankit Ganeshpurkar

4.1 INTRODUCTION

Natural products are an essential element in the development of human civilisation. Mankind uses a variety of plant parts, extracts and phytoconstituents for the treatment of numerous diseases and ailments. Various traditional medicinal systems, treaties and folklores have emphasised the importance of phytomedicines. Phytomedicines are plants and their originated products with some healing and therapeutic properties (Petrovska, 2012). The plants are a rich source of a variety of secondary metabolites produced for signalling, defence and other functions. Alkaloids, glycosides, flavonoids, tannins and saponins are abundantly found in flora across the globe, especially in flowering plants. Alkaloids are one of the significant classes which have numerous uses along with medicinal benefits. These are small molecular nitrogenous natural products with basic nature. The compounds bear heterocyclic rings along with a carbon skeleton with at least one nitrogen. These are classified as true, pseudo and proto alkaloids and other classifications based on their phytochemical and therapeutic use (Kurek, 2019). The various nitrogenous rings, viz. pyrrolidine, tropane, pyrrolizidine, piperidine, quinolizidine, indolizidine, pyridine, isoquinoline, thiazole, quinoline, indole, imidazole, purines and others, constitute various alkaloidal skeleton. An estimate suggests that about 25% of higher plants contain alkaloids and several thousand types have been identified to date (Dewick Paul, 2009). Alkaloids are present in almost every plant part, including bark, leaf, root, stem and flowers. Various plant families, viz. Ranunculaceae, Solanaceae, Papaveraceae and Amaryllidaceae, are predominantly rich in alkaloids (Cushnie et al, 2014). The alkaloids show a broad spectrum of biological activities, including hypoglycaemic, anti-inflammatory, antimalarial, cardioprotective, anti-cancer, antibacterial, antiasthma, vasodilatory, antifibrosis, anti-tussive actions, etc. (Adamski et al, 2020). The alkaloids, such as codeine caffeine (central nervous system [CNS] stimulant), (anti-tussive), morphine (painkillers), quinine (antimalarial), sanguinarine (antibacterial), galantamine and physostigmine (anti-Alzheimer agents), vinblastine and vincristine (anti-cancer) are used in modern medicine.

Alkaloids are one of the critical components of treatments for various CNS diseases and disorders, viz. Alzheimer's disease (AD), schizophrenia, depression

DOI: 10.1201/9781003389781-4

and others. Alteration and imbalance in neurotransmitters lead to the path for the pathogenesis of various neurological disorders (Ressler and Nemeroff, 2000). The alkaloids bear chemical structures like indole, tyrosine, imidazole and phenyl ethyl amine, which resemble the substrate precursor for neurotransmitters such as dopamine, serotonin, adrenaline and acetylcholine (ACh). Hence, they have a greater CNS effect compared to other secondary plant metabolites (Debnath et al, 2018). Vast preclinical and clinical studies have been conducted on alkaloids for their neuroprotective effects. Some important plant extracts rich in alkaloids are listed in Table 4.1.

Neurodegeneration is a detrimental phenomenon involving neuronal damage and ultimately death. The progression involves various physical, toxic, environmental and infection-related causes. Excitotoxicity, mitochondrial dysfunction, toxin buildup, abnormal protein aggregation, impaired voltage-sensitive sodium and

TABLE 4.1
Neuroprotective Plant Extract Rich in Alkaloids and Their Biological Activities.

S. No	Source of Plant Extract	Major Alkaloidal Constituents	Dose(s)/Species	Activity	References
1	*Glaucium corniculatum*, Papaveraceae	Allocryptopine	497 μg/mg, PC12 cells	Neuroprotection	(Dolanbay et al, 2021)
2	*Lupinus mutabilis Lindl.*, Fabaceae	Sparteine	13 mg/kg s.c, Wistar rats	Anticonvulsant	(Villalpando-Vargas et al, 2020)
3	*Salsola oppositofolia*, Chenopodiaceae	*N*-methylisosalsoline	16.30 μg/mL, cholinesterase inhibition assay	AChE inhibition Anti-AD	(Tundis et al, 2009)
4	*P. somniferum*, Papaveraceae	Protopine	3.92 mg/kg p.o., Sprague-Dawley rats	Focal cerebral ischemia	(Xiao et al, 2007)
5	*Gelsemium elegans*, Gelsemiaceae	Koumine	0.056–7 mg/kg s.c., Wistar rats	Diabetic neuropathy	(Ling et al, 2014)
6	*Cytisus laborinum L.*, Fabaceae	Cytisine and 5-bromocytisine	1 mg/kg s.c., Sprague-Dawley rats	Anti-Parkinsonian	(Abin-Carriquiry et al, 2008)
7	*Ephedra sp.*, Ephedraceae	Ephedrine	1.5 mg/kg i.p., Sprague-Dawley rat pups	Ischemic brain injury	(Chen et al, 2009)
8	*Coptis chinensis*, Ranunculaceae	Berberine	40 mg/kg p.o., mice	Anti-Huntington's disease	(Jiang et al, 2015)
9	*Cinchona sp.*, Rubiaceae	Quinine	10 mg/kg p.o., mice	Cocaine addiction	(Huertas et al, 2015)
10	*Tabernanthe iboga*, Apocynaceae	Ibogaine	20 mg/kg i.p., Wistar rats	Anti-addiction	(Paškulin et al, 2006)

calcium channel, impaired gamma amino butyric acid (GABA)-mediated inhibition lysosomal dysfunction, membrane damage, DNA damage, hampered axonal transport and oxidative stress are some of the factors leading to neurodegeneration (Vajda, 2002b). Neuroprotection results in the rescue, recovery or regeneration of the nervous system, its cells and its function (Vajda, 2002a). The neuroprotective mechanism involved the prevention of aggregation of misfolded proteins, improper mitochondrial functioning, modulation of the functioning of various ion channels and reducing Ca^{2+}, glutamate and other toxin-mediated excitotoxicity. The prevention of apoptosis mediated via caspases inhibition, amyloid-β (Aβ) peptide disaggregation, Bcl-2 family modulation, p53 gene inhibition and heat-shock proteins activation also renders neuroprotection (Papaliagkas et al, 2007).

The current chapter aims to describe the numerous neuroprotective actions of alkaloids and their synthetic derivatives by various pathways, including antioxidant, brain-derived neurotrophic factor (BDNF), kinase, receptor inhibition, modulation of synthesis and metabolism of the neurotransmitter, enzymatic inhibition and anti-inflammatory activity.

4.2 NEUROPROTECTIVE ALKALOIDS

4.2.1 Atropine

Atropine (1) is a tropane alkaloid obtained from plants of the Solanaceae family and possesses a neuroprotective effect (Kohnen-Johannsen and Kayser, 2019). Atropine sulphate, when used in combination with ketamine, displayed an antiepileptic effect along with neuroprotection in the status epilepticus model induced by soman gas in rats. The administration of atropine sulphate (10 mg/kg i.p.) reduced neutrophil granulocyte infiltration and glial activation in mice, resulting in the elimination of neuroinflammation (Dhote et al, 2012). Toluene disrupts cerebral cortex and striatal cholinergic neurotransmission leading to cognitive impairments in Sprague-Dawley rats. Atropine (2 mg/kg i.p.) improved the antioxidant defence of the brain by elevation of glutathione (GSH) levels and reduced malondialdehyde (MDA) levels leading to improved memory and cognitive functions in toluene-induced neurodegeneration (Abdel-Salam et al, 2020). Interestingly, some contrasting results were also reported, viz. atropine administered at the doses of 10 and 100 mg/kg in mice and rats for seven days, respectively, resulting in recognition memory deficit evaluated by Y-maze and novel object recognition test. Further, the histopathological findings confirmed neural cell death by the presence of large nuclear fragmentation in the frontal cortex and distortion of the granular layer that led to neurodegeneration. The findings raised the question of using atropine as a neuroprotective in clinical studies (Enye et al, 2017; Olawepo et al, 2017). Similarly, scopolamine (2), another tropane alkaloid clinically used as an anticholinergic medication for postoperative nausea and vomiting, worsens the psychotic symptoms and reduces gastric motility. Recently, scopolamine attracted more attention due to its neurodegenerative effect and close monitoring is suggested while dealing with scopolamine (Riad and Hithe, 2021). The chemical structures of the neuroprotective alkaloids are presented in Figure 4.1.

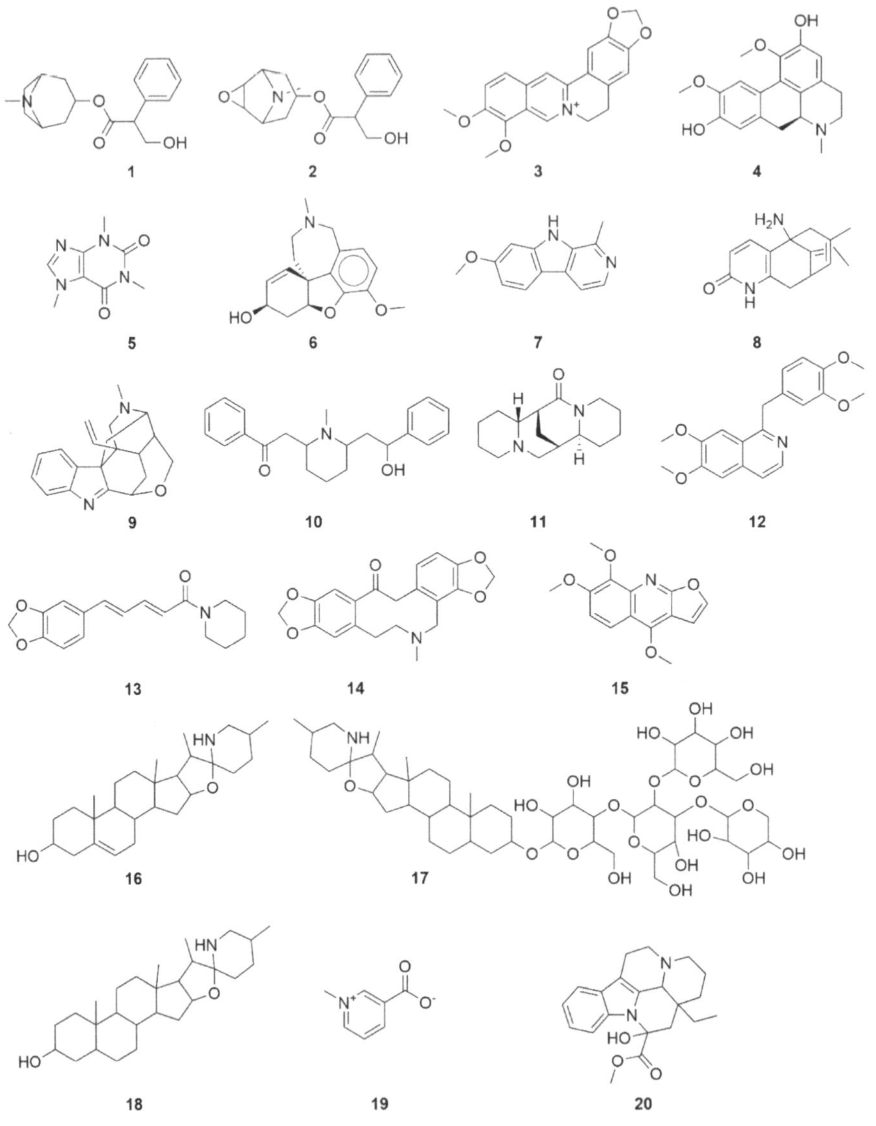

FIGURE 4.1 Structure of neuroprotective alkaloids.

4.2.2 Berberine

Berberine (3) is an isoquinoline alkaloid found in several plants, with its highest concentration reported in *Coptidis rhizoma*. The oral administration of berberine at a dose of 200 mg/kg easily crosses the blood-brain barrier (BBB) with its bioavailability after 12 h of administration in the brain. Such findings promoted the exploration of the neuroprotective potential of berberine (Tan et al, 2013). Besides the antioxidant activity, berberine is profoundly used to treat several neurodegenerative diseases. Berberine has displayed anti-ageing activity in *Drosophila melanogaster,* which is mediated via the AMP-activated protein kinase (AMPK) pathway and oxidative stress modulation (Xu et al, 2017). The treatment of berberine (25 and 100 mg/kg) in a transgenic TgCRND8 mice model of AD reduced glycogen synthase kinase (GSK)-3β expression and Aβ plaque formation. The neuroprotective effect in N2a mouse neuroblastoma cells was exerted by suppressing amyloid precursor protein (APP) and tau hyperphosphorylation mediated through Akt/GSK-3β signalling (Durairajan et al, 2012).

The administration of mercury, aluminium, cadmium and fluoride for 90 days in rats in varying doses (20–50 mg/kg) releases free radicals and pro-inflammatory mediators that induce neurotoxicity and cell death through Aβ and tau hyperphosphorylation. Berberine neutralised the free radicals activity resulting in a reduction of oxidative stress, which further mediated the anti-inflammatory response and anticholinergic activity against heavy metals-induced neurotoxicity (Abdel Moneim, 2015; Hussien et al, 2018). Zhou et al explained the role of berberine in neuroprotection through *in vitro* and *in vivo* stroke models. In the middle cerebral artery occlusion (MCAO) model of stroke, berberine immensely reduced infarct volume. An increased release of cytochrome c (Cyt c) is a prominent feature of mitochondrial dysfunction. In oxygen-glucose deprivation (OGD) induced cellular injury in PC12 cells, berberine (1–4 μM) significantly improved cell survival and prevented apoptosis by reducing mitochondrial Cyt c release rendering neuroprotection (Zhou et al, 2008). Berberine also provides neuroprotective action via modulation of the mitogen-activated protein kinase (MAPK) signalling pathway in diabetic neuropathy by decreasing mRNA expression of neuritin, p38 and c-Jun N-terminal kinase in streptozocin (STZ)-induced diabetic neuropathy in rats (Zhou et al, 2016). The mechanism of actions of major alkaloids is presented in Figure 4.2.

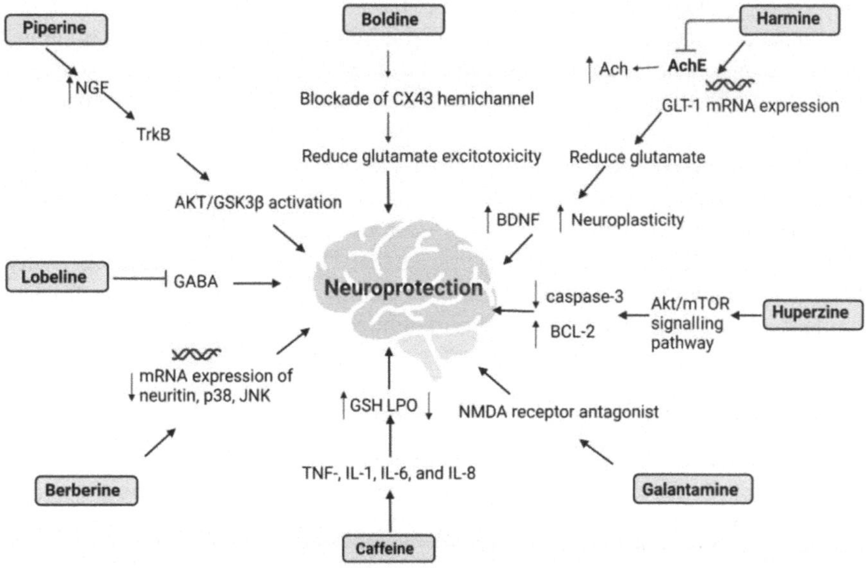

FIGURE 4.2 Neuroprotective mechanism of various alkaloids.

4.2.3 Boldine

Boldine (4) is an aporphine alkaloid obtained from the plant *Peumus boldus Molina* belonging to the family Monimiaceae. The research on boldine was started in 1950 due to its high abundance in Chilean boldo trees. The majority of studies reported that boldine is a strong antioxidant and responsible for its neuroprotective actions (de Lima et al, 2017; Konrath et al, 2008; Yi et al, 2017). Boldine has a cytoprotective action and is widely screened against numerous animal models of cancer and infectious diseases. It attenuates synaptic failure and mitochondrial dysfunctions in the cellular model of primary hippocampal neurons and HT22 cells. In this study, intracellular Ca^{+2} levels were restored to normal levels on treatment with boldine in the presence of Aβ oligomers. Boldine also decreased Aβ aggregation, restored mitochondrial membrane potential by maintaining mitochondrial Ca^{+2} levels and reduced lipid peroxidation, which improved neuronal viability (Toledo et al, 2021). Boldine has been extensively studied for the modulation of neuroinflammation and oxidative stress for neuroprotective actions. In the MCAO-induced stroke model and pentylenetetrazol-induced epilepsy model of rats and mice, respectively, boldine improved the behavioural parameters via nuclear factor kappa B (NFκβ) inhibition, reduction in the MDA levels and elevating antioxidant enzyme levels, thus proven to be effective in stroke and epilepsy through its neuroprotective action (de Lima et al, 2017).

4.2.4 Caffeine

Caffeine (5) is the most widely used constituent by humans in their day-to-day life, obtained from *Coffea arabica* belonging to the family Rubiaceae. Its products have gained popularity globally due to central stimulatory actions. Higher doses of caffeine cause inhibition of GABA and affect the mood of animals and humans in a dose-dependent manner. The regular normal dose of caffeine for humans is limited to 750 mg/kg. The intravenous administration of caffeine (10 mg/kg) attenuated cerebral ischemia/reperfusion (I/R) injury in rats by modulation of oxidative stress and inflammatory mediators viz tumor necrosis factor-alpha (TNF-α), interleukin (IL)-1, IL-6 and IL-8 secretion and reduced the injury (Sun et al, 2013). Caffeine at a dose of 10 mg/kg p.o. restored the striatal dopamine levels in 6-hydroxydopamine (OHDA)-induced lesioned rats and protected the neurons (Machado-Filho et al, 2014). In 1-methyl-4-phenyl-1,2,3,6-tetra-hydropyridine (MPTP)-subjected C57 BALB/c mice, the striatal dopaminergic neurons were protected by caffeine (30 mg/kg p.o) which diminished the MPTP activity in the brain of mice (Bagga et al, 2016).

4.2.5 Galantamine

Galantamine (6) is a tetracyclic alkaloid obtained from *Galanthus nivalis* belonging to the family Amaryllidaceae. Galantamine has gained popularity in research for its BBB permeability and availability in the brain due to its low polarity index. Galantamine was the first compound from plants to treat cognitive deficits, dementia and AD. It was found that galantamine is a strong neuroprotective agent and has a therapeutic effect on multiple neurological disorders such as Parkinson's disease

(PD), AD, stroke, epilepsy and Huntington's disease. Elevated kynurenine in the hippocampus showed psychotic, depressive episodes and cognitive decline. The intraprefrontal cortex (PFC) infusion of galantamine, alpha 7 nicotinic acetylcholine receptors (α7nAChRs) positive modulator, improved the cognitive deficit in kynurenine-induced cognitive deficit in rats (Alexander et al, 2012). N-methyl-D-aspartate (NMDA)-induced neurotoxicity is one of the plausible causes of AD. Galantamine (5 μM) and also in combination with memantine protects rat cortical neurons. The study showed that the synergistic effect of combinatorial therapy is much more effective in AD as compared to memantine and prevents neurodegeneration and neurotoxicity (Lopes et al, 2013). Besides the combination therapy for AD, galantamine was also explored in the form of chitosan-loaded nanoparticles for AD through its antioxidant activity. The rat brain homogenate suggested that intranasal galantamine nanoparticles were found to be effective in AD and were nontoxic for 30 days of administration, which makes it a safe therapy (Kandil et al, 2021). Kainic acid, a strong neurotoxin, is well known for neuronal degeneration in the CA3 and CA4 regions of the rat brain. Kainic acid majorly affects the entorhinal area of the brain which results in cognitive loss and AD. Galantamine was tested against the intrahippocampal kainic acid (0.4 μg/2 μl)-induced mitochondrial dysfunction and cholinesterase overactivity leading to a cognitive deficit in Wistar rats. Galantamine (2.5 mg/kg p.o) potentiates the protective effects of rofecoxib and caffeic acid. Mechanistically, galantamine balanced the oxidative stress reflected in increased antioxidant status and reduced inflammatory mediators along with inhibitory activity on NFκβ (Kumar et al, 2011).

4.2.6 Harmine

Harmine (7) is a β-carboline alkaloid obtained from *Peganum harmala,* Zygophyllaceae. Harmine and its derivatives gained popularity worldwide due to their broad spectrum of pharmacological activities. Several *in silico, in vitro and in vivo* studies were performed for the exploration of its activities and elucidation of the mechanism. The array of multifactorial activities of harmine strongly prevents neurodegeneration. It attenuates ethanol-induced neurodegeneration and sodium nitrite-induced memory deterioration. The protective mechanism was due to the modulation of acetylcholinesterase (AChE) activity and oxidative stress, which directly inhibits the deterioration of mitochondrial functions (Biradar et al, 2013). Glutamate-induced excitotoxicity affects neuronal events are a well-known cause of neurodegeneration. Glutamate transporters 1 (GLT-1) have been documented previously for their role in stroke, AD and cerebral ischemia. Harmine was tested on a four-vessel occlusion model of global cerebral ischemia (GCI) characterised by excessive glutamate release from neurons. Harmine (30 mg/kg i.p.) attenuated glutamate excitotoxicity by elevation of GLT-1 expression leading to reduce glutamate accumulation in the synaptic cleft. The increased GLT-1 mRNA expression by harmine would be a possible mechanism of neuroprotection in cerebral ischemia (Sun et al, 2014). Harmine was evaluated in the learning and memory deficit induced by the traumatic brain injury (TBI) model. TBI significantly causes glutamate excitotoxicity and inflammatory cytokine elevations. Harmine (30 mg/kg i.p.) effectively protects memory functions in rats. It also reduced the inflammatory mediators stimulated by astrocytes in the hippocampus and improved GLT-1 expression. The study suggested that harmine attenuates inflammation which may synergistically reduce extracellular glutamate and generate cognition-enhancer effects (Zhong et al, 2015). Moreover, harmine attenuated rotenone-induced parkinsonism in rats mediated by behavioural and antioxidant effects and was found to improve striatal dopamine levels (El Madani et al, 2016). It is well established that BDNF and tyrosine-protein kinase (trkB) signalling role in cognitive functions. STZ induces diabetes and also produces cognitive deficits in rats via activation of the NLR family pyrin domain containing 3 (NLRP3) and might be responsible for numerous neurological disorders. Harmine ameliorates the learning memory and cognitive effects via inhibition of NLRP3. Notably, NLRP3 inflammasome inhibition can upregulate BDNF expression by suppressing IL-1β, resulting in neuroprotective effects (Liu et al, 2020). Further, the *in vitro* evaluation of harmine on PC12 neuronal cells also showed neuroprotection. Harmine-mediated regulated α-synuclein homeostasis through positive modulation of the ubiquitin-proteasome system in PC12 neuronal cells (Cai et al, 2019). Clinical studies using proteomics also supported that harmine promotes vesicle formation, axonal transport and neurotrophin signalling pathway for neuroprotection (Karmirian et al, 2021).

4.2.7 Huperzine

Huperzine (8) is a constituent of *H. serrata (Thunb.) Trevis.* (Lycopodiaceae). In China, *Huperzia serrata* has been used for centuries to treat a variety of diseases. Previously conducted studies on the brain pharmacokinetics of huperzine found that high BBB permeability, multi-target therapeutics actions and good cerebral concentration make it a potential lead for brain illnesses (Chen et al, 2019). Huperzine showed a protective role on SH-SY5Y cells. *In vivo* study indicated that the huperzine treatment increased antioxidant activity and restored the cognitive functions impaired by scopolamine (Dang et al, 2023). It has been reported that eight weeks of treatment of huperzine in patients with AD alleviates cognitive symptoms. The neuroprotective effects of huperzine result from the inhibition of AChE, increased BDNF expression, NMDA antagonism and reactive oxidative species (ROS) neutralisation, leading to neuronal survival (Gul et al, 2019). Huperzine provided protection for NSC34, neuroblastoma spinal motor cells, against cell death induced by staurosporine, thapsigargin and H_2O_2. The results significantly improved by 17.15%, 35.0% and 10.38% when the cultures were pre-treated with huperzine (10 μM) for 2 h (Hemendinger et al, 2008). Glutamate intoxication insults reduce cell viability, increase caspase-3 activity, a marker for apoptosis, and reduce neurotrophin expression reflected by reduced BDNF levels. Huperzine alleviated glutamate intoxication in mouse HT22 cells and reduced oxidative stress, lactate dehydrogenase (LDH) and improved cell viability. Phosphoinositide 3-kinase (PI3K)/Akt/mTOR signalling pathway is involved in neuroprotective action of huperzine. Huperzine reversed glutamate toxicity by activating BDNF/TrkB that further activates the PI3K/Akt signalling pathway, subsequently contributing to neuronal survival (Mao et al, 2016). Huperzine mediates neuroprotective effect by an anti-inflammatory mechanism in the MCAO rat model of transient focal cerebral ischemia. Huperzine (0.1 mg/kg) improved cognitive functions and reduced infarction size. It inhibited the activity of NFκβ and AChE and decreased the overexpression of proinflammatory mediators. The increased synaptic ACh concentration rendered cholinergic anti-inflammatory effect (Wang et al, 2008).

4.2.8 Koumine

Koumine (9) is an indole alkaloid with anti-inflammatory and analgesic properties. It has shown an anti-allodynic effect in rats against STZ-induced neuropathy with increased sensory nerve conduction velocity and reduced damage to axonal myelin structure. Koumine was effective at a dose of 7 mg/kg with repetitive administration (Ling et al, 2014). In another study, strychnine mediated blocked of mRNA expression of 3α-hydroxysteroid oxidoreductase (3α-HSOR) was increased by Koumine in primary spinal neuronal culture. The radioligand binding study indicated that the alkaloid occupied the strychnine binding site at the glycine receptor and probably acted as an orthosteric agonist. It activated the 3α-HSOR/allopregnanolone pathway producing an anti-allodynic effect (Shoaib et al, 2019). Another independent study also confirmed the induction of 3α-HSOR by koumine in a neuropathic pain model (Qiu et al, 2015). In chronic constriction injury (CCI)-induced neuropathy, koumine reduced the astroglia activation and pro-inflammatory cytokine levels. It also induced autophagy which was evident by the LC3-II/I ratio reduction and reduced P62 expression that inhibited apoptosis. The increased expression of Bcl-xl with decreased Bax expression and caspase-3 cleavage was also observed. Further, koumine decreased the astrocyte-mediated neuroinflammation induced by lipopolysaccharide (LPS) and enhanced autophagy in primary astrocyte culture (Jin et al, 2018). Koumine also reduced the M1 microglial polarisation and Notch-*recombination signal-binding protein Jκ* signalling in diabetic neuropathic pain induced by STZ in rats (Jin et al, 2021). In the postoperative pain model, koumine displayed reduced allodynia and thermal hyperalgesia. It reduced the expression of cytokines and inhibition of astrocytes and microglial activation, resulting in neuroprotection (Xiong et al, 2017).

4.2.9 Lobeline

Lobeline (10), a pyridine alkaloid, has displayed a broad spectrum of activity. Lobeline displayed neuroprotective activity in the pilocarpine-induced seizure in mice. It resulted in reduced DNA damage and cleavage induced by pilocarpine in the cerebral cortex and hippocampus. The free radical level was also found to be reduced along with an increased catalase (CAT) activity rendering additional neuroprotection (da Costa e Silva et al, 2018). In MPTP-induced neurotoxicity, lobeline has displayed a reduced motor deficit in mice. The tyrosine hydroxylase immunohistochemistry of substantia nigra and striatum showed reduced immunoreactivity indicating the neuroprotective mechanism of lobeline (Li et al, 2014). Similarly, the neuroprotective effect of lobeline was also established in methamphetamine-induced toxicity in the striatal region of the rat. It reversed a decreased level of synaptosomal, membrane-associated and vesicular monoamine transporter 2 with an increase in dopamine level (Eyerman and Yamamoto, 2005).

4.2.10 17 Oxo Sparteine

17 Oxo Sparteine (11) has shown a neuroprotective effect in PC12 cells and hip pocampal neurons against toxicity caused by soluble Aβ oligomers. However, the neuroprotection was completely reversed by α-bungarotoxin. The alkaloid increased the frequency of spontaneous calcium transients and Akt phosphorylation. It indicated that the neuroprotection was mediated by the activation of nAChR (Gavilan et al, 2019).

4.2.11 Papaverine

Papaverine (12), a poppy alkaloid, is obtained from opium and has anti-spasmodic properties. However, in MPTP/LPS-induced neurodegeneration in mice, papaverine led to reduced inflammasome activation, thereby suppressing mature IL-1β expression. Additionally, it inhibited NF-κβ and enhanced the activity of the cAMP-response element binding protein (CREB), providing neuroprotection (Leem et al, 2021). The *in vitro* study on BV2 microglial cells challenged with LPS showed the induction of proinflammatory factors (IL-1β, TNF-α, inducible nitric oxide synthase (iNOS) and cyclooxygenase-2 (COX-2)). The treatment with papaverine resulted in the suppression of Il1rn, Socs3, Nos2 and Ptgs2 genes and reduced expression of IL-1β and TNF-α (Dang et al, 2016). In another study, the level of proinflammatory cytokines and nitric oxide (NO) was reduced after treatment with papaverine in LPS-challenged BV2 microglial cells. It also enhanced cAMP level and CREB phosphorylation. Further, a similar neuroprotective effect of papaverine was observed in the LPS-induced neuroinflammation and MPTP-induced PD model in mice leading to reduced dopaminergic neuronal cell death (Lee et al, 2019). In another independent study on MPTP-induced PD model in rats, papaverine ameliorated microglia/astrocyte activation in the striatum and substantia nigra. Further, α-synuclein expression and aggregation were also reduced, rendering neuroprotection (Leem et al, 2020). In the TBI model, papaverine reduced the receptor for advanced glycation end products (RAGE) positive glial cells and the number of NF-κB positive neurons and glia. It also increased NeuN-positive cells and reduced neuroinflammation and apoptosis (Saglam et al, 2021). Interestingly, papaverine afforded neuroprotection in sepsis-induced critical illness neuropathy. It significantly attenuated levels of lactic acid, MDA TNF-α, C-reactive protein, IL-6 and HMGB1, along with improved neuronal conduction observed in the electrophysiological study (Solmaz et al, 2022). In quinolinic acid-induced excitotoxicity, papaverine increased cAMP and NAD^+ levels in cortical neuronal cells. It restored mitochondrial membrane potential and upregulated the expression of BDNF, CREB and synaptic proteins. It also decreased ROS and caspase 3/7 levels rendering neuroprotection (Bhat et al, 2021).

4.2.12 Piperine

Piperine (13) is a piperidine alkaloid obtained from black pepper and other species of the Piperaceae family. It is popular for its neuroprotective effects due to its high permeability and low toxicity. In the MPTP-induced PD model, piperine displayed a neuroprotective effect without any decrease in tyrosine hydroxylase-positive cells in the substantia nigra. Additionally, piperine reduced activated microglia, IL-1β expression and oxidative stress. The balance of Bcl-2/Bax also indicated the anti-apoptotic property (Yang et al, 2015). In another study, the pretreatment with piperine exerted a neuroprotective effect against glutamate-induced toxicity and apoptosis in hippocampal neurons (Fu et al, 2010). The striatal lesions were also decreased on the treatment with piperine, along with reduced levels of TNF-α and iNOS in the 6-OHDA-induced PD model of Wistar rats (Correia et al, 2015). Piperine showed neuroprotection against permanent MCAO injury in rats. It causes inhibition of Bax with upregulated Bcl-2 expression. The apoptotic proteins, Caspase-3, Caspase-9 and Cyt-c, were downregulated upon treatment (Hua et al, 2019). In the transient MCAO model, piperine suppressed mitochondrial dysfunctioning and prevented Cyt c release and caspase-3 activation (Kaushik et al, 2021). Piperine also reduced demyelination and immune cell infiltration as well as astrocyte and microglia activation in experimental autoimmune encephalomyelitis models. It also reduced IL-1β, TNF-α and iNOS and enhanced HO-1, IL-10, Nrf2 and MBP expression. The anti-apoptotic effect of piperine was established through reduced caspase-3 expression along with increased BDNF and NeuN-expressing cells (Nasrnezhad et al, 2021). In the kainic acid-induced excitotoxicity model, piperine reduced the level of matrix metalloproteinase (MMP) 9, an enzyme responsible for nerve growth factor (NGF) degradation. The treatment also increased the expression of pro-NGF processing enzyme MMP-7 along with NGF and NGF-activated receptor TrkA in the rat hippocampus. This indicated that piperine has a neuroprotective effect (Hsieh et al, 2022). Piperine also mitigated corticosterone-induced neurotoxicity, decreased ROS and reversed the reduced expression of BDNF mRNA in cultured rat pheochromocytoma (PC12) cells (Mao et al, 2012).

4.2.13 Protopine

Protopine (14) is an opium poppy alkaloid with benzylisoquinoline ring in the structure. Protopine prevented cell death of PC12 cells from H_2O_2 with an increase in the level of superoxide dismutase (SOD), CAT and glutathione peroxidase (GP). It reversed the raised intracellular Ca^{+2} concentration and altered mitochondrial membrane potential along with the anti-apoptotic effect produced by the downregulation of caspase-3 expression (Xiao et al, 2008). Protopine also decreased the infarction area and serum LDH level in the MCAO model in rats. The neuroprotective activity was indicated by increased serum SOD activity and decreased Ca^{+2} level and terminal dUTP nick end labelling (TUNEL)-positive cells in the affected neuronal cells (Xiao et al, 2007).

4.2.14 Skimmianine

Skimmianine (15) is a furoquinoline alkaloid usually present in plants of the Rutaceae family. In LPS-activated BV-2 microglia, the treatment of skimmianine increased the phosphorylation of NF-κβ/p65 and Iκβα proteins and interfered with NF-κβ binding. Further, the increased MAO-2 expression was observed in HT-22 cells activated by microglia-conditioned media due to skimmianine treatment indicating its neuroprotective potential (Ogunrinade et al, 2023).

4.2.15 Solasodine

Solasodine (16) is an anti-oxidant glycoalkaloid present in Solanum species. In I/R-injury rat model, solasodine showed a marked decrease in neuronal damage with a significant increase in SOD, CAT, GSH and total thiols (Sharma et al, 2014). It also reduced glutamate-induced excitotoxicity of PC12 cells. However, the alkaloid did not prevent cell death due to glucose deprivation and mitochondrial damage (García-Pupo et al, 2016).

4.2.16 Tomatine and Tomatidine

Tomatine (17) and tomatidine (18) are the steroidal alkaloids present in tomatoes. Tomatidine has shown neuroprotective activity by alleviating injury oxygen-glucose deprivation followed by reperfusion (OGD/R) in mice cortical neurons and N_2a cells. It induced autophagy by increasing the number of lysosomes rather than through autophagosome formation. The proteolytic activity and levels of Cathepsin B and D were raised in the cells. It also increased the expression and nuclear translocation of transcription factor EB (TFEB) (Ahsan et al, 2020). In another study, tomatine and tomatidine reduced glutamate-induced toxicity on SH-SY5Y cells and rendered neuroprotection. The mitochondrial membrane potential also remained unaltered in AGS, Caco-2 and SH-SY5Y cells after the treatment. Further, the ROS levels were decreased in SH-SY5Y cells (Taveira et al, 2014). Tomatine has shown a neuroprotective effect against H_2O_2 insult on SH-SY5Y cells. It raised anti-oxidant enzyme levels and BNDF expression in the cells along with the downregulation of Bax and activity of caspase-3 and 9 (Huang et al, 2014).

4.2.17 Trigonelline

Trigonelline (19) is a pyridine alkaloid obtained from fenugreek seed, coffee, garden peas, etc. Trigonelline has a neuroprotective effect against ischemia induced by MCAO in rats. It has shown a reduction in the elevated nitrite and MDA levels relieving oxidative stress. It inhibited reduced GSH-mediated myeloperoxidase expression in the cortex of the brain (Pravalika et al, 2019). In the 6-OHDA-induced PD model, trigonelline prevented neuronal death and apoptosis in substantia nigra pars compacta and restored MDA level (Mirzaie et al, 2016). In the chronic d-galactose model, trigonelline displayed a marked reduction in advanced glycation end product (AGEs) levels and oxidative stress along with a significant reduction in AChE levels leading to neuroprotection (Chowdhury et al, 2018). In OGD/R-induced hippocampal neurons, trigonelline treatment suppressed caspase-3 activity and bax expression along with the induction of bcl-2 expression. It induced PI3K/Akt signalling leading to neuroprotection in hippocampal neuronal injury (Qiu et al, 2020). Trigonelline also reversed the LPS-induced behavioural and memory deficits and restored levels of anti-oxidant enzymes and AChE in the hippocampus and cortex. The levels of TNF-α and IL-6 were alleviated along with the upregulation of BDNF, indicating the promising neuroprotective property (Chowdhury et al, 2018). Similar results were found in another study where trigonelline ameliorated LPS-induced hippocampal oxidative stress and neuroinflammation and reduced expression of NF-κβ, toll-like receptor 4 (TLR4), and TNF-α (Khalili et al, 2018). Trigonelline was also found to be effective against diabetic neuropathy with the restoration of the impaired motor and sensory nerve conduction and downregulation of glucagon-like peptide 1 protein and phosphorylated p38 MAPK protein in plasma and sciatic nerve, respectively (Zhou and Zhou, 2012). In $A\beta_{1-40}$-induced toxicity in rats, trigonelline has shown improved mitochondrial membrane potential and lowered hippocampal MDA, protein carbonyl and LDH levels. It also significantly ameliorated hippocampal levels of TNF-α, IL-6, glial fibrillary acidic protein (GFAP), S100b and Cox2 (Fahanik-Babaei et al, 2019).

4.2.18 Vincamine

Vincamine (20) is an indole-containing monoterpenoid alkaloid present in *Vinca minor.* Vincamine has displayed neuroprotection against $A\beta_{25-35}$ induced toxicity and reduced apoptosis in PC12 cells. It activates PI3K/Akt pathway and upregulates SOD (Han et al, 2017). In another study, vincamine resulted in 50% clearance of brain iron content, reducing oxidative stress (Fayed, 2010). Vindeburnol, a vincamine derivative, reduced the astrocyte activation and demyelination in the cerebellum of C57BL/6 mice treated immunised with myelin oligodendrocyte glycoprotein (MOG_{35-55}). Locus coeruleus also appeared to have lower astrocyte activation along with attenuated tyrosine hydroxylase-positive neuronal hypertrophy (Polak et al, 2012).

The neuroprotective mode of action of other alkaloids is presented in Table 4.2.

4.3 SYNTHETIC AND SEMI-SYNTHETIC DERIVATIVES OF NEUROPROTECTIVE ALKALOIDS

The synthetic and semi-synthetic derivatives of alkaloids along with the neuroprotective mechanism, are presented in Table 4.3.

4.4 FUTURE PROSPECTS

Alkaloids are indeed an interesting class of plant secondary metabolites synthesised from amino acids with a wide spectrum of medicinal properties. Being widely distributed in the plant kingdom, they are well used in traditional medicine in the form of extracts. The advancement in modern technology has been helpful in the isolation of these alkaloids in their pure form. The alkaloids have shown various CNS activities with neuroprotection as one of the crucial properties. Neurodegeneration is an unavoidable phenomenon related to ageing, which involves dysfunctional cellular machinery leading the cell death. The plausible cause would be membrane damage, DNA damage, dysfunctional mitochondria and hampered cellular pathways such as axonal transport and protein degradation machinery. Neuroprotection is important

TABLE 4.2
Alkaloids along with Their Neuroprotective Mechanism.

Alkaloid	Dose(s)	Model	Mode of Action(s)	References
Aloperine 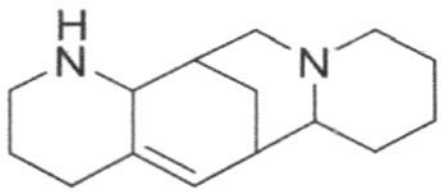	25, 50, and 100 mg/l	OGD/R	↑ Mitochondrial membrane potential, inhibited intracellular-free Ca^{+2}, ↓ ROS, ↑ SOD, CAT and GP	(Ma et al, 2015)
Capsaicin 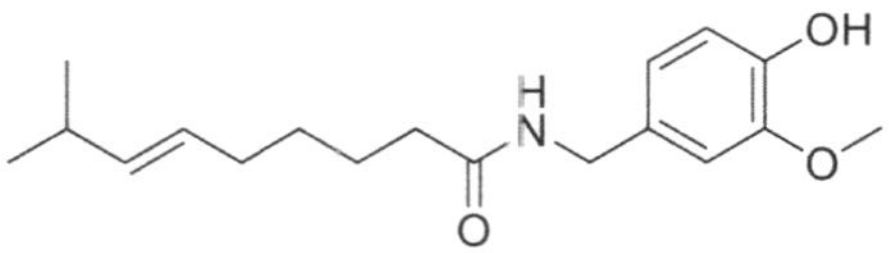	10 mg/kg	Cold water stress in rats	↑ Synapsin I and PSD93 levels prevent tau hyperphosphorylation through the reversal of suppression of protein phosphatase 2A	(Jiang et al, 2013)
	20 and 30 mg/kg	Male Sprague–Dawley rats	↑ Amyloidogenic APP processing, ↑ membrane-bound APP	(Pákáski et al, 2009)
	0.2 and 2.0 mg/kg	Hypoxic-ischemic neonatal brain injury model	↓ Infarction volume and prevent apoptosis	(Khatibi et al, 2011)
	3–30 µM/L	Hypoxia-reoxygenation-induced primary rat hippocampal neuron	↑ Akt phosphorylation, ↓ apoptosis	(Guo et al, 2008)
Coptisine 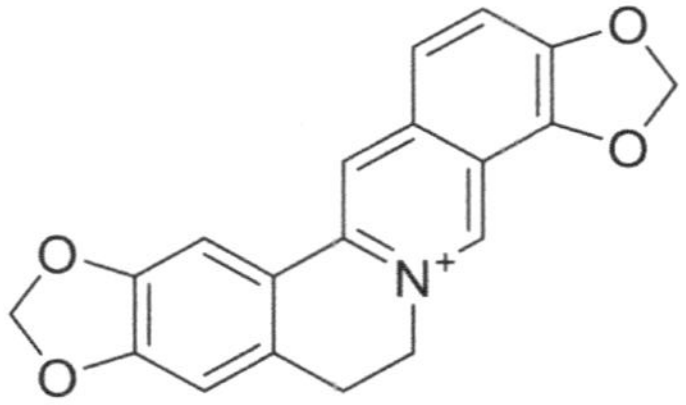	0.1–40 µM	tert-butyl hydroperoxide-induced apoptosis	↓ Apoptosis, ↓ thioredoxin-interacting protein (TXNIP) gene, apoptosis ↓ signal-regulating kinase	(Friedemann et al, 2015)

(Continued)

TABLE 4.2 *(Continued)*
Alkaloids along with Their Neuroprotective Mechanism.

Alkaloid	Dose(s)	Model	Mode of Action(s)	References
Cryptolepine 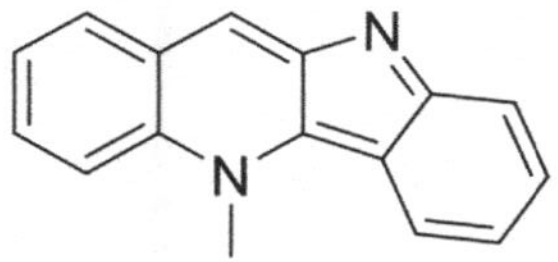	2.5–20 µM	IL-1β-stimulated SK-N-SH neuroblastoma	↓ Production of IL-6, PGE2 and TNFα, inhibition of NF-κβp65 nuclear translocation	(Olajide et al, 2013a)
		LPS-induced neuroinflammation in rat microglia	↓ NF-KappaB inhibited p38 and MAPKAPK2 phosphorylation in the microglia	(Olajide et al, 2013b)
Cystine 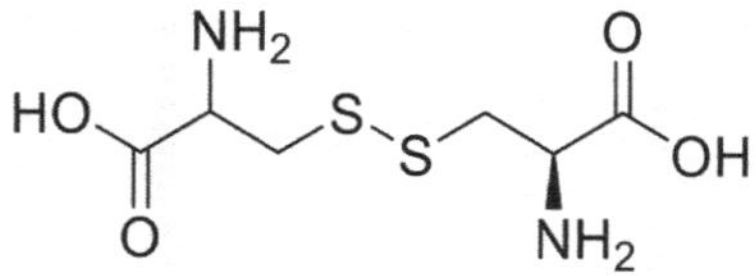	0–400 µM	NMDA-injured cortical neurons	↓ GluN2B-containing NMDA receptors and modulate Bcl-2	(Li et al, 2013)
Isocorynoxeine 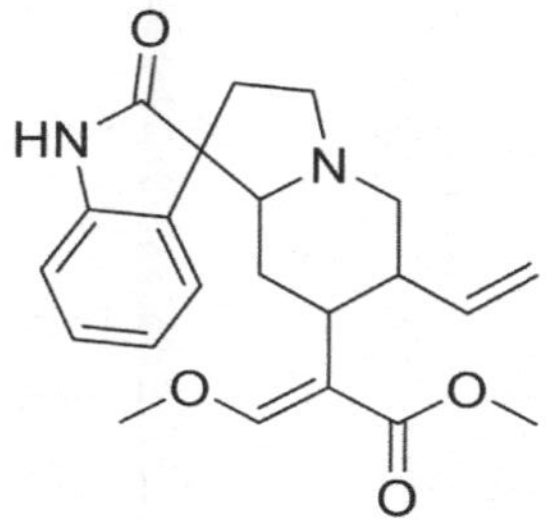	100 µM	Glutamate-induced HT22 cells	↓ Cell death	(Qi et al, 2015)

(Continued)

TABLE 4.2 *(Continued)*
Alkaloids along with Their Neuroprotective Mechanism.

Alkaloid	Dose(s)	Model	Mode of Action(s)	References
Oxysophocarpine 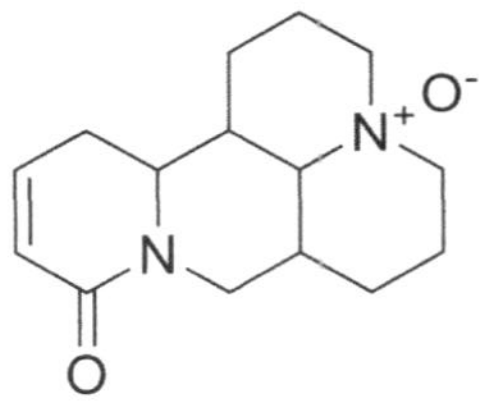	1, 2 and 5 μM/L	OGD/R-insulted neonatal rat	↓ Neuronal death, ↓ morphologic impairment	(Zhu et al, 2014)
Racemoside A 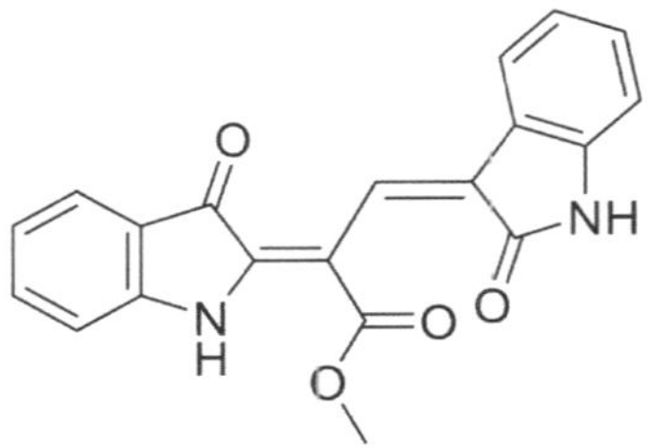	10 μM	$A\beta_{25-35}$-induced SH-SY5Y cell damage	↓ Cell death	(Liu et al, 2013)

TABLE 4.3
Synthetic and Semi-Synthetic Derivatives of Neuroprotective Alkaloids.

Alkaloidal Derivative	Dose(s)	Model(s)	Mode of Action(s)	References
N-(3,4-dimethoxyphenethyl)-2-((1,3,7-trimethyl-2,6-dioxo-2,3,6,7-tetrahydro-1H-purin-8-yl)thio)acetamide 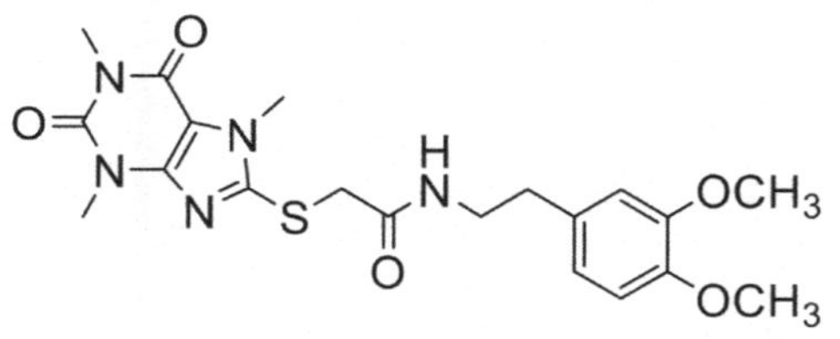	100 µM	6-OHDA-induced toxicity in SH-SY5Y cell	↑ Cell viability	(Kasabova-Angelova et al, 2020)
N′-(2-hydroxybenzylidene)-3-(1,3,7-trimethyl-2,6-dioxo-2,3,6,7-tetrahydro-1H-purin-8-ylthio)propanehydrazide 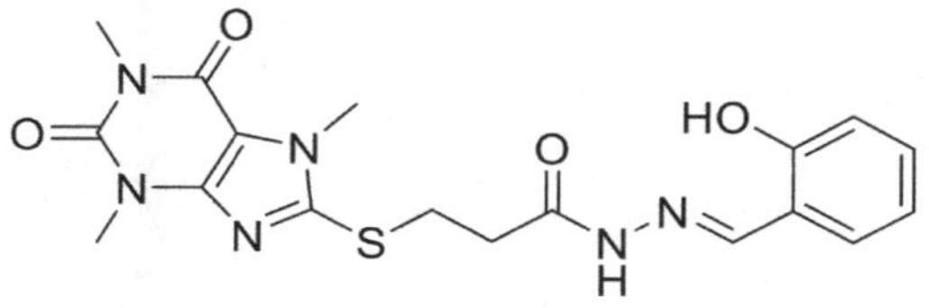	100 µM	Rat brain synaptosomes and mitochondria	↑ Cell viability, ↑ GSH, ↓ MDA, ↓ oxidative stress	(Kondeva-Burdina et al, 2022)

(Continued)

TABLE 4.3 *(Continued)*
Synthetic and Semi-Synthetic Derivatives of Neuroprotective Alkaloids.

Alkaloidal Derivative	Dose(s)	Model(s)	Mode of Action(s)	References
Vinpocetine 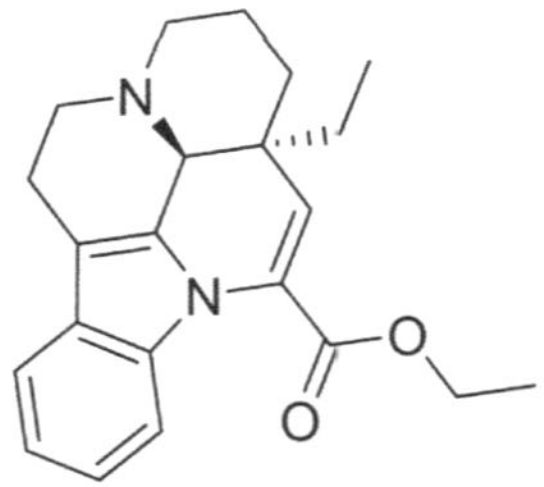	1–50 µM	Glutamate excitotoxicity on primary cortical neurons	Inhibition of voltage-dependent Na^+ channel and interaction with glutamate receptors	(Bönöczk et al, 2000; Tárnok et al, 2008)
	1–50 µM	Primary cortical neurons	↓ Glutamate excitotoxicity, reduction of inner mitochondrial membrane potential	100
	–	Microglial cell M2 in OGD ischemic model OGD-induced damage BV2 cells	Inhibits phosphodiesterase (PDE) 1-B	(Zang et al, 2021)
	1–100 µM	Sodium azide-induced hypoxia in chick embryo cerebral hemisphere neuronal culture	↓ Neuronal death	(Krieglstein and Rischke, 1991)
	3 mg/kg	Permanent MCAO model Primary cortical cell	↓ Infraction volume ↓ LDL level, ↓ viability ↓ Transient NMDAR, transient N-methyl-D-aspartate, ↓ veratridine-induced excitotoxicity	(Dézsi et al, 2002)
Cis-apovincaminic acid 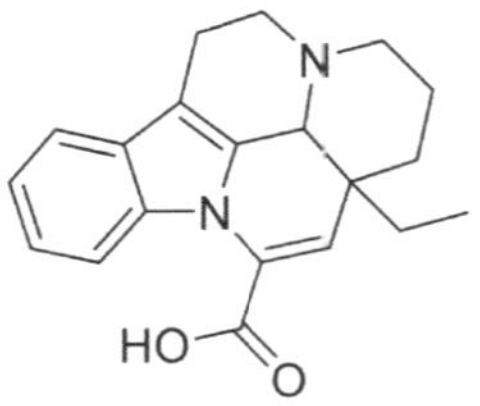	10 mg/kg	NMDA-induced neurodegeneration in rats	Alleviate entorhinal NMDA lesions and microglia activation	(Nyakas et al, 2009)

for the prolonged survival of neuronal cells, especially in the case of neurodegenerative diseases. Alkaloids displayed neuroprotection through various routes and mechanisms, viz. minimising DNA and cellular damage by neutralising ROS, preventing the release of pro-inflammatory mediators, activating anti-apoptotic pathways and decreasing the expression of inflammatory markers. The bioavailability and BBB permeability of the alkaloids are still a major concern. The semi-synthetic and synthetic derivatives with suitable physicochemical properties for bioavailability could be a plausible avenue for the development of new lead molecules. However, there is still a need for the development of such derivatives, with a few available to date.

REFERENCES

Abdel-Salam, Omar M. E., Amany Ameen Sleem, Yasser Ashry Khadrawy, and Fatma A. Morsy. 2020. "Prevention of toluene-induced brain neurodegeneration by atropine and neostigmine." *Journal of Basic Pharmacology and Toxicology* 4 (1): 1–9.

Abdel Moneim, Ahmed E. 2015. "The neuroprotective effect of berberine in mercury-induced neurotoxicity in rats." *Metabolic Brain Disease* 30: 935–942.

Abin-Carriquiry, J. Andrés, Gustavo Costa, Jessika Urbanavicius, Bruce K. Cassels, Marco Rebolledo-Fuentes, Susan Wonnacott, and Federico Dajas. 2008. "In vivo modulation of dopaminergic nigrostriatal pathways by cytisine derivatives: Implications for Parkinson's disease." *European Journal of Pharmacology* 589 (1–3): 80–84.

Adamski, Z., L. L. Blythe, L. Milella, and S. A. Bufo. 2020. "Biological activities of alkaloids: From toxicology to pharmacology." *Toxins (Basel)* 12 (4): 210. https://doi.org/10.3390/toxins12040210, https://www.ncbi.nlm.nih.gov/pubmed/32224853.

Ahsan, Anil, Yanrong Zheng, Shijia Ma, Mengru Liu, Ming Cao, Yue Li, Wanqing Zheng, Xinyu Zhou, Minhang Xin, Wei-Wei Hu, Zhong Chen, and Xiangnan Zhang. 2020. "Tomatidine protects against ischemic neuronal injury by improving lysosomal function." *European Journal of Pharmacology* 882: 173280. https://doi.org/10.1016/j.ejphar.2020.173280. https://www.sciencedirect.com/science/article/pii/S0014299920303721.

Alexander, Kathleen S., Hui-Qiu Wu, Robert Schwarcz, and John P. Bruno. 2012. "Acute elevations of brain kynurenic acid impair cognitive flexibility: Normalization by the alpha7 positive modulator galantamine." *Psychopharmacology* 220: 627–637.

Bagga, Puneet, Anup N. Chugani, and Anant B. Patel. 2016. "Neuroprotective effects of caffeine in MPTP model of Parkinson's disease: A 13C NMR study." *Neurochemistry International* 92: 25–34.

Bhat, Abid, Vanessa Tan, Benjamin Heng, Sharron Chow, Salundi Basappa, Musthafa M. Essa, Saravana B. Chidambaram, and Gilles J. Guillemin. 2021. "Papaverine, a phosphodiesterase 10a inhibitor, ameliorates quinolinic acid-induced synaptotoxicity in human cortical neurons." *Neurotoxicity Research* 39 (4): 1238–1250.

Biradar, S. M., Hanumanthachar Joshi, and K. C. Tarak. 2013. "Cerebroprotective effect of isolated harmine alkaloids extracts of seeds of Peganum harmala L. on sodium nitrite-induced hypoxia and ethanol-induced neurodegeneration in young mice." *Pakistan Journal of Biological Sciences: PJBS* 16 (23): 1687–1697.

Bönöczk, Péter, Balázs Gulyás, Vera Adam-Vizi, András Nemes, Egon Kárpáti, Béla Kiss, Margit Kapás, Csaba Szántay, István Koncz, Tibor Zelles, and Adam Vas. 2000. "Role of sodium channel inhibition in neuroprotection: Effect of vinpocetine." *Brain Research Bulletin* 53 (3): 245–254. https://doi.org/10.1016/S0361-9230(00)00354-3, https://www.sciencedirect.com/science/article/pii/S0361923000003543.

Cai, Cui-Zan, He-Feng Zhou, Ning-Ning Yuan, Ming-Yue Wu, Simon Ming-Yuen Lee, Jiao-Yan Ren, Huan-Xing Su, Jin-Jian Lu, Xiu-Ping Chen, and Min Li. 2019. "Natural alkaloid harmine promotes degradation of alpha-synuclein via PKA-mediated ubiquitin-proteasome system activation." *Phytomedicine* 61: 152842.

Chen, Siyuan, Nong Xiao, and Xiaoping Zhang. 2009. "Effect of combined therapy with ephedrine and hyperbaric oxygen on neonatal hypoxic–ischemic brain injury." *Neuroscience Letters* 465 (2): 171–176.

Chen, Yifan, Gang Cheng, Rongfeng Hu, Shengqi Chen, Wenjie Lu, Song Gao, Hongmei Xia, Bin Wang, Chaojie Sun, and Xiangjiang Nie. 2019. "A nasal temperature and pH dual-responsive in situ gel delivery system based on microemulsion of huperzine A: Formulation, evaluation, and in vivo pharmacokinetic study." AAPS PharmSciTech 20: 1–12.

Chowdhury, Amrita A., Nitin B. Gawali, Vipin D. Bulani, Pankaj S. Kothavade, Snehal N. Mestry, Padmini S. Deshpande, and Archana R. Juvekar. 2018. "In vitro antiglycating effect and in vivo neuroprotective activity of trigonelline in d-galactose induced cognitive impairment." *Pharmacological Reports* 70 (2): 372–377. https://doi.org/10.1016/j.pharep.2017.09.006, https://www.sciencedirect.com/science/article/pii/S1734114017301846.

Chowdhury, Amrita A., Nitin B. Gawali, Renuka Munshi, and Archana R. Juvekar. 2018. "Trigonelline insulates against oxidative stress, proinflammatory cytokines and restores BDNF levels in lipopolysaccharide induced cognitive impairment in adult mice." *Metabolic Brain Disease* 33 (3): 681–691. https://doi.org/10.1007/s11011-017-0147-5.

Correia, Alyne Oliveira, Abílio Augusto Pimentel Cruz, Arôdo Tenório Ribeiro de Aquino, Joanisson Rubens Gomes Diniz, Karizia Bianca Ferreira Santana, Pedro Ivo Martins Cidade, Jaine Dantas Peixoto, Daniel Luna Lucetti, Maria Elizabeth Pereira Nobre, and Giovany Michely Pinto da Cruz. 2015. "Neuroprotective effects of piperine, an alkaloid from the Piper genus, on the Parkinson's disease model in rats." *Journal of Neurology and Therapeutics* 1 (1): 1–18.

Cushnie, T. P. Tim, Benjamart Cushnie, and Andrew J. Lamb. 2014. "Alkaloids: An overview of their antibacterial, antibiotic-enhancing and antivirulence activities." *International Journal of Antimicrobial Agents* 44 (5): 377–386. https://doi.org/10.1016/j.ijantimicag.2014.06.001, https://www.sciencedirect.com/science/article/pii/S0924857914001885.

da Costa e Silva, Liana Dantas, Patrícia Pereira, Gabriela Gregory Regner, Fernanda Brião Menezes Boaretto, Cleonice Hoffmann, Pricila Pflüger, Lucas Lima da Silva, Luiza Reinhardt Steffens, Ana Moira Morás, Dinara Jaqueline Moura, and Jaqueline Nascimento Picada. 2018. "DNA damage and oxidative stress induced by seizures are decreased by anticonvulsant and neuroprotective effects of lobeline, a candidate to treat alcoholism." *Metabolic Brain Disease* 33 (1): 53–61. https://doi.org/10.1007/s11011-017-0130-1.

Dang, Thu Kim, Min Hong, Vui Thi Dao, Duong Thuy Nguyen, Khanh Van Nguyen, Hai Thanh Nguyen, Sana Ullah, Hiep Tuan Tran, and Sun Yeou Kim. 2023. "Neuroprotective effects of total alkaloids fraction of Huperzia serrata on scopolamine-induced neurodegenerative animals." *Phytotherapy Research* 37 (1): 140–150.

Dang, Yalong, Yalin Mu, Kun Wang, Ke Xu, Jing Yang, Yu Zhu, and Bin Luo. 2016. "Papaverine inhibits lipopolysaccharide-induced microglial activation by suppressing NF-κB signaling pathway." *Drug Design, Development and Therapy* 10: 851–859. https://doi.org/10.2147/DDDT.S97380, https://www.tandfonline.com/doi/abs/10.2147/DDDT.S97380.

de Lima, Neila Maria R., Emerson de O. Ferreira, Mara Yone S. D. Fernandes, Francisco Arnaldo V. Lima, Kelly Rose T. Neves, Marta Regina S. do Carmo, and Geanne M. de Andrade. 2017. "Neuroinflammatory response to experimental stroke is inhibited by boldine." *Behavioural Pharmacology* 28 (2): 223–237.

Debnath, Bikash, Waikhom Somraj Singh, Manik Das, Sanchari Goswami, Mahesh Kumar Singh, Debasish Maiti, and Kuntal Manna. 2018. "Role of plant alkaloids on human health: A review of biological activities." *Materials Today Chemistry* 9: 56–72.

Dewick Paul, M. 2009. *Medicinal Natural Products*. 3rd ed. Chichester: Wiley.

Dézsi, László, Istvánné Kis-Varga, József Nagy, Zsolt Komlódi, and Egon Kárpáti. 2002. "Neuroprotective effects of vinpocetine in vivo and in vitro. Apovincaminic acid derivatives as potential therapeutic tools in ischemic stroke." *Acta Pharmaceutica Hungarica* 72 (2): 84–91.

Dhote, Franck, Pierre Carpentier, Laure Barbier, André Peinnequin, Valérie Baille, Fabien Pernot, Guy Testylier, Claire Beaup, Annie Foquin, and Fréderic Dorandeu. 2012. "Combinations of ketamine and atropine are neuroprotective and reduce neuroinflammation after a toxic status epilepticus in mice." *Toxicology and Applied Pharmacology* 259 (2): 195–209.

Dolanbay, Serap Nigdelioglu, Fatma Gonca Kocanci, and Belma Aslim. 2021. "Neuroprotective effects of allocryptopine-rich alkaloid extracts against oxidative stress-induced neuronal damage." *Biomedicine & Pharmacotherapy* 140: 111690.

Durairajan, Siva Sundara Kumar, Liang-Feng Liu, Jia-Hong Lu, Lei-Lei Chen, Qiuju Yuan, Sookja K. Chung, Ling Huang, Xing-Shu Li, Jian-Dong Huang, and Min Li. 2012. "Berberine ameliorates β-amyloid pathology, gliosis, and cognitive impairment in an Alzheimer's disease transgenic mouse model." *Neurobiology of Aging* 33 (12): 2903–2919. https://doi.org/10.1016/j.neurobiolaging.2012.02.016, https://www.sciencedirect.com/science/article/pii/S0197458012001613.

El Madani, Mevidette A., R. M. A. Elsalam, Amina S. Attia, S. M. El-Shenawy, and Mahmoud S. Arbid. 2016. "Neuropharmacological effects of naringenin, harmine and adenosine on parkinsonism induced in rats." *Scholars Research Library* 8 (5): 45–57.

Enye, Linus Anderson, Edem Ekpenyong Edem, Ishola Olakunle Azeez, Mary Essien Umoh, Achonwa-Njemanze Nneoma, and Abimbola John Ajayi. 2017. "Spatial memory, motor coordination, cerebellar and hippocampal histoarchitectural changes following atropine administration to adult mice." International Biological and Biomedical Journal 3 (3): 125–132.

Eyerman, David J, and Bryan K Yamamoto. 2005. "Lobeline attenuates methamphetamine-induced changes in vesicular monoamine transporter 2 immunoreactivity and monoamine depletions in the striatum." *Journal of Pharmacology and Experimental Therapeutics* 312 (1): 160–169.

Fahanik-Babaei, Javad, Tourandokht Baluchnejadmojarad, Farnaz Nikbakht, and Mehrdad Roghani. 2019. "Trigonelline protects hippocampus against intracerebral Aβ (1–40) as a model of Alzheimer's disease in the rat: Insights into underlying mechanisms." *Metabolic Brain Disease* 34: 191–201.

Fayed, A. H. 2010. "Brain trace element concentration of rats treated with the plant alkaloid, vincamine." Biological Trace Element *Research* 136 (3): 314–319. https://doi.org/10.1007/s12011-009-8550-3.

Friedemann, Thomas, Udo Schumacher, Yi Tao, Alexander Kai-Man Leung, and Sven Schröder. 2015. "Neuroprotective activity of coptisine from *Coptis chinensis* (Franch)." *Evidence-Based Complementary and Alternative Medicine* 2015: 827308. https://doi.org/10.1155/2015/827308.

Fu, Min, Zhao-Hui Sun, and Huan-Cong Zuo. 2010. "Neuroprotective effect of piperine on primarily cultured hippocampal neurons." *Biological and Pharmaceutical Bulletin* 33 (4): 598–603.

García-Pupo, L., A. Zaldo-Castro, V. Exarchou, J. E. Tacoronte-Morales, L. Pieters, W. Vanden Berghe, Y. Nuñez-Figueredo, and R. Delgado-Hernández. 2016. "In vitro neuroprotective and anti-inflammatory activities of natural and semi-synthetic spirosteroid analogues." *Molecules* 21 (8). https://doi.org/10.3390/molecules21080992.

Gavilan, Javiera, Daniela Mennickent, Oscar Ramirez-Molina, Sergio Triviño, Claudia Perez, Tiare Silva-Grecchi, Pamela A. Godoy, Jose Becerra, Luis G. Aguayo, Gustavo Moraga-Cid, Victoria San Martin, Gonzalo E. Yevenes, Patricio A. Castro, Leonardo

Guzman, and Jorge Fuentealba. 2019. "17 Oxo sparteine and lupanine, obtained from Cytisus scoparius, exert a neuroprotection against soluble oligomers of amyloid-β toxicity by nicotinic acetylcholine receptors." *Journal of Alzheimer's Disease* 67: 343–356. https://doi.org/10.3233/JAD-180945, https://content.iospress.com/articles/journal-of-alzheimers-disease/jad180945.

Gul, Amara, Jehan Bakht, and Farah Mehmood. 2019. "Huperzine-A response to cognitive impairment and task switching deficits in patients with Alzheimer's disease." *Journal of the Chinese Medical Association* 82 (1): 40–43.

Guo, Shi-Yin, Guo-Ping Yang, De-Jian Jiang, Feng Wang, Tao Song, Xing-He Tan, and Zhen-Qiu Sun. 2008. "Protection of capsaicin against hypoxia–reoxygenation-induced apoptosis of rat hippocampal neurons." *Canadian Journal of Physiology and Pharmacology* 86 (11): 785–792. https://doi.org/10.1139/y08-083, https://cdnsciencepub.com/doi/abs/10.1139/Y08-083.

Han, J., Q. Qu, J. Qiao, and J. Zhang. 2017. "Vincamine alleviates amyloid-β 25-35 peptides-induced cytotoxicity in PC12 cells." Pharmacognosy Magazine 13 (49): 123–128. https://doi.org/10.4103/0973-1296.196309.

Hemendinger, Richelle A., Edward J. Armstrong, Rafal Persinski, Julianne Todd, Jean-Luc Mougeot, Franklin Volvovitz, and Jeffrey Rosenfeld. 2008. "Huperzine A provides neuroprotection against several cell death inducers using in vitro model systems of motor neuron cell death." *Neurotoxicity Research* 13: 49–61.

Hsieh, Ting-Yang, Yi Chang, and Su-Jane Wang. 2022. "Piperine provides neuroprotection against kainic acid-induced neurotoxicity via maintaining NGF signalling pathway." *Molecules* 27 (9): 2638. https://www.mdpi.com/1420-3049/27/9/2638.

Hua, Shiyao, Jiayue Liu, Yiwei Zhang, Juan Li, Xinhui Zhang, Lin Dong, Yunsheng Zhao, and Xueyan Fu. 2019. "Piperine as a neuroprotective functional component in rats with cerebral ischemic injury." *Food Science & Nutrition* 7 (11): 3443–3451.

Huang, S. L., H. B. He, K. Zou, C. H. Bai, Y. H. Xue, J. Z. Wang, and J. F. Chen. 2014. "Protective effect of tomatine against hydrogen peroxide-induced neurotoxicity in neuroblastoma (SH-SY5Y) cells." Journal *of* Pharmacy *and* Pharmacology 66 (6): 844–854. https://doi.org/10.1111/jphp.12205.

Huertas, Adriana, William D. Wessinger, Yuri V. Kucheryavykh, Priscila Sanabria, Misty J. Eaton, Serguei N. Skatchkov, Legier V. Rojas, Gerónimo Maldonado-Martínez, and Mikhail Y. Inyushin. 2015. "Quinine enhances the behavioral stimulant effect of cocaine in mice." *Pharmacology Biochemistry and Behavior* 129: 26–33.

Hussien, Hend M., Aml Abd-Elmegied, Doaa A. Ghareeb, Hani S. Hafez, Hany E. A. Ahmed, and Nehad Abd El-Moneam. 2018. "Neuroprotective effect of berberine against environmental heavy metals-induced neurotoxicity and Alzheimer's-like disease in rats." *Food and Chemical Toxicology* 111: 432–444.

Jiang, Wenxiao, Wenjie Wei, Marta A. Gaertig, Shihua Li, and Xiao-Jiang Li. 2015. "Therapeutic effect of berberine on Huntington's disease transgenic mouse model." *PloS One* 10 (7): e0134142.

Jiang, Xia, Lin-Wei Jia, Xiao-Hong Li, Xiang-Shu Cheng, Jia-Zhao Xie, Zhi-Wei Ma, Wei-Jie Xu, Yue Liu, Yun Yao, and Lai-Ling Du. 2013. "Capsaicin ameliorates stress-induced Alzheimer's disease-like pathological and cognitive impairments in rats." *Journal of Alzheimer's Disease* 35 (1): 91–105.

Jin, Gui-Lin, Li-Mian Hong, Hai-Ping Liu, Rong-Cai Yue, Zu-Cheng Shen, Jian Yang, Ying Xu, Hui-Hui Huang, Yi Li, Bo-Jun Xiong, Yan-Ping Su, and Chang-Xi Yu. 2021. "Koumine modulates spinal microglial M1 polarization and the inflammatory response through the notch-RBP-Jκ signaling pathway, ameliorating diabetic neuropathic pain in rats." *Phytomedicine* 90: 153640. https://doi.org/10.1016/j.phymed.2021.153640, https://www.sciencedirect.com/science/article/pii/S0944711321001835.

Jin, Gui-Lin, Rong-Cai Yue, Sai-Di He, Li-Mian Hong, Ying Xu, and Chang-Xi Yu. 2018. "Koumine decreases astrocyte-mediated neuroinflammation and enhances autophagy, contributing to neuropathic pain from chronic constriction injury in rats." *Frontiers in Pharmacology* 9: 989.

Kandil, Lamia Said, Ragwa M. Farid, Safaa S. ElGamal, and Amira Sayed Hanafy. 2021. "Intranasal galantamine/chitosan complex nanoparticles elicit neuroprotection potentials in rat brains via antioxidant effect." *Drug Development and Industrial Pharmacy* 47 (5): 735–740.

Karmirian, Karina, Lívia Goto-Silva, Juliana Minardi Nascimento, Marcelo N. Costa, José Alexandre Salerno, Isis Moraes Ornelas, Bart Vanderborght, Daniel Martins-de-Souza, and Stevens Rehen. 2021. "Proteomic changes induced by harmine in human brain organoids reveal signaling pathways related to neuroprotection." *bioRxiv* 2021.06.16.448740.

Kasabova-Angelova, Alexandra, Magdalena Kondeva-Burdina, Javor Mitkov, Maya Georgieva, Virginia Tzankova, and Alexander Zlatkov. 2020. "Neuroprotective and MAOB inhibitory effects of a series of caffeine-8-thioglycolic acid amides." *Brazilian Journal of Pharmaceutical Sciences* 56.

Kaushik, Pooja, Mubashshir Ali, Mohd Salman, Heena Tabassum, and Suhel Parvez. 2021. "Harnessing the mitochondrial integrity for neuroprotection: Therapeutic role of piperine against experimental ischemic stroke." *Neurochemistry International* 149: 105138. https://doi.org/10.1016/j.neuint.2021.105138, https://www.sciencedirect.com/science/article/pii/S0197018621001844.

Khalili, Mohsen, Mitra Alavi, Elham Esmaeil-Jamaat, Tourandokht Baluchnejadmojarad, and Mehrdad Roghani. 2018. "Trigonelline mitigates lipopolysaccharide-induced learning and memory impairment in the rat due to its anti-oxidative and anti-inflammatory effect." *International Immunopharmacology* 61: 355–362. https://doi.org/10.1016/j.intimp.2018.06.019, https://www.sciencedirect.com/science/article/pii/S1567576918302704.

Khatibi, Nikan H., Vikram Jadhav, Shelton Charles, Jeffrey Chiu, John Buchholz, Jiping Tang, and John H. Zhang. 2011. "Capsaicin pre-treatment provides neurovascular protection against neonatal hypoxic-ischemic brain injury in rats." *Acta Neurochirurgica. Supplement* 111: 225–230.

Kohnen-Johannsen, Kathrin Laura, and Oliver Kayser. 2019. "Tropane alkaloids: Chemistry, pharmacology, biosynthesis and production." *Molecules* 24 (4): 796. https://www.mdpi.com/1420-3049/24/4/796.

Kondeva-Burdina, Magdalena, Javor Mitkov, Iva Valkova, Lily Peikova, Maya Georgieva, and Alexander Zlatkov. 2022. "Quantitative structure–neurotoxicity assessment and in vitro evaluation of neuroprotective and MAO-B inhibitory activities of series N′-substituted 3-(1,3,7-trimethyl-xanthin-8-ylthio)propanehydrazides." *Molecules* 27 (16): 5321. https://www.mdpi.com/1420-3049/27/16/5321.

Konrath, Eduardo L., Katiane Santin, Melissa Nassif, Alexandra Latini, Amélia Henriques, and Christianne Salbego. 2008. "Antioxidant and pro-oxidant properties of boldine on hippocampal slices exposed to oxygen–glucose deprivation in vitro." *Neurotoxicology* 29 (6): 1136–1140.

Krieglstein, Josef, and Ralf Rischke. 1991. "Vinpocetine increases the neuroprotective effect of adenosine in vitro." *European Journal of Pharmacology* 205 (1): 7–10. https://doi.org/10.1016/0014-2999(91)90762-F, https://www.sciencedirect.com/science/article/pii/001429999190762F.

Kumar, Anil, Atish Prakash, and Deeksha Pahwa. 2011. "Galantamine potentiates the protective effect of rofecoxib and caffeic acid against intrahippocampal kainic acid-induced cognitive dysfunction in rat." *Brain Research Bulletin* 85 (3–4): 158–168.

Kurek, Joanna. 2019. *Alkaloids: Their Importance in Nature and Human Life*. BoD–Books on Demand.

Lee, Yu-Young, Jin-Sun Park, Yea-Hyun Leem, Jung-Eun Park, Do-Yeon Kim, Youn-Hee Choi, Eun-Mi Park, Jihee Lee Kang, and Hee-Sun Kim. 2019. "The phosphodiesterase 10 inhibitor papaverine exerts anti-inflammatory and neuroprotective effects via the PKA signaling pathway in neuroinflammation and Parkinson's disease mouse models." *Journal of Neuroinflammation* 16 (1): 246. https://doi.org/10.1186/s12974-019-1649-3.

Leem, Y. H., J. S. Park, J. E. Park, D. Y. Kim, and H. S Kim. 2021. "Papaverine exerts neuroprotective effect by inhibiting NLRP3 inflammasome activation in an MPTP-induced microglial priming mouse model challenged with LPS." *Biomolecules & Therapeutics* 29 (3): 295–302. https://doi.org/10.4062/biomolther.2021.039.

Leem, Yea-Hyun, Jin-Sun Park, Jung-Eun Park, Do-Yeon Kim, Jihee Lee Kang, and Hee-Sun Kim. 2020. "Papaverine inhibits α-synuclein aggregation by modulating neuroinflammation and matrix metalloproteinase-3 expression in the subacute MPTP/P mouse model of Parkinson's disease." *Biomedicine & Pharmacotherapy* 130: 110576. https://doi.org/10.1016/j.biopha.2020.110576, https://www.sciencedirect.com/science/article/pii/S0753332220307691.

Li, Chao-Yue, Li-Ming Zhao, Xi-Wen Shi, and Jia-Dong Zhang. 2014. "Lobeline shows protective effects against MPTP-induced dopaminergic neuron death and attenuates behavior deficits in animals." *Experimental and Therapeutic* Medicine 7 (2): 375–378. https://doi.org/10.3892/etm.2013.1413.

Li, Yu-Jiao, Qi Yang, Kun Zhang, Yan-Yan Guo, Xu-Bo Li, Le Yang, Ming-Gao Zhao, and Yu-Mei Wu. 2013. "Cytisine confers neuronal protection against excitotoxic injury by down-regulating GluN2B-containing NMDA receptors." *NeuroToxicology* 34: 219–225. https://doi.org/10.1016/j.neuro.2012.09.009, https://www.sciencedirect.com/science/article/pii/S0161813X12002215.

Ling, Qian, Ming Liu, Min-Xia Wu, Ying Xu, Jian Yang, Hui-Hui Huang, and Chang-Xi Yu. 2014. "Anti-allodynic and neuroprotective effects of koumine, a benth alkaloid, in a rat model of diabetic neuropathy." *Biological and Pharmaceutical Bulletin* 37 (5): 858–864.

Liu, Ding-Quan, Shui-Chun Mao, Hai-Yan Zhang, Xiao-Qing Yu, Mei-Tang Feng, Bin Wang, Li-Hua Feng, and Yue-Wei Guo. 2013. "Racemosins a and b, two novel bisindole alkaloids from the green alga Caulerpa racemosa." *Fitoterapia* 91: 15–20.

Liu, Peifang, Hui Li, Yueqiu Wang, Xiaolin Su, Yang Li, Meiling Yan, Lan Ma, and Hui Che. 2020. "Harmine ameliorates cognitive impairment by inhibiting NLRP3 inflammasome activation and enhancing the BDNF/TrkB signaling pathway in STZ-induced diabetic rats." *Frontiers in Pharmacology* 11: 535.

Lopes, João P., Glauco Tarozzo, Angelo Reggiani, Daniele Piomelli, and Andrea Cavalli. 2013. "Galantamine potentiates the neuroprotective effect of memantine against NMDA-induced excitotoxicity." *Brain and Behavior* 3 (2): 67–74.

Ma, Ning-Tian, Ru Zhou, Ren-Yuan Chang, Yin-Ju Hao, Lin Ma, Shao-Ju Jin, Juan Du, Jie Zheng, Cheng-Jun Zhao, Yang Niu, Tao Sun, Wei Li, Kazuo Koike, Jian-Qiang Yu, and Yu-Xiang Li. 2015. "Protective effects of aloperine on neonatal rat primary cultured hippocampal neurons injured by oxygen–glucose deprivation and reperfusion." *Journal of Natural Medicines* 69 (4): 575–583. https://doi.org/10.1007/s11418-015-0928-2.

Machado-Filho, João Ananias, Alyne Oliveira Correia, Anyssa Brilhante Aires Montenegro, Maria Elizabeth Pereira Nobre, Gilberto Santos Cerqueira, Kelly Rose Tavares Neves, Maria da Graça Naffah-Mazzacoratti, Esper Abrão Cavalheiro, Gerly Anne de Castro Brito, and Glauce Socorro de Barros Viana. 2014. "Caffeine neuroprotective effects on 6-OHDA-lesioned rats are mediated by several factors, including pro-inflammatory cytokines and histone deacetylase inhibitions." *Behavioural Brain Research* 264: 116–125.

Mao, Qing-Qiu, Zhen Huang, Siu-Po Ip, Yan-Fang Xian, and Chun-Tao Che. 2012. "Protective effects of piperine against corticosterone-induced neurotoxicity in PC12 cells." *Cellular and Molecular Neurobiology* 32: 531–537.

Mao, Xiao-Yuan, Hong-Hao Zhou, Xi Li, and Zhao-Qian Liu. 2016. "Huperzine A alleviates oxidative glutamate toxicity in hippocampal HT22 cells via activating BDNF/TrkB-dependent PI3K/Akt/mTOR signaling pathway." *Cellular and Molecular Neurobiology* 36: 915–925.

Mirzaie, M., M. Khalili, Z. Kiasalari, and M. Roghani. 2016. "Neuroprotective and antiapoptotic potential of trigonelline in a striatal 6-hydroxydopamine rat model of Parkinson's disease." *Neurophysiology* 48 (3): 176–183. https://doi.org/10.1007/s11062-016-9586-6.

Nasrnezhad, Reza, Sohrab Halalkhor, Farzin Sadeghi, and Fereshteh Pourabdolhossein. 2021. "Piperine improves experimental autoimmune encephalomyelitis (EAE) in Lewis rats through its neuroprotective, anti-inflammatory, and antioxidant effects." *Molecular Neurobiology* 58 (11): 5473–5493. https://doi.org/10.1007/s12035-021-02497-5.

Nyakas, Csaba, Klára Felszeghy, Róbert Szabó, Jan N. Keijser, Paul G.M. Luiten, Zsolt Szombathelyi, and Károly Tihanyi. 2009. "Neuroprotective effects of vinpocetine and its major metabolite cis-apovincaminic acid on NMDA-induced neurotoxicity in a rat entorhinal cortex lesion model." *CNS Neuroscience & Therapeutics* 15 (2): 89–99. https://doi.org/10.1111/j.1755-5949.2009.00078.x, https://onlinelibrary.wiley.com/doi/abs/10.1111/j.1755-5949.2009.00078.x.

Ogunrinade, Folashade A., Victoria U. Iwuanyanwu, Satyajit D. Sarker, and Olumayokun A. Olajide. 2023. "Neuroprotection by skimmianine in lipopolysaccharide-activated BV-2 microglia." *Molecules* 28 (3): 1317.

Olajide, Olumayokun A., Harsharan S. Bhatia, Antonio C. P. de Oliveira, Colin W. Wright, and Bernd L. Fiebich. 2013a. "Anti-neuroinflammatory properties of synthetic cryptolepine in human neuroblastoma cells: Possible involvement of NF-κB and p38 MAPK inhibition." *European Journal of Medicinal Chemistry* 63: 333–339. https://doi.org/10.1016/j.ejmech.2013.02.004, https://www.sciencedirect.com/science/article/pii/S0223523413000901.

______. 2013b. "Inhibition of neuroinflammation in LPS-activated microglia by cryptolepine." *Evidence-Based Complementary and Alternative Medicine* 2013: 459723. https://doi.org/10.1155/2013/459723.

Olawepo, A, A. O. Ishola, M. S. Ajao, O. J. Olayemi, and L. A. Olayaki. 2017. "Atropine exposure in adolescence predispose to adult memory loss in Wistar rats." *International Journal of Biological and Chemical Sciences* 11 (5): 1937–1947.

Pákáski, Magdolna, Marietta Hugyecz, Péter Sántha, Gábor Jancsó, Annamária Bjelik, Ágnes Domokos, Zoltán Janka, and János Kálmán. 2009. "Capsaicin promotes the amyloidogenic route of brain amyloid precursor protein processing." *Neurochemistry International* 54 (7): 426–430. https://doi.org/10.1016/j.neuint.2009.01.012, https://www.sciencedirect.com/science/article/pii/S0197018609000308.

Papaliagkas, V., A. Anogianaki, G. Anogianakis, and G. Ilonidis. 2007. "The proteins and the mechanisms of apoptosis: A mini-review of the fundamentals." *Hippokratia* 11 (3): 108–113.

Paškulin, Roman, Polona Jamnik, Marko Živin, Peter Raspor, and Borut Štrukelj. 2006. "Ibogaine affects brain energy metabolism." *European Journal of Pharmacology* 552 (1–3): 11–14.

Petrovska, B. B. 2012. "Historical review of medicinal plants' usage." Pharmacognosy *Reviews* 6 (11): 1–5. https://doi.org/10.4103/0973-7847.95849.

Polak, P. E., S. Kalinin, D. Braun, A. Sharp, S. X. Lin, and D. L. Feinstein. 2012. "The vincamine derivative vindeburnol provides benefit in a mouse model of multiple sclerosis: Effects on the locus coeruleus." *Journal of Neurochemistry* 121 (2): 206–16. https://doi.org/10.1111/j.1471-4159.2012.07673.x.

Pravalika, Kanta, Deepaneeta Sarmah, Harpreet Kaur, Kanchan Vats, Jackson Saraf, Madhuri Wanve, Kiran Kalia, Anupom Borah, Dileep R. Yavagal, Kunjan R. Dave, and Pallab Bhattacharya. 2019. "Trigonelline therapy confers neuroprotection by reduced glutathione mediated myeloperoxidase expression in animal model of ischemic stroke." *Life Sciences* 216: 49–58. https://doi.org/10.1016/j.lfs.2018.11.014, https://www.sciencedirect.com/science/article/pii/S0024320518307203.

Qi, Wen, Jiahong Sun, James W Simpkins, and Dan Yuan. 2015. "Isolation and identification of twelve metabolites of isocorynoxeine in rat urine and their neuroprotective activities in HT22 cell assay." *Planta Medica* 81 (01): 46–55.

Qiu, Hong-Qiang, Ying Xu, Gui-Lin Jin, Jian Yang, Ming Liu, Su-Ping Li, and Chang-Xi Yu. 2015. "Koumine enhances spinal cord 3α-hydroxysteroid oxidoreductase expression and activity in a rat model of neuropathic pain." *Molecular Pain* 11: s12990-015-0050-1.

Qiu, Zhengguo, Kefeng Wang, Chao Jiang, Yuqiang Su, Xiaoying Fan, Jing Li, Sha Xue, and Li Yao. 2020. "Trigonelline protects hippocampal neurons from oxygen-glucose deprivation-induced injury through activating the PI3K/Akt pathway." *Chemico-Biological Interactions* 317: 108946. https://doi.org/10.1016/j.cbi.2020.108946, https://www.sciencedirect.com/science/article/pii/S0009279719311901.

Ressler, Kerry J, and Charles B Nemeroff. 2000. "Role of serotonergic and noradrenergic systems in the pathophysiology of depression and anxiety disorders." *Depression and Anxiety* 12 (S1): 2–19.

Riad M., and Candice C. Hithe. 2023. Scopolamine. In *StatPearls* [Internet]. StatPearls Publishing. https://www.ncbi.nlm.nih.gov/books/NBK554397/.

Saglam, Emre, Selim Zırh, Canan Cakir Aktas, Sevda Fatma Muftuoglu, and Burcak Bilginer. 2021. "Papaverine provides neuroprotection by suppressing neuroinflammation and apoptosis in the traumatic brain injury via RAGE- NF-<kappa>B pathway." *Journal of Neuroimmunology* 352: 577476. https://doi.org/10.1016/j.jneuroim.2021.577476, https://www.sciencedirect.com/science/article/pii/S0165572821000035.

Sharma, Tejas, Vishal Airao, Nimesh Panara, Devendra Vaishnav, Vishavas Ranpariya, Navin Sheth, and Sachin Parmar. 2014. "Solasodine protects rat brain against ischemia/reperfusion injury through its antioxidant activity." *European Journal of Pharmacology* 725: 40–46. https://doi.org/10.1016/j.ejphar.2014.01.005, https://www.sciencedirect.com/science/article/pii/S0014299914000065.

Shoaib, Rana Muhammad, Jing-Yang Zhang, Xiao-Fang Mao, and Yong-Xiang Wang. 2019. "Gelsemine and koumine, principal active ingredients of Gelsemium, exhibit mechanical antiallodynia via spinal glycine receptor activation-induced allopregnanolone biosynthesis." *Biochemical Pharmacology* 161: 136–148. https://doi.org/10.1016/j.bcp.2019.01.014, https://www.sciencedirect.com/science/article/pii/S0006295219300206.

Solmaz, Volkan, Mahmut Kaya, Fatma Betul Uslu, Ozum Atasoy, and Oytun Erbaş. 2022. "Papaverine has therapeutic potential for sepsis-induced neuropathy in rats, possibly via the modulation of HMGB1-RAGE axis and its antioxidant prosperities." *Journal of Investigative Surgery* 35 (1): 7–13. https://doi.org/10.1080/08941939.2020.1809751.

Sun, Lingyan, Xia Tian, Lingshan Gou, Xin Ling, Ling Wang, Yan Feng, Xiaoxing Yin, and Yi Liu. 2013. "Beneficial synergistic effects of concurrent treatment with theanine and caffeine against cerebral ischemia–reperfusion injury in rats." *Canadian Journal of Physiology and Pharmacology* 91 (7): 562–569.

Sun, Piyun, Shuyan Zhang, Yanyan Li, and Lihua Wang. 2014. "Harmine mediated neuroprotection via evaluation of glutamate transporter 1 in a rat model of global cerebral ischemia." *Neuroscience Letters* 583: 32–36.

Tan, Xiang-Shan, Jing-Yi Ma, Ru Feng, Chao Ma, Wen-Jing Chen, Yu-Peng Sun, Jie Fu, Min Huang, Chi-Yu He, and Jia-Wen Shou. 2013. "Tissue distribution of berberine and its metabolites after oral administration in rats." *PloS One* 8 (10): e77969.

Tárnok, K., E. Kiss, P. G. M. Luiten, C. Nyakas, K. Tihanyi, K. Schlett, and U. L. M. Eisel. 2008. "Effects of vinpocetine on mitochondrial function and neuroprotection in primary cortical neurons." *Neurochemistry International* 53 (6): 289–295. https://doi.org/10.1016/j.neuint.2008.08.003, https://www.sciencedirect.com/science/article/pii/S0197018608001320.

Taveira, Marcos, Carla Sousa, Patrícia Valentão, Federico Ferreres, João P. Teixeira, and Paula B. Andrade. 2014. "Neuroprotective effect of steroidal alkaloids on glutamate-induced toxicity by preserving mitochondrial membrane potential and reducing oxidative stress." *The Journal of Steroid Biochemistry and Molecular Biology* 140: 106–115. https://doi.org/10.1016/j.jsbmb.2013.12.013, https://www.sciencedirect.com/science/article/pii/S0960076013002860.

Toledo, Juan P., Eduardo J. Fernández-Pérez, Ildete L. Ferreira, Daniela Marinho, Nicolas O. Riffo-Lepe, Benjamin N. Pineda-Cuevas, Luis F. Pinochet-Pino, Carlos F. Burgos, A. Cristina Rego, and Luis G. Aguayo. 2021. "Boldine attenuates synaptic failure and mitochondrial deregulation in cellular models of Alzheimer's disease." *Frontiers in Neuroscience* 15: 617821.

Tundis, Rosa, Federica Menichini, Filomena Conforti, Monica R. Loizzo, Marco Bonesi, Giancarlo Statti, and Francesco Menichini. 2009. "A potential role of alkaloid extracts from Salsola species (Chenopodiaceae) in the treatment of Alzheimer's disease." *Journal of Enzyme Inhibition and Medicinal Chemistry* 24 (3): 818–824.

Vajda, F. J. 2002a. "Neuroprotection and neurodegenerative disease." Journal *of* Clinical Neuroscience 9 (1): 4–8. https://doi.org/10.1054/jocn.2001.1027.

Vajda, Frank JE. 2002b. "Neuroprotection and neurodegenerative disease." *Journal of Clinical Neuroscience* 9 (1): 4–8.

Villalpando-Vargas, Fridha, Laura Medina-Ceja, Anne Santerre, and Edgar A. Enciso-Madero. 2020. "The anticonvulsant effect of sparteine on pentylenetetrazole-induced seizures in rats: A behavioral, electroencephalographic, morphological and molecular study." *Journal of Molecular Histology* 51: 503–518.

Wang, Zhi-Fei, Juan Wang, Hai-Yan Zhang, and Xi-Can Tang. 2008. "Huperzine A exhibits anti-inflammatory and neuroprotective effects in a rat model of transient focal cerebral ischemia." *Journal of Neurochemistry* 106 (4): 1594–1603.

Xiao, Xianghua, Juntian Liu, Jingwen Hu, Tianxia Li, and Yuanhui Zhang. 2007. "Protective effect of protopine on the focal cerebral ischaemic injury in rats." *Basic & Clinical Pharmacology & Toxicology* 101 (2): 85–89.

Xiao, Xianghua, Juntian Liu, Jingwen Hu, Xiuping Zhu, Hua Yang, Chaoyun Wang, and Yuanhui Zhang. 2008. "Protective effects of protopine on hydrogen peroxide-induced oxidative injury of PC12 cells via Ca2+ antagonism and antioxidant mechanisms." *European Journal of Pharmacology* 591 (1): 21–27. https://doi.org/10.1016/j.ejphar.2008.06.045, https://www.sciencedirect.com/science/article/pii/S0014299908006286.

Xiong, Bo-Jun, Ying Xu, Gui-Lin Jin, Ming Liu, Jian Yang, and Chang-Xi Yu. 2017. "Analgesic effects and pharmacologic mechanisms of the Gelsemium alkaloid koumine on a rat model of postoperative pain." *Scientific Reports* 7 (1): 14269. https://doi.org/10.1038/s41598-017-14714-0.

Xu, Zhifang, Wei Feng, Qian Shen, Nannan Yu, Kun Yu, Shenjun Wang, Zhigang Chen, Seiji Shioda, and Yi Guo. 2017. "Rhizoma coptidis and berberine as a natural drug to combat aging and aging-related diseases via anti-oxidation and AMPK activation." *Aging and Disease* 8 (6): 760.

Yang, Wei, Yu-Hua Chen, Hao Liu, and Hong-Dang Qu. 2015. "Neuroprotective effects of piperine on the 1-methyl-4-phenyl-1,2,3,6-tetrahydropyridine-induced Parkinson's disease mouse model." International Journal of Molecular Medicine 36 (5): 1369–1376. https://doi.org/10.3892/ijmm.2015.2356.

Yi, Chenju, Pascal Ezan, Paola Fernandez, Julien Schmitt, Juan C. Saez, Christian Giaume, and Annette Koulakoff. 2017. "Inhibition of glial hemichannels by boldine treatment reduces neuronal suffering in a murine model of Alzheimer's disease." *Glia* 65 (10): 1607–1625.

Zang, Jiankun, Yousheng Wu, Xuanlin Su, Tianyuan Zhang, Xionglin Tang, Dan Ma, Yufeng Li, Yanfang Liu, Ze'an Weng, Xuanzhuo Liu, Chi Kwan Tsang, Anding Xu, and Dan Lu. 2021. "Inhibition of PDE1-b by vinpocetine regulates microglial exosomes and polarization through enhancing autophagic flux for neuroprotection against ischemic stroke." *Frontiers in Cell and Developmental Biology* 8. https://doi.org/10.3389/fcell.2020.616590, https://www.frontiersin.org/articles/10.3389/fcell.2020.616590.

Zhong, Zeqi, Yuan Tao, and Hui Yang. 2015. "Treatment with harmine ameliorates functional impairment and neuronal death following traumatic brain injury." *Molecular Medicine Reports* 12 (6): 7985–7991.

Zhou, Ji-Yin, and Shi-Wen Zhou. 2012. "Protection of trigonelline on experimental diabetic peripheral neuropathy." *Evidence-Based Complementary and Alternative Medicine* 2012: 164219. https://doi.org/10.1155/2012/164219.

Zhou, Jiyin, Xiaohuang Du, Min Long, Zuo Zhang, Shiwen Zhou, Jianyun Zhou, and Guisheng Qian. 2016. "Neuroprotective effect of berberine is mediated by MAPK signaling pathway in experimental diabetic neuropathy in rats." *European Journal of Pharmacology* 774: 87–94.

Zhou, Xi-Qiao, Xiao-Ning Zeng, Hui Kong, and Xiu-Lan Sun. 2008. "Neuroprotective effects of berberine on stroke models in vitro and in vivo." *Neuroscience Letters* 447 (1): 31–36.

Zhu, Qing-Luan, Yu-Xiang Li, Ru Zhou, Ning-Tian Ma, Ren-Yuan Chang, Teng-Fei Wang, Yi Zhang, Xiao-Ping Chen, Yin-Ju Hao, and Shao-Ju Jin. 2014. "Neuroprotective effects of oxysophocarpine on neonatal rat primary cultured hippocampal neurons injured by oxygen-glucose deprivation and reperfusion." *Pharmaceutical Biology* 52 (8): 1052–1059.

5 Molecular Mechanism of Action of Phytoconstituents in Neuropsychiatric Disorders

Urmila Aswar, Rashmi Patil, Likhit Akotkar, and Atmaram Pawar

5.1 INTRODUCTION

Neuropsychiatric disorders are highly prevalent and significantly have increased the global disease burden. They are cumulatively complex diseases with a poorly described mechanism, and research into them is progressing slowly. Approximately 970.1 million individuals worldwide suffer from mental problems, a number that has significantly increased since the global pandemic in 2020. Stress, major depressive disorder (MDD), anxiety, psychosis, schizophrenia (SZ), bipolar disorder (BD), insomnia and obsessive-compulsive disorder (OCD) are some of the most common neuropsychiatric disorders (Figure 5.1). Drawing a distinction between "disorder" and "normality" of a person's wellness is a crucial part of classifying and identifying mental diseases (Sotille et al, 2022). Recently phytomedicines have gained immense demand as a prophylactic as well as therapeutic use for numerous neuropsychiatric disorders. Herbs are becoming more and more popular all over the world due to their effectiveness in treating sickness due to their multitarget mechanisms and the drawbacks of pharmaceutical treatments in terms of side effects and expense. Current chapter will discuss the various phytoconstituents studied clinically and preclinically for neuropsychiatric disorders and emphasis is also given to understand their mechanism of actions.

5.2 PHYTOCONSTITUENTS FOR STRESS-RELATED DISORDERS

Stress is a divisive condition caused by stressors, putting at risk the homeostasis necessary for survival and well-being. The stress system is a complex array of molecular, cellular and neuroendocrine components. According to physiological theory, the sympathetic-adrenomedullary axis (SAM axis) and adrenergic system are immediately activated in the body's response to stress, followed by the hypothalamus-pituitary-adrenal axis (HPA axis), which are the main players in stress (Figure 5.2).

DOI: 10.1201/9781003389781-5

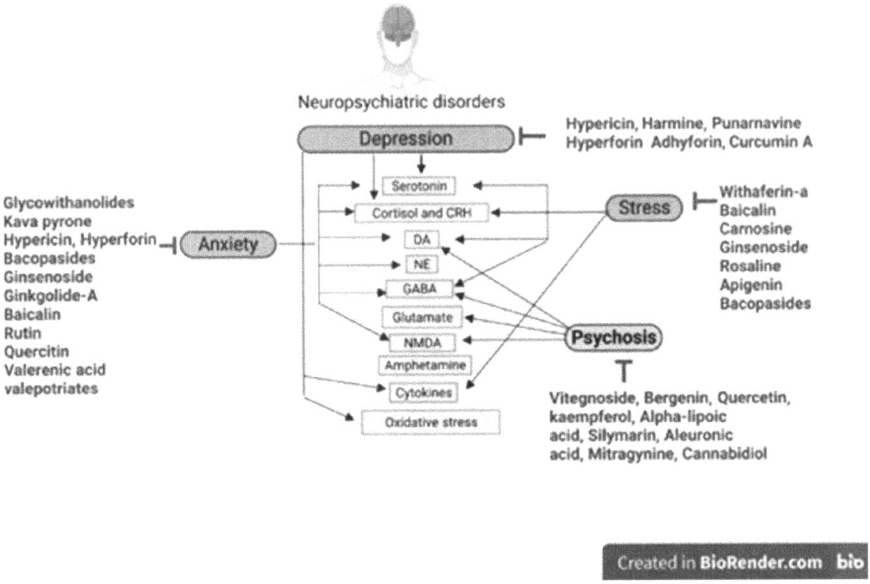

FIGURE 5.1 Diagrammatic representation of various phytoconstituents and their role in major neuropsychiatric disorders (stress, anxiety, depression and psychosis). Abbreviations: corticotropin-releasing hormone (CRH), dopamine (DA), nor-epinephrine (NE), gamma-aminobutyric acid (GABA), n-methyl-d-aspartate (NMDA).

It affects the cognitive, reward and fear systems in the brain, and centrally it produces structural and functional alterations that result in neuropsychiatric diseases such as anxiety, depression and sleeplessness (Musazzi et al, 2018; Patriquin and Mathew, 2017; Yang et al, 2015). The cytokine pathway, the neurotransmitter pathway (dopamine, serotonin) and the neuroendocrine pathway (cortisol) are the three primary mechanisms implicated in stress. The body releases arginine, vasopressin, corticotropin-releasing hormone (CRH), pro-opiomelanocortin-derived peptides, neuropeptide Y, activating locus coeruleus (LC) and autonomic norepinephrine centres in the brainstem known as allostasis for survival and adaptation to the changing internal and external environment. The thyroid hormone axis, growth, reproductive, gastrointestinal, cardiorespiratory, metabolic and immunologic systems are all affected at the periphery. Many neuropathological and emotional disorders, including depression, SZ, post-traumatic stress disorder and cognitive impairment, may be caused by long-term chronic stress. By enhancing sleep, food and exercise, as well as integrating meditation and some herbs into our food, medicine or diet can reduce stress.

The families of plants, Solanaceae, Araliaceae, Lamiaceae, Crassulaceae, Asteraceae and Scrophulariaceae are the most frequently utilised as anti-stress remedies around the world. The most popular herbs for stress management include *Withania somnifera* (WS), *Rosmarinus officinalis, Panax ginseng, Rhodiola rosea* L., *Matricaria recutita* L., *Bacopa monnieri, Convolvulus pluricaulis, Eleutherococcus senticosus, Schisandra chinensis* and *Centella asiatica* L. WS,

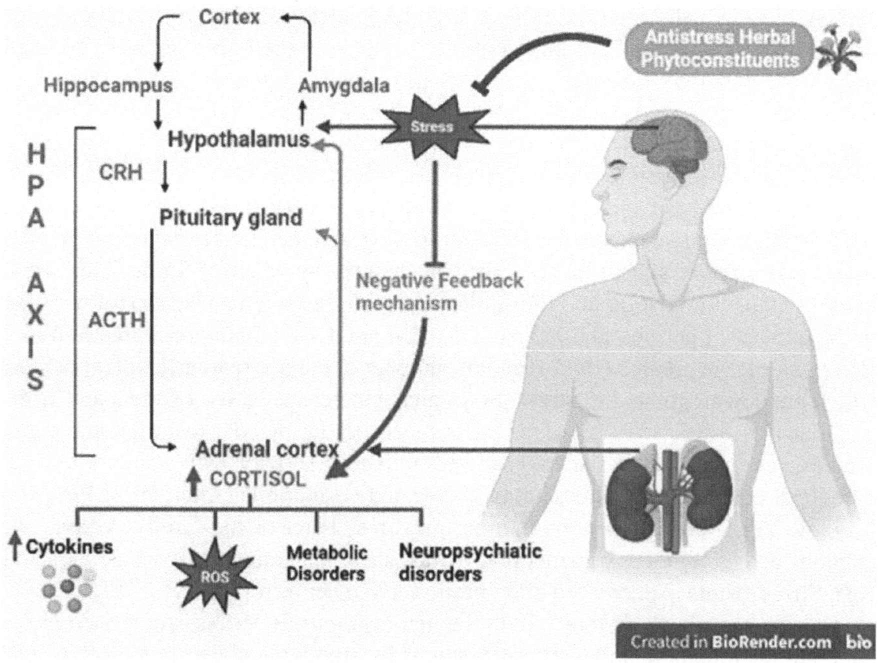

FIGURE 5.2 Diagrammatic representation of stress pathway, representing HPA axis dysregulation. Stress activates a complex network of major endocrine hormones known as the HPA axis. Hypothalamus secretes CRH into the bloodstream and increases the activity of the sympathetic nervous system, signalling the pituitary gland to secrete adrenocorticotropic hormone. ACTH released into the bloodstream travels down to the adrenal cortex, resulting in adrenal glands to secrete glucocorticoids like cortisol. Cortisol has receptors in almost all cell types and helps our body cope with the immediate stress by negative feedback mechanism that impact back on the function of the hypothalamus and pituitary. In case of extreme persistent stress, it overloads the HPA axis and it continues to release hormones on accelerated basis leading to desensitisation and stop recognising signals to stop producing cortisol. It results in non-functioning of negative feedback loop leading to various metabolical, neuropsychological and inflammatory disorders. Introduction of herbal phytoconstituents results in blockage of stress response by modulation of the HPA axis through corticotrophin-releasing hormone and cortisol. Abbreviations: Hypothalamic-pituitary-adrenal (HPA) axis, corticotropin-releasing hormone (CRH), adrenocorticotropic hormone (ACTH) and reactive oxygen species (ROS).

also known as ashwagandha, reduces stress symptoms primarily via modulating the HPA axis, SAM axis, gamma-aminobutyric acid (GABA)ergic and serotonergic pathways, as well as other stress-related pathways. The active phytochemical, withaferin-A, withanolide acts mainly through the suppression of cortisol and corticosterone which get elevated during stress-induced HPA axis dysregulation. A recent clinical trial involving 60 healthy adults between the ages of 18 and 75 with below 20 value on the perceived stress scale (PSS) was studied for adaptogenic and anxiolytic effects of ashwagandha root extract (Salve et al, 2019). PSS is a 14-item instrument to measure stress. For eight weeks, they received either ashwagandha extract 125 mg, ashwagandha extract 300 mg or identical placebo twice daily for eight weeks as a capsule. A significant improvement in the PSS scale was observed in 58 volunteers (who completed the trial) with a marked drop in blood cortisol levels as well as improved sleep pattern. Another clinical study focused on treating an obese adult population with 300 mg of ashwagandha root extract (>5% withanolides) twice a day for 10 weeks. This has led to decreased levels of cortisol in the blood, body weight and basic metabolic index, as well as decreased body fat. A preclinical study in rats showed decreased corticosterone levels in the hippocampus in the stress model (Baitharu et al, 2013). The other mechanism by which WS alleviates stress is by immune stimulation and anti-inflammatory effects. This leads to an increase in the number of peripheral T cells (CD3, CD4 and CD8 populations), interleukin-2 (IL-2), interferon gamma (INF-γ) and the number of polymorphonuclear leukocytes, which are downregulated in stress. Preclinical studies using WS extracts and withanolides-rich extracts (2.7%) on stress induced in rats by cold restraint stress and forced swimming-induced stress, showed antistress activity with increased white blood cell counts, neutrophils, lymphocytes and eosinophils (Anju, 2011; Puri and Purohit, 2016). *Panax ginseng,* another antistress plant of the family Araliaceae, contains active phytoconstituents that reduce stress including ginsenoside Rb1 and Rg3. Stress results in decreased brain brain-derived neurotrophic factor (BDNF) levels altering heat shock protein – 70 in the hippocampus. BDNF is synthesised in the endoplasmic reticulum and plays its functions by binding on the tropomyosin receptor kinase B receptor further activating the signal transduction cascade to promote neuroplasticity, neuronal survival and differentiation (Lee and Kim, 2010).

Clinical study of Korean red ginseng (KRG) in two high-stress occupations, that is, nurses and firefighters aged 20–60 years were included in the study (Baek et al, 2019). The stress levels of these volunteers and cognition were assessed using the Beck Depression Inventory and the Stress Response Inventory. Chronic stress in volunteers had raised levels of cortisol and catecholamine, adrenaline and reduced triglycerides. Treatment with KRG for 2 g/day of KRG (LAX-101) or placebo for 6 weeks showed reduced levels of cortisol and epinephrine and normalised triglycerides. This is assumed to be the possible mechanism for central nervous system (CNS) stabilisation. Further, KRG was found to contain Rb1, 6.44 mg/g; Rb2, 2.25 mg/g; Rc, 2.68 mg/g; Rd, 0.50 mg/g; Re, 2.08 mg/g; Rf, 0.89 mg/g; Rg1, 3.21 mg/g; Rg2, 0.29 mg/g; Rg3, 0.18 mg/g. Treatment with Rb1 also increases the levels of thermal shock protein 70 and BDNF that have neuroprotective and anti-stress properties (Kim et al, 2013). A pre-clinical study consisting of total *ginseng panax* saponins (TSPG) was studied on chronic unpredictable mild stress (CUMS)-induced stress in rats. They were treated with TSPG (50 mg/kg/day, 100 mg/kg/day) during the entire modelling period. TSPG was found to improve depressive behaviour and to exhibit neuroprotection. TSPG inhibited pro-inflammatory cytokines, as well as P38 and Jun N terminal kinase (JNK) protein phosphorylation in the hippocampus, indicating significant antidepressant and antistress activity. *Camellia sinensis* L. commonly called green tea from the Theaceae family. The active phytoconstituent, L-theanine is thenon-proteinogenic amino acid, passes freely through the blood-brain barrier, acts by reduction of presynaptic glutamate release, increased inhibitory neurotransmitter GABA, and enhanced glycine and dopamine release (Kakuda et al, 2008;

Shen et al, 2019; Yamada et al, 2007). Clinical study on fifth-year college students ($n = 20$) of the school of pharmaceutical sciences, practising outside the university at the hospital or a drug store, for 11 weeks, where they were administered theanine (200 mg, twice a day, after breakfast and lunch). These were assessed for anxiety using a state anxiety inventory test. At the end of the study, treatment with theanine supplementation was found to be beneficial in suppressing psychosocial stress. It was assumed that theanine improved the HPA axis (Unno et al, 2013).

5.3 PHYTOCONSTITUENTS USED FOR ANXIETY

People who experience anxiety have difficulty with their sleep habits, their ability to focus and their ability to interact with others. Anxiety disorders like panic disorder are characterised by recurrent, unprovoked panic attacks that leave sufferers feeling extremely anxious, approaching doom, dizzy and short of breath (Markowitz et al, 1989). Agoraphobia, generalised anxiety disorder (GAD), social phobia and phobias of animals, heights, elevators, blood or injections are examples of other anxiety disorders (Ballenger, 2007; Craske et al, 1989; Magee et al, 1996; Schweizer, 1995). Due to the varied character of anxiety disorders, concomitant conditions including stress, diabetes, depression or hereditary factors could contribute to their aetiology (Amray et al, 2019). Serotonin (5-hydroxytryptamine; 5-HT), GABA, noradrenaline (NE), adrenocorticotropic hormone (ACTH), CRH, neuropeptide-Y and cholecystokinin (CCK), main excitatory amino acid, glutamate, N-methyl-D aspartate (NMDA), alpha-amino-3-hydroxy-5-methylisoxazole propionate (AMPA) and kainic acid are a few neurotransmitters involved in the development of anxiety. Serotonergic dysfunction leads to an aberrant 5-HT release and/or reuptake regulation or abnormal 5-HT signal response. A key stress-response mechanism that causes anxiety is the LC noradrenergic system that releases epinephrine during stress and works in tandem with the corticotropin-releasing factor (CRF) system. CRF neurons are found in the paraventricular nucleus, where they mediate the HPA axis's activity in response to stress, which raises plasma cortisol levels and increases CRF. The serotonergic and noradrenergic systems are interconnected and together play a prime role in the regulation of mood and anxiety (Bremner et al, 2010). The inhibitory neurotransmitter, GABA influences the level of dopamine, 5-HT and NMDA. Thus, NMDA antagonists and non-NMDA type receptors can reduce excitatory glutamate receptor activity and possess anxiolytic effects (LeDoux, 1994)). Alternately, increasing GABA neurotransmission can reduce the output of the excitatory system. Thus, for effective control of the behavioural and physiological responses related to anxiety, a balance between NE, 5-HT, glutamate receptor-mediated excitement and GABA receptor-mediated inhibition should exist. Selective 5-HT reuptake inhibitors (SSRI) and 5-HT-norepinephrine reuptake inhibitors are the first-line treatments for anxiety.

An alarming amount of people are becoming overly dependent on synthetic pharmaceuticals to treat or relieve anxiety disorders, yet there are still many people who prefer herbal remedies. Important phytochemicals acting on the CNS are being identified, including saponins, alkaloids, polyphenols, triterpenes, essential oils, fatty acids and flavonoids. The phytoconstituents from families of herbs – Solanaceae, Piperaceae, Hypericaceae, Plantaginaceae, Passifloraceae, Araliaceae, Lamiaceae,

Caprifoliaceae, Turneraceae, Elaeocarpaceae, Lamiaceae, Asteraceae and Apiaceae are commonly used as anti-anxiety remedies. According to the European Medicines Agency (EMA) and the National Agency of Sanitary Surveillance (ANVISA), *Withania somnifera* L., *Piper methysticum, Hypericum perforatum* L., *Bacopa monnieri* L., *Passiflora incarnata* L. and *Panax ginseng* are safe herbal remedies. *Ocimum sanctum* L. also known as Tulsi contains eugenol (70%) as a major constituent. Clinical study carried out in patients suffering from generalised anxiety disorder were selected and given the ethanolic extract of *Ocimum sanctum* L. supplement (500 mg capsule, twice daily) for 60 days. On completion of the study, reduced stress, attenuated anxiety and negated depression attributing to the regulation of hypothalamohypophyseal-adrenocortical axis were noted (Bhattacharyya et al, 2008). OciBest, an extract of *Ocimum sanctum* L, having ociglycoside-I (hydroxychavicol glucoside/4-allyl-1-O-β-D-glucopyronosyl-2-hydroxybenzene; >0.1% w/w), rosmarinic acid (>0.2% w/w) and triterpene acids (>2.5% w/w) was tested in 158 patients in a randomised, double-blind trial which exhibited antistress effects (Saxena et al, 2012). Eugenol treatment in rats exposed to restrain for 4 hours attenuated ulcer index and reduced cortisol without affecting NE and amelioration of HPA axis (Garabadu et al, 2011). The most commonly active phytoconstituents present in *Melissa officinalis* L. (Lamiaceae) also known as lemon herb are rosmarinic acid, ursolic acid and oleanolic acid is native to Europe and Asia and is responsible for its antianxiety effect. Clinical investigation in patients with anxiety and sleep disorder, supplemented with dried powder of aerial part of *Melissa officinalis* L. (3 g) for 56 days showed improvement in anxiety score (Haybar et al, 2018). The author suggested its antianxiety effects attributed to its serotonergic activity, reduced cortisol level and its GABA transaminase inhibitory effect. The preclinical study in rats has proven that *Melissa officinalis* L. act by inhibiting GABA transaminase and shows promising anxiolytic effect (Bathala et al, 2012). *Hypericum perforatum* L. (Saint John's Wort (SJW)) containing active phytoconstituents hypericin and hyperforin is majorly used in anxiety disorder. Hypericin inhibits monoamine oxidase-A (MAO-A) and monoamine oxidase-B (MAO-B) enzymes and regulates dopamine levels. Hyperforin elevates the extracellular levels of 5-HT, dopamine, NE, GABA and L-glutamate. *Passiflora incarnata* L. (Passion flower) contain active phytoconstituents such as orientin, isoorientin, vitexin, isovitexin and chrysin shows antianxiety effect by its action on benzodiazepine and GABA receptors (Table 5.1).

5.4 PHYTOCONSTITUENTS USED FOR DEPRESSION

Depression is a neuropsychiatric disorder characterised by anxiety, aggression, dementia and emotional imbalance. According to the World Health Organization, depression is considered the most prevalent among all neuropsychiatric disorders. Worldwide 322 million people are suffering from depression (Vasha et al, 2023). Lack of diagnosis makes depression a more severe form, resulting in loneliness and suicidal tendencies. Various studies have reported complications and comorbidities associated with depression such as cardiovascular changes, constipation, colitis, erectile dysfunctions and thyroid problems. In severe cases, it may lead to MDD, Alzheimer's disease and diabetes. Various synthetic drugs such as tricyclic

TABLE 5.1
Summary of Phytoconstituents Reported for Various Neuropsychiatric Disorders and Their Mechanism of Action.

S. No	Antipsychotic	Biological Source-Family	Molecular Mechanism	References
		Antipsychotic		
1.	Vitegnoside	*Vitex negundo* (Lamiaceae)	 Antagonizes dopaminergic activity.	(Abidin et al, 2015; Ahmed et al, 2014)
2.	Bergenin	*Securinega virosa* (Euphorbiaceae)	 Antidopaminergic activity on the melatonin MT1 and MT2 and promotes sleep onset.	(Magaji et al, 2015)
3.	Alkaloid, saponin glycoside, steroids, reducing sugar, resin and flavonoid	*Albizia zygia* (DC) (Leguminosae-Mimosoideae)	 Enhance glutamate and dopamine release in the limbic striatal regions.	(Amoateng et al, 2017; Kumbol et al, 2018)
4.	Quercetin, kaempferol	*Lonchocarpus cyanescens* (Fabaceae)	 Inhibition of hyperactivation caused by dopamine.	(Arowona et al, 2014)
5.	Saponins	*Crassocephalum bauchiense* (Asteraceae)	 Blockade of dopamine D-2 receptors and GABAergic activation	(Taïwe et al, 2012)
6.	Flavonoids, essential oils	*Ocimum sanctum* (Lamiaceae)	 Decreases dopamine levels and reduces oxidative stress.	(Sharma et al, 2016)
7.	Alpha lipoic acid	Spinach/*Spinacia oleracea* (Amaranthaceae), brocolli, tomato	 Restore monoamines by modulation of 5HT3 receptor.	(Perera et al, 2011)
8.	Silymarin	*Silybum marianum* (L.) Gaertn (Asteraceae)	 Block the reuptake of 5-HT into presynaptic 5-HT neurons.	(Karcı and Celik, 2020)
9.	Aleuronic acid	*Valeriana Officinalis* L. (Valerianaceae)	 GABA A receptor modulator.	(Karcı and Celik, 2020; Kochlamazashvili et al, 2010)

(Continued)

TABLE 5.1 ***(Continued)***
Summary of Phytoconstituents Reported for Various Neuropsychiatric Disorders and Their Mechanism of Action.

S. No	Antipsychotic	Biological Source-Family	Molecular Mechanism	References
10.	Morin	*Morus alba* (Moraceae)	 Attenuate neurochemical change and reduce ROS.	(Ben-Azu et al, 2018)
11.	Mitragynine	*Mitragyna speciosa* (Rubiaceae)	 Inhibit dopamine-induced contractile response.	(Vijeepallam et al, 2016)
12.	Cannabidiol	*Cannabis sativa* (Cannabaceae)	 Activates 5-HT1A and glutamate receptor (NMDA), inhibit FAAH enzyme.	(Xiang et al, 2022)
		Antidepressants		
13.	Curcumin Chrysin Arctigenin Glycyrrhizic acid	*Curcuma longa* L. (Zingiberaceae) *Apis dorsata* (Apidae) *Carthamus tinctorius* L. (Asteraceae) *Glycyrrhiza glabra* L. (Fabaceae)	 ↓IDO/TDO	(Borges Filho et al, 2016; Seo et al, 2015; Temml et al, 2013; Wang et al, 2018)
14.	Eugenol Antiepilepsirine Palmatine Gallic acid β-Carotene Icariin Fisetin	*Syzygium aromaticum* (Myrtaceae) *Albizzia julibrissin* (Fabaceae) *Berberis aristate* (Berberidiaceae) *Emblica officinalis* (Phylantheaceae) *Daucus carota subsp.* sativus (Apiaceae) *Epimedium grandiflorum* (Berberidaceae) *Fragaria ananassa* (Rosaceae)	 ↓MAO-A	(Tao et al, 2005)
15.	Cinnamic aldehyde Resveratrol Salvianolic acid B Ferulic acid Quercetin	*Cinnamomum verum* (Lauraceae) *Muscadine grapes* (Vitaceae) *Salvia miltiorrhiza* (Danshen)	 ↓PG-E2 ↓ TNF-α and IL-Iβ; ↑ Bcl-2 expression; ↓ NF-κB gene expression.	(Yao et al, 2015)

(Continued)

TABLE 5.1 ***(Continued)***
Summary of Phytoconstituents Reported for Various Neuropsychiatric Disorders and Their Mechanism of Action.

S. No	Antipsychotic	Biological Source-Family	Molecular Mechanism	References
		(Lamiaceae) *Pavetta owariensis* (Rubiaceae) *Coccinia indica* (Cucurbitaceae)	↓ ACTH and CORT; ↑ CAT, SOD and GSH peroxidase activities; ↓ MDA; ↓ IL-6, IL-1β and TNF-α; ↓ NLRP3 inflammasome activation. ↓ IL-1β, IL-6 and TNF-α; ↓ ACTH and CORT; ↓ NF-κB activation; ↓ nNOS expression; ↑ IL-10; GR protein expression. ↑ GSH; ↑ SOD and CAT activity; ↑ 5-HT; ↓ Glu; ↓ IL-1β and TNF-α.	
		Antistress		
16.	Withaferin-A, withanolide	*Withania somnifera* (Solanaceae)	 Reduction of cortisol levels.  Increase upregulates receptor population in corpus striatum.	(Archana and Namasivayam, 1998; Jamalludin and Manshoor, 2022)
17.	Asiatic acid, madecassic acid, asiaticoside, madecassoside and brahmic acid	*Centella asiatica* L. (Apiaceae)	 Decrease oxidative stress.	(Zakaria et al, 2023; Kalshetty et al, 2012)
18.	Baicalin and wogonin	*Scutellaria baicalensis* (Lamiaceae)	 Antioxidative activity, interact with GABA receptor, reduce serum levels of IL-2, corticosterone.	(Lee et al, 2007)
19.	Carnosine, carnosic acid, carnosal, rosmanol, epirosmanol	*Rosmarinus officinalis* (Lamiaceae)	 Modulate sympathetic nervous system and HPA axis, activate cholinergic activity (AchE activity) in PC12 cells via phosphorylation of ERK1/2.	(Kayashima et al, 2020)

(Continued)

TABLE 5.1 *(Continued)*
Summary of Phytoconstituents Reported for Various Neuropsychiatric Disorders and Their Mechanism of Action.

S. No	Antipsychotic	Biological Source-Family	Molecular Mechanism	References
20.	Ginsenoside Rb1 and Rg3	*Panax ginseng* (Araliaceae)	Increased ACTH and corticosterone.	(Huang et al, 2019)
21.	L-Theanine	*Camellia sinensis* (Theaceae)	Reduce presynaptic glutamate release, increase inhibitory neurotransmitter GABA and enhance glycine and dopamine release.	(Sarris et al, 2019; Unno et al, 2013)
22.	Rosalin (WS® 1375), Rodiolin, rosin, rosavin, rosarin and rosiridin	*Rhodiola rosea* (Crassulaceae)	Normalise cortisol synthesis.	(Emilija and Quintela, 2022)
23.	Apigenin	*Matricaria recutita* (Asteraceae)	Affinity for benzodiazepine receptors, binds to GABA receptors, reduces the levels of the stress-induced ACTH.	(Ebrahimi et al, 2022)
24.	Rosmarinic acid	*Melissa officinalis* (Lamiaceae)	Inhibits GABA transaminase activity and slow the degradation of GABA.	(Wang et al, 2022)
		Antianxiety		
25.	Rosmarinic acid triterpenoids, ursolic acid oleanolic acid	*Melissa officinalis* (Lamiaceae)	Inhibitor of rat brain GABA transaminase. Inhibited the oxidative stress.	(Heydari et al, 2019)
26.	Triterpenes, derivatives of flavones	*Cimicifuga pregula* L. Nutt (Ranunculaceae)	Act in the hypothalamus vasomotor centre, dopaminergic action.	(Fajemiroye et al, 2016)
27.	Linalool and linalyl acetate	*Lavandula angustifolia* Mill. (Lamiaceae)	Anxiolytic-like effect through 5-HT1A receptors.	(López et al, 2017)

(Continued)

TABLE 5.1 ***(Continued)***
Summary of Phytoconstituents Reported for Various Neuropsychiatric Disorders and Their Mechanism of Action.

S. No	Antipsychotic	Biological Source-Family	Molecular Mechanism	References
28.	Hypericin, hyperforin	*Hypericum perforatum* L. (Hypericaceae)	Selective inhibitor of MAO-A and MAO-B Inhibition of 5-HT, NE and DA uptake; antagonist of NMDA receptors; moderate interactions with the GABA A receptor, Suppression of the release of interleukin.	(Oliveira et al, 2016)
29.	Linalool and β-pinene	*Litsea glaucescens* (Lauraceae)	Interaction with the serotonergic 5-HT1A receptors, a2- and b-adrenoceptors and dopaminergic receptors D1.	(Guzmán-Gutiérrez et al, 2015)
30.	Methyl isoeugenol and oleanolic acid	*Pimenta pseudocaryophyllus* (Gomes), *ChamguavaL. R. Landrum* (Myrtaceae)	Acts on 5-HT1A receptor.	(Fajemiroye et al, 2016)
31.	Mesembrine, mesembranol, mesembrenol and mesembrenone	*Sceletium tortuosum* (Aizoaceae)	Inhibitory effects on 5-HT reuptake, binds to gamma butyric acid (GABA).	(Carpenter et al, 2016)
32.	Bacopasides	*Bacopa monnieri* (Scrophulariaceae)	Lower susceptibility to stress even before exposure.	(Calabrese et al, 2008; Sathyanarayanan et al, 2013)
33.	Benzo-flavonoids	*Passiflora incarnate* (Passifloraceae)	Acts on benzodiazepine and GABA receptors, MAO enzyme inhibiting properties.	(Dhawan et al, 2001)
34.	Eleutherosides, sennosides	*Eleutherococcus senticosus* (Araliaceae)	HPA axis regulation.	(Lee and Rhee, 2017)

(Continued)

TABLE 5.1 ***(Continued)***
Summary of Phytoconstituents Reported for Various Neuropsychiatric Disorders and Their Mechanism of Action.

S. No	Antipsychotic	Biological Source-Family	Molecular Mechanism	References
35.	Ethanol extract of flower	*Echium amoenum* (Boraginaceae)	Reduce oxidative stress and elevation of catecholamine levels.	(Nouri et al, 2019)
36.	Ginsenoside Rb1	*Panax ginseng* (Araliaceae)	Modulate monoamine neurotransmitter system, regulate the expression of neurotrophic factors, regulate HPA axis function, anti-inflammatory action.	(Bui et al, 2022)
37.	Hydroalcoholic extract	*Salvia reuterana* (Lamiaceae)	Acetylcholinesterase inhibitor, GABA receptor agonist.	(Rabbani et al, 2005)
38.	Ginkgolide-A	*Ginkgo biloba* (Ginkgoaceae)	Regulate dopamine, 5-HT and inflammatory glial-derived proteins.	(Deravi, 2022)
39.	Baicalin, baicalein	*Scutellaria lateriflora* (Lamiaceae)	$GABA_A$ agonist	(Awad et al, 2003)
40.	Valerenic acid, valepotriates	*Valeriana officinalis* L. (Caprifoliaceae)	$GABA_A$ agonist	(Murphy et al, 2010)
41.	Mother tincture	*Turnera aphrodisiaca* (Passifloraceae)	GABA agonist	(Kumar et al, 2008)
42.	Riparin III	*Aniba riparia* (Lauraceae)	Monoamine inhibition	(Sousa et al, 2004)
43.	Hydroethanolic extract of the aerial part	*Aloysia polystachya* (Verbenaceae)	Selective 5-HT reuptake inhibition.	(Carmona et al, 2019)
44.	Clary (aromatic essential oil)	*Salvia sclarea* (Lamiaceae)	Modulation of the dopaminergic pathway.	(Seol et al, 2010)
45.	Root tincture	*Artemisia vulgaris* (Asteraceae)	$GABA_A$ receptor modulation.	(Hernandez-Leon et al, 2017; Khan et al, 2016)

(Continued)

TABLE 5.1 ***(Continued)***
Summary of Phytoconstituents Reported for Various Neuropsychiatric Disorders and Their Mechanism of Action.

S. No	Antipsychotic	Biological Source-Family	Molecular Mechanism	References
46.	Rutin	*Camellia sinensis* (L.) Kuntze (Theaceae)	$GABA_A$ receptors modulation.	(Hernandez-Leon et al, 2017)
47.	Quercetin	*Chenopodium album (Amaranthaceae)*	Modulation of GABAergic and serotonergic system.	(Cárdenas-Rodríguez et al, 2014)
48.	Kaempferol, quercetin alkaloid and flavonoid	*Sansevieria trifasciata* (Asparagaceae) *Ginkgo biloba* (Ginkgoaceae) *Elaeocarpus ganitrus* (Elaeocarpaceae)	Modulation of molecular pathways such as NF-kB, PI3k/AKT, MAPK, Bcl2, Caspase 3 and VEGF.	(Kim et al, 2013)
49.	Honokiol	*Magnolia obovata* (Magnoliaceae)	Effects on GABA-benzodiazepine receptor, inhibitory activity on inflammatory mediator.	(Ham et al, 2020)
50.	Eugenol, 1,8-cineole, 1,8-bisabolene	*Ocimum sanctum* (Lamiaceae)	Regulation of HPA axis Cortical sparing effect.	(Bhattacharyya et al, 2008)
51.	Linalool and linalyl acetate	*Lavandula angustifolia* Mill. (Lamiaceae)	Effect on 5-HT1A receptor.	(Caputo et al, 2018)
52.	Apigenin	*Matricaria recutita* (Asteraceae)	Ligand for the central benzodiazepine receptors.	(Viola et al, 1995)
53.	Linalool, γ-terpinene, α-pinene	*Coriandrum sativum* (Apiaceae)	MAO-B inhibition.	(Hosseini et al, 2021)

Abbreviations: Corticotropin-releasing hormone (CRH), dopamine (DA), nor-epinephrine (NE), 5-hydroxytryptamine (5-HT), gamma-aminobutyric acid (GABA), N-methyl-D-aspartate (NMDA), monoamine oxidase (MAO), hypothalamic–pituitary–adrenal (HPA) axis, corticotropin-releasing hormone (CRH), adrenocorticotropic hormone (ACTH), reactive oxygen species (ROS), acetylcholinesterase (AchE), fatty acid amide hydrolase (FAAH), tryptophan-2,3-dioxygenase (TDO), indole 2,3 amine dioxygenase (IDO), superoxide dismutase (SOD), catalase (CAT), tumour necrosis factor-alpha (TNF-α), interleukin-1 beta (IL-1β), nuclear factor kappa b (NF-kb), mitogen-activated protein kinase (MAPK).

antidepressants, SSRI, monoamine oxidase inhibitors and newer drugs like brexanolone and esketamine which are quick-acting have become the choice for therapy. However, they are associated with many side effects such as cardiovascular disorder and suicidal ideations. Other complications include constipation and urine retention, causing discomfort to the patients and decreased responsiveness to the therapy (Aswar et al, 2017; Ghoneim and O'Hara, 2016).

Various *in vitro* and *in vivo* studies were reported for the antidepressant use of plant extracts such as *Boophone disticha* L., *Leonotis leonurus* L. and *Mentha longifolia* L., *Bacopa monieri* L., *Dittrichia viscosa* L. (Alkahtani et al, 2022; Arrieta-Báez et al, 2022; Badaoui et al, 2022; Murlanova et al, 2022). The researchers have studied several phytoconstituents in context with their molecular mechanism with the aid of animal behaviour, computational molecular docking, molecular dynamics, enzymatic studies, network pharmacology, patch clamp technique, electrophysiology and radioligand binding assays, some of them are discussed below.

5.4.1 Monoamine Oxidase A Inhibitors (MAOI-A)

Monoamine oxidase is widely distributed in the periphery as well as the CNS and is involved in the metabolism of various monoamines. It is present in two isoforms, namely MAO-A and MAO-B. 5-HT, norepinephrine and dopamine have specific selectivity with MAO-A, which is the key enzyme for their metabolism (Myburg et al, 2022). Numerous preclinical and clinical studies suggested increased expressions of MAO in chronic stress-induced rats and depressed patients (Domschke et al, 2008; Stefanovic et al, 2016). It was found that under the stress conditions activation of kruppel-like factor 11 upregulates the MAO-A activity, which further rapidly breaks down the monoamines such as 5-HT and dopamine resulting in MDD (Harris et al, 2015). A number of phytoconstituents have been evaluated and reported as MAO-A inhibitors which attenuate depression-like behaviour in rats. Plants like SJW, with active constituent hypericin which is an MAO-A inhibitor, have been used from ancient times to treat MDD (Assadi et al, 2011; Suzuki et al, 1984). Many polyherbal formulations contain phytoconstituents such as harmine (*Peganum harmala* L.), punarnavine (*Boerhaavia diffusa* L.) (Dhingra and Valecha, 2014) hyperforin and adhyforin (*Hypericum perforatum* L.) act via modulation of MAO activity (Jensen et al, 2001). *Curcuma longa* L. contain active phytoconstituents curcumin A and curcumin B. *In vitro* study of the MAO-A activity showed that curcumin A possess inhibitory effect on it with a docking score of –9.78 kcal/mol by *in silico* molecular simulation method (Aswar et al, 2020; Nazemi et al, 2019). Further *in silico* and *in vivo* evaluation by isolation of ethanolic extract of *Piper longum* L. a piperine alkaloid and its derivative antiepilepsirine showed antidepressant effect. Piperine competitively inhibits MAO-A with Ki 19.0 ± 0.9 μM. The antidepressant activity was further confirmed by reduced immobility in tail suspension and force swim tests (Li et al, 2007). Another preclinical study in CUMS model in mice where administration of water extract of *Gastrodia elata* Blume (0.5 mg/kg) and palmatine (0.5 mg/kg) has shown to reduce the MAO-A and increased 5-HT in the hippocampus. It was supported by behavioural study, sucrose preference test and open field test (Dhingra and Bhankher, 2014).

5.4.2 HPA Axis and Anti-Neuroinflammatory Phytoconstituents

The activation of inflammatory cascade induces overexpression of CRH and elevates cortisol levels in hypothalamus altering the behavioural function such as mood, cognition and locomotor activity resulting in neuroinflammation and associated depression (Baune, 2017; Yang et al, 2014). In addition, the glucocorticoid receptors (GCs) present in the hippocampus play a major role in cognition, mood and emotion. In a preclinical study on zebrafish, a reserpine-induced stress model was used where ethanolic extract of *Centella asiatica* L. ameliorated depression-like behaviour. Active constituents, asiatic acid and madecassic acid modulated HPA axis activity leading to reduced cortisol release thereby restoring the monoamines in zebrafish brains (Zakaria et al, 2023). Resveratrol shows myriad of beneficial effects such as anti-inflammatory, antioxidant and HPA axis modulation and has been extensively studied for its antidepressant effect. *In vivo* study conducted by Ge et al (2013) in CUMS induced depression in rats, when administered with 15 mg/kg resveratrol elucidated antidepressant activity. The mRNA expression of corticotropin-releasing hormone (CRH) in the hippocampus remained unaltered, but peripheral regulation of HPA axis caused reduced serum cortisol levels. Improvement in behavioural models such as increased sucrose preference, reduced immobility in Forced Swim Test and reduction of lipid peroxidation supports its antidepressant effect (Ge et al, 2013). Another phytoconstituent salidroside present in *Rhodiola rosea* was studied in olfactory bulbectomy-induced rats. Administration of salidroside (20, 40 mg/kg p.o) for 2 weeks remarkably reduced the tumour necrosis factor-alpha (TNF-α) and IL-1 beta (IL-1β) levels and markedly increased the GC expression in the hippocampus. Salidroside showed its antidepressant effect by reduction of CRH in the hypothalamus and serum cortisol levels (Yang et al, 2014). Phytoconstituents are multifunctional in nature which not only target a single disease but also attenuate the comorbidities associated with it. Depression associated with diabetes worsens the diabetic symptoms and blood glucose. The activation of inflammatory mediators leads to HPA axis activation and oxidative stress, thus making diabetes depression a vicious cycle (Patil and Aswar, 2021). A potent flavonoid quercetin is richly present in apples, broccoli and ginkgo biloba. In a preclinical study on diabetic depressed rats when administered (100 mg/kg i.p.), balanced the HPA axis and reduced blood glucose levels (Demir et al, 2016). The important and key regulator for neuroinflammation associated with depression is nuclear factor kappa-light-chain-enhancer of activated B cells (NF-κB) which regulates TNF-α, IL-6, IL-1 and IL-10 levels. Due to its direct correlation with depression, various polyphenols were studied for its inhibitory activity. The polyphenols are ferulic acid, chrysin and hesperidin (Borges Filho et al, 2016; Zheng et al, 2019). It was found that stress activates the glycogen synthase kinase-3 beta (GSK3β) which further stimulates inflammatory cascade by increasing NF-κB activity leading to the activation of nucleotide-binding domain, leucine-rich–containing family, pyrin domain–containing-3 inflammasome (NLRP-3). They also stimulate caspase 1 activity which helps in the formation of immature to mature form of IL-1β in the hippocampus. Baicalin a polyphenol obtained from the *Scutellaria baicalensis Georgi,* phosphorylate the GSK3β and further activate NF-κB. Baicalin attenuates depressive symptoms by inhibition of GSK3β/NF-κB/NLRP3 pathway which makes baicalin as a strong anti-neuroinflammatory and neuroprotective phytoconstituent (Zeng et al, 2018).

5.4.3 Inhibition of Kynurenine Pathway

Pathogenesis of depression and MDD mainly involves tryptophan metabolism and kynurenine pathway (indole 2,3 amine dioxygenase (IDO)/tryptophan-2,3-dioxygenase (TDO)). Tryptophan is a precursor protein for the 5-HT synthesis, and under normal physiological conditions, the tryptophan hydroxylase converts the tryptophan into 5-HT. TDO and IDO are the enzymes involved in the metabolism of tryptophan in the kynurenine pathway. Due to stress, the immune activation stimulates the IDO and TDO activity which further rapidly converts tryptophan into kynurenine resulting in the formation of the neurotoxin called quinolinic acid. Such events alter the 5-HT synthesis and induce its deficiency in the brain leading to mood alteration and depressive symptoms (Qin et al, 2018). Glycyrrhizic acid *(Glycyrrhiza glabra)* inhibits high-mobility group box 1 (HMGB1) danger-associated protein-mediated quinolinic acid formation and reduces kynurenine formation and improves the 5-HT balance in the rat hippocampus (Wang et al, 2018). Another active phytoconstituent cajaninstilbene acid is isolated from pigeon pea leaves. Cajaninstilbene acid when administered to mice at doses of 15 and 30 mg/kg showed antidepressant effect in mice via tryptophan metabolism attributed to its action by mTOR. It regulates 5-HT formation and has been proved to play an important role in cognition, synaptic plasticity and neural protein synthesis (Zhang et al, 2020).

A number of hypotheses have been developed and explored on target-oriented therapy for depression but very few succeeded in proper management of major depressive episodes. The neurotrophic hypothesis proposes the involvement of BDNF in depression. Zhaoxiang Ren performed a study to determine the action of dihydromyricetin *(Ampelopsis grossedentata)* lipopolysaccharide-induced MDD. As results suggested that the dihydromyricetin acts via GSK3B which further phosphorylates and improves the cAMP response element-binding protein (CREB) activity to regulate the transcription of BDNF in hippocampal tissue and cultured cells of the hippocampus (Ren et al, 2018). *Crocus sativus* L. commonly known as saffron also regulates the expression of BDNF when administered for 21 days at a dose of 160 mg/kg i.p (Ghasemi et al, 2015). The phytoconstituents such as piperine, tetrandrine, eugenol, baicalin, chrysin, naringenin, curcumin, oleanolic acid, ginsenoside Rg1 and genipin were found to increase the expression of BDNF (Bahramsoltani et al, 2015).

5.5 PHYTOCONSTITUENTS USED FOR PSYCHOSIS

The clinically evident characteristics of the major five domains, such as delusions, hallucinations, disorganised speech, abnormal catatonic behaviour and decreased emotional expressiveness, make up psychosis (Reus et al, 2016). Other symptoms include a lack of comprehension and poor social skills (Turner et al, 2018). It is divided into dementia praecox and manic depression (BD and SZ). Trauma, sensory and sleep deprivation, drugs including cannabis, lysergic acid diethylamide (LSD), amphetamines and very occasionally MDD are other factors that might cause psychosis (Griswold et al, 2021; Turner et al, 2018). Rapid sedation, antipsychotic medicines and cognitive behavioural therapy (CBT) can all be used to treat psychosis. A severe mental illness called SZ is linked to abnormal neural circuitry. It contributes to the global prevalence of 26 million populations (Koyanagi et al, 2022). According to researchers, the lack of understanding of the pathogenesis of SZ is primarily due to multiple origins connected to metabolic, genetic and environmental factors, as well as auto-immune and abnormalities in neurotransmission in the brain (Dhindsa and Goldstein, 2016; Rodrigues-Amorim et al, 2018). Additionally, SZ is linked to a variety of brain disorders that cause an excess of presynaptic dopamine (DA) production in the striatum (Grace and Gomes, 2019; McCutcheon et al, 2019). The pathophysiology of SZ is significantly influenced by neurotrophins, including BDNF, nerve growth factor (NGF), neurotrophin-3 (NT-3) and neurotrophin-4/5 (NT-4/5) (Shoval and Weizman, 2005). Around 1%–3% of people have BD, formerly known as manic-depressive psychosis. It is a recurring, severe and life-threatening mental illness (Belmaker, 2004). It is the primary cause of disability, cognitive impairment and functional impairment, along with rising suicide rates. Unbalance in monoaminergic neurotransmitter systems, such as the serotonergic, noradrenergic and dopaminergic, is a hallmark of BD, a heritable psychiatric illness. Neurotrophic molecules BDNF, which are essential for brain plasticity and are evidently decreased in BD (Grande et al, 2010; Machado-Vieira et al, 2014). Antidepressant therapy, electroconvulsive therapy, lithium supplementation, psychosocial psychotherapy and CBT are used to manage BD.

Psychosis is a mental illness that has an impact on a patient's overall well-being in terms of his thoughts, perception, cognition, emotion, etc. It is stigmatised or believed to be the result of God's curse, witchcraft or taboo in various regions, such as South Africa and Asia. The disease involves a strong conviction that it can only be cured by conventional techniques, such as using local plants and following antiquated rituals. The Leguminosae family, which also includes the Fabaceae, Apocynaceae, Liliaceae, Amaranthaceae, Rubiaceae and Lamiaceae, is home to the majority of the anti-psychotic herbs used around the world. The different active components from these plants, their reported scientific research in animal models and their mechanisms of action are all taken into consideration in this chapter. Additionally, an attempt will be made to examine their experience in clinical trials and their prospects.

Several plants, including *Piper retrofractum, Thevetia peruviana* L., *Euphorbia neriifolia* L., *Vitex negundo, Cannabis sativa* and *Brassica juncea*, have been linked to the treatment of psychotic episodes, according to a study conducted in Bangladesh. At the University of Cologne in Germany, 20 individuals with SZ received cannabidiol (CBD), a substance derived from the *Cannabis sativa* plant. It has been discovered that CBD lessens anxiety brought on by psychosis. The traditional Indian medical system, Ayurveda, is extensively utilised to treat psychosis since it is a powerful and efficient system of therapy. *Acorus calamus*, found in Brahmyadiyoga, is utilised as an anti-psychotic ayurvedic drug. Since ancient times, Chinese herbal medicine has been a cornerstone in the treatment of psychoses. Between 300 and 100 BC, "The Canon of Internal Medicine" was compiled. In the past, it has been discovered that extracts from Datura metal, Parasiticus on Coriaria and its host Coriaria can stop psychotic episodes. A Japanese herbal medicine called Yokukansan (TJ-54) is well-known for having ferulic acid, glycyrrhizin and Saikosaponin A-rich components. Using the brief psychiatric rating scale (BPRS), clinical global impression scale-severity (CGI-S) and positive and negative syndrome scale (PANSS), the medication was examined clinically in an open-label trial. As a result of TJ-54's ability to modify 5HT2A receptor activation as well as its anti-glutaminergic capability, the results demonstrated a considerable improvement with TJ-54 monotherapy (Miyaoka et al, 2013).

The catalepsy caused by haloperidol and the stereotypy caused by amphetamine can both be reduced by methanolic preparations of *Vitex negundo* L. roots. DA-induced contractions of the vas deferens can also be suppressed. Its anti-dopaminergic action was demonstrated by this research. The neuroprotective compounds found in *Vitex negundo* are called vitegnoside, negundoside, agunside, flavones, luteolin and casticin. However, further in-depth research is needed to determine the phytochemical that gives *Vitex negundo* its anti-psychotic properties. The dopamine theory of psychosis is currently out of date because recent research has shown that psychosis involves more than only the D2 receptor. Targets for innovative medication include the function of 5-HT (1 and 2A) receptors and glutamate receptors with D2 antagonism. Similarly, the plants/phytochemicals used by traditional healers are known to act with myriad effects on different receptors, neurotransmitters and channels as well. *Cannabis sativa* contains an alkaloid known as the cannabinoid. According to research, it is the first clinical experiment to examine potential antipsychotic effects of CBD. Nineteen patients received 800 mg of CBD as monotherapy for 4 weeks, and the impact on the PANSS and BPRS was examined. The effects of CBD were found to be equivalent to amisulpride, the standard treatment; moreover, CBD was devoid of any extrapyramidal side effects as observed with amisulpride. Many more clinical trials were conducted in later years and suggested the antipsychotic effect of CBD at a dose ranging from 800 to 1000 mg for 4 weeks. CBD is known to act via 5HT1A receptor and GPR55 (novel cannabinoid receptor) activation. The GPR55 is present abundantly in CNS, its activation improves psychotic symptoms and cognitive impairments, as well as neuroinflammation (Dyck et al, 2021). Caryophyllene, obtained from the same plant, is also demonstrated to prevent positive and negative symptoms in patients at a dose of 0.4 mg/kg to about 2 mg/kg by acting cannabinoid receptor type 2 receptors (Anavi-Goffer et al, 2016). Stephalidine (SPD) is another important constituent of Chinese herbal medicine *Stephania intermedia*. SPD is found to exhibit a D2 antagonism effect while having D1 agonistic activity. It stimulated apomorphine (specific D1) induced rotations while inhibiting amphetamine (specific D2) induced rotations in the 6-OHDA model. Dual actions of (-)-stepholidine on dopamine receptor subtypes after substantia nigra lesion. Moreover, it inhibited prepulse inhibition and increased the hindlimb retraction time indicating its D2 antagonistic effect thereby exhibiting an antipsychotic effect with low potency to induce EPS (Ellenbroek et al, 2006). Baicalin is a polyphenol present in plants such as *Scutellaria baicalensis*, found to alleviate the negative symptoms associated with SZ when given as an adjunct to olanzapine. The study was carried out on 45 patients, and the treatment was given for 24 weeks. The treatment attenuated PANSS scores and showed marked improvements in cognitive skills, including attention/vigilance (AV), working memory (WM) and brief visuospatial memory test (BVMT). Other phytoconstituents such as stigmasterol (a phytosterol), tutin (sesquiterpene), glycine, an amino acid and kavapyrone from kava are also found to be effective in the treatment of psychoses. The detailed information of other phytoconstituents reported antipsychotic activity is presented in (Table 5.1).

5.6 PHYTOCONSTITUENTS USED FOR OCD

In neurology practice, OCD is a persistent and incapacitating mental illness. It is diagnosed almost as frequently as asthma and diabetes mellitus, making it the fourth most prevalent mental disorder. Despite the several pharmacological classes that are available for the treatment of OCD, achieving full remission is still difficult (Stein, 2002). Over the past ten years, research on herbal medicines has increased. OCD is characterised by recurrent, ego-diatomic ritualistic thoughts (compulsions) coupled with presumably intentional acts (compulsions). It frequently co-occurs with significant depression, and it is regarded as an anxiety condition that is resistant to benzodiazepines. It has a 1%–3% lifetime prevalence rate. The only medications that consistently work in treating OCD patients are strong SSRIs. Pharmacotherapy for OCD has been studied extensively, but research into efficient herbal remedies for OCD has only recently begun. Plants that are used to treat depression and anxiety may be a potential therapeutic approach for OCD. Evidence suggests that the ayurvedic medicine of India known as shankhpushpi, which contains *Bacopa monnieri* L. Cyanoglucoside from *Colocasia esculenta* and *Convolvulus pluricaulis*, may be effective in treating OCD. 5-HT is believed to play a part in controlling anxiety. It is suggested that OCD patients may have 5-HT receptors that are relatively understimulated, and patients with OCD benefit from taking SSRI. It follows that an ethanolic extract of *Bacopa monnieri* may have a similar effect to an SSRI or some other facilitative impact on serotonergic neurotransmission (Ayati et al, 2020).

A number of therapeutic properties, including anxiolytic, neuroprotective, antioxidant, analgesic, immunomodulatory, antibacterial, antidiabetic and cardioprotective effects, have been linked to this herb in ancient literature. Numerous bioactive phytoconstituents, including the alkaloid (convolamine), flavonoid (kaempferol) and phenolics (scopoletin, β-sitosterol and ceryl alcohol), have been linked to the neuroprotective capabilities of these herbs. The neuro-pharmacological profile such as neuroprotective, nootropic and neuro-modulatory effects of *Convolulus prostratus* has been highlighted by numerous research teams. The literature shows mainly extracts used for the treatment of OCD in pre-clinical as well as clinical setup. The various extracts reported are methanolic extract of *Benincasa hispida* (doses: 200, 400, 600 mg/kg) tested on male Swiss albino mice which have influence on 5-HT reuptake and also shows the presence of tryptophan in the extract which is an important precursor of 5-HT in the serotonergic neurons and may be enhancing the biosynthesis of 5-HT to facilitate the anti-compulsive effect (Girdhar et al, 2010), CBD (15, 30, 60 mg/kg; tested on male mice) shows facilitation of CB1 receptor-mediated neurotransmission, suggesting the involvement of the endocannabinoid system in the pathophysiology of OCD (Casarotto et al, 2010). Curcumin demonstrated increased

5-HT levels and decreased dopamine in treated rats. Quinpirole-induced OCD was tested using an open field consisting of four objects of different shape and colour. The vehicle-treated rat demonstrated obsession towards two objects, the effects of curcumin (10 mg/kg) were comparable to that of paroxetineperoxatine (block the 5-HT reuptake transporter; SERT) and thus increase the concentration of (synaptic 5-HT). Curcumin also reduced the level of dopamine in quinpirole-treated rats. The study demonstrated anti-OCD effect of curcumin due to increased 5-HT, 5-HT1 receptors and reduction of dopamine (Chimakurthy and Murthy, 2010). The plant *Crocus sativus* L. (saffron) is grown in several nations, including Iran, India, Italy, Spain and Greece. The well-known spice saffron is obtained from the red stigmas of the flowers. Crocins, picrocrocin and safranal are the major components of saffron. The peculiar water-soluble carotenoids are called as crocins, which are glucosyl esters of crocetin. In folk medicine, the stigmas of *Crocus sativus* L. are used as an anticatarrhal, eupeptic, expectorant and emmenagogue, among other things (Girdhar et al, 2010). Crocin is tested in OCD at dose of 30 and 50 mg/kg, i.p., which were administered to rats followed by treatment with 1-(3-Chlorophenyl)piperazine hydrochloride (mCPP). 1,3-Chlorophenyl pyprizine hydrochloride induced excessive cell grooming is a model of OCD; it is a non-selective 5-HT receptor agonist with more affinity towards 5-HT 2C and 5-HT 2B receptor that results in excessive cell grooming. Crocin significantly reduced the duration of grooming, no. of grooming, indicating anti-OCD effect due to its effect on the 5-HT2 receptor (Georgiadou et al, 2012). Further, crocin was studied in clinical trials on 50 patients with OCD aged 18–60 years who received crocin (15 mg) and were evaluated using the Yale-Brown obsessive-compulsive scale (Y-BOCS). The efficacy of saffron and its active constituents on different mental disorders might be due to its effect on monoamine neurotransporters, including 5-HT, dopamine and norepinephrine as well as 5-HT receptors (Shafiee et al, 2018).

Bottle gourd is the popular name for *Lagenaria siceraria* (Molina) Standley (LS), also known as *Lagenaria leucantha Rusby* belonging to the family Cucurbitaceae. LS fruits have been utilised for centuries for their cardioprotective, cardiotonic, general tonic, aphrodisiac and diuretic qualities. It is also used to treat pectoral cough, asthma and other bronchial problems, as well as discomfort, ulcers and fever. Modern phytochemical screening techniques have revealed the presence of flavone C-glycosides, fucosterol, campesterol and triterpenoid cucurbitacins B, D, G and H in it. The methanolic crude extract of LS (25 and 50 mg/kg i.p.) demonstrated reduced marble burring behaviours indicating anti-OCD effects. OCD is frequently characterised by compulsive behaviour. LS may decrease serotonergic neurotransmission similarly to an SSRI or in some other way, making more 5-HT readily available to other nerve cells (Prajapati et al, 2011). Originally from Europe, North Africa and Asia, St. John's wort is a flowering plant in the Hypericaceae family. Chemically, hypericum extracts contain a variety of biologically active and complex substances, such as naphthodianthrones (rutin, hyperoside, isoquercitrin, quercitrin and pseudorutin), phloroglucinol derivatives (hyperforin, adhyperforin), flavonoids (rutin, hyperoside, isoquercitrin, quercitrin and quercetin). Hyperforin (3%) and hypericin (0.3%–0.5%) concentrations are typically used to standardise it. Twelve individuals with OCD were treated for at least a year in an open-label trial by Taylor and

Kobak, with a fixed dose of 450 mg of 0.3% hypericin twice daily (extended-release formulation) for a period of 12 weeks. The Yale-Brown obsessive-compulsive scale (YBOCS) score was found to decrease, indicating a considerable improvement in OCD (Taylor and Kobak, 2000).

5.7 PHYTOCONSTITUENTS FOR INSOMNIA

The common sleep problem known as insomnia is characterised by issues with falling asleep, remaining asleep and waking up early in the morning. It is among the most widespread sleep disorders in the world and significantly lowers one's quality of life. One of the most common neuropsychiatric disorders, it is estimated that 10%–15% of the general population and 30%–60% of the elderly are affected by it. A number of major health problems, such as cardiovascular disease, diabetes, obesity, mental disorders and cognitive dysfunction are all made more likely by persistent insomnia. The physiology of sleep involves neurotransmitters such as orexins A and B, adenosine, glycine, ACh, 5-HT and melatonin. NA, histamine, DA, glutamate and GABA are among those that have a substantial impact on the process. Both pharmaceutical and natural approaches can be used to treat insomnia. Because of the possibility for tolerance and dependence toward conventional drugs as well as their numerous side effects, the surveys favour complementary and alternative treatment (Siegel, 2004). There have been investigations on 16 medicinal plants used as a single herb and 13 polyherbal blends. The herbs *Valeriana officinalis* L., *Matricaria chamomilla* L., *Viola odorata* L. and *Passiflora incarnata* L. are the most popular and effective for treating insomnia. The active phytoconstituents for treatment of insomnia are valeneomerins A-D, mauritine-A, mucronine-D, nummularine-D, sativanine-A and sativanine-B, frangulanine, nummularine-B mucronine, apigenin, luteolin, rutin, herniarin, umbelliferone, trans-caffeic, protocatechuic, gentisic, p-hydroxybenzoic, 4-hydroxyphenylacetic, trans- and cis-coumaric acids quercetin, kaempferol, vitexin, iso-vitexin. They act mainly through modulation of the GABAergic system (Sarris et al, 2019).

5.8 CONCLUSION

Based on the literature, herbs are regularly consumed worldwide for various neuropsychiatric disorders. These herbs majorly belong to the family Solanaceae, Linaceace, Plantaginaceae, Epipacea, Convolvuaceae, Ariariliacae, Hyperiaceae, Passifloracea, Pipreciacae and Zingiberacae. The phytoconstituents present in herbs act through diverse mechanisms such as some act by increasing neurotransmitters, for example, theanine, curcumin, resveratrol, ursolic acid, etc. While others act on receptors such as cannabinoid, quercetin, cyanoglucoside, etc., certain phytoconstituents are reported as anti-inflammatory and HPA axis modulators, for example, withanoloids, rosmarinic acid, ursolic acid, salidroside, baicalin profoundly reduces cortisol and ameliorate HPA axis. Taking into account the side effects associated with pharmaceutical drugs used in neuropsychological disorders, these phytoconstituents can act as alternative and complementary drugs and provide hope for the discovery of new drugs.

ACKNOWLEDGEMENTS

We are also thankful to Ms Dhanashree Javere and Ms Pradnya Bagad, Department of Pharmacology, Poona College of Pharmacy, Bharati Vidyapeeth (Deemed to be University), Erandwane, Pune-411038, India for carrying the literature survey for the manuscript.

CONFLICT OF INTEREST

The authors declare no conflict of interest.

REFERENCES

Abidin, Lubna, Aftab Ahmad, Shokat Rasool Mir, Mohd Mujeeb, and Shah Alam Khan. 2015. "Ethnobotany, phytochemistry and pharmacological potential of Vitex negundo l. (five-leaved chaste tree): An updated review." *Journal of Coastal Life Medicine* 3 (10): 826–833.

Ahmed, Md, Kabidul Azam, and Md Nur. 2014. "Traditional knowledge and formulations of medicinal plants used by the traditional medical practitioners of Bangladesh to treat schizophrenia like psychosis." *Schizophrenia Research and Treatment* 2014: 679810.

Alkahtani, Jawaher, Mohamed S Elshikh, Yheni Dwiningsih, Muthaiyan Ahalliya Rathi, Rengasamy Sathya, and P Vijayaraghavan. 2022. "In-vitro antidepressant property of methanol extract of Bacopa monnieri." *Journal of King Saud University-Science* 34 (8): 102299.

Amoateng, Patrick, Dorcas Osei-Safo, Kennedy Kwami Edem Kukuia, Samuel Adjei, Obed Awintuma Akure, Constance Agbemelo-Tsomafo, Shirley Nyarko Adu-Poku, and Kenneth Yaw Agyeman-Badu. 2017. "Psychotropic effects of an alcoholic extract from the leaves of Albizia zygia (Leguminosae-Mimosoideae)." *Evidence-Based Complementary and Alternative Medicine* 2017: 9297808. https://doi.org/10.1155/2017/9297808.

Amray, Afshan N, Khurram Munir, Nusrat Jahan, Fatima B Motiwala, and Sadiq Naveed. 2019. "Psychopharmacology of pediatric anxiety disorders: A narrative review." *Cureus* 11 (8): e5487.

Anavi-Goffer, Sharon, Andrew J Irving, and Ruth A Ross. 2016. "Modulation of l-α-lysophosphatidylinositol/GPR55 MAP kinase signalling by CB2 receptor agonists: Identifying novel GPR55 inhibitors." *Journal of Basic and Clinical Physiology and Pharmacology* 27 (3): 303–310.

Anju, L. 2011. "Adaptogenic and anti-stress activity of Ocimum sanctum in mice." *Research Journal* of *Pharmaceutical*, *Biological* and *Chemical* Sciences 2: 670–678.

Archana, R, and A Namasivayam. 1998. "Antistressor effect of Withania somnifera." *Journal of Ethnopharmacology* 64 (1): 91–93.

Arowona, Ismot T, Mubo A Sonibare, and Solomon Umukoro. 2014. "Antipsychotic property of solvent-partitioned fractions of Lonchocarpus cyanescens leaf extract in mice." *Journal of Basic and Clinical Physiology and Pharmacology* 25 (2): 235–240.

Arrieta-Báez, Daniel, Mayra Beatriz Gómez-Patiño, Noé Jurado Hernández, Lilian Mayagoitia-Novales, Ana María Dorantes-Barrón, and Rosa Estrada-Reyes. 2022. "Antidepressant-like effects of a methanol extract of Leonotis nepetifolia in mice." *Natural Product Research* 36 (23): 6170–6176.

Assadi, Assad, Mohammad Reza Zarrindast, Abolghasem Jouyban, and Morteza Samini. 2011. "Comparing of the effects of hypericin and synthetic antidepressants on the expression of morphine-induced conditioned place preference." *Iranian Journal of Pharmaceutical Research: IJPR* 10 (3): 619.

Aswar, Manoj, Mangesh Bhalekar, Akshata Trimukhe, and Urmila Aswar. 2020. "Self-microemulsifying drug delivery system (SMEDDS) of curcumin attenuates depression in olfactory bulbectomized rats." *Heliyon* 6 (8): e04482.

Aswar, Urmila, Shilpa Chepurwar, Sumit Shintre, and Manoj Aswar. 2017. "Telmisartan attenuates diabetes induced depression in rats." *Pharmacological Reports* 69 (2): 358–364.

Awad, R, JT Arnason, V Trudeau, C Bergeron, JW Budzinski, BC Foster, and Zul Merali. 2003. "Phytochemical and biological analysis of skullcap (Scutellaria lateriflora l.): A medicinal plant with anxiolytic properties." *Phytomedicine* 10 (8): 640–649.

Ayati, Zahra, Jerome Sarris, Dennis Chang, Seyed A Emami, and Roja Rahimi. 2020. "Herbal medicines and phytochemicals for obsessive–compulsive disorder." *Phytotherapy Research* 34 (8): 1889–1901.

Badaoui, Majda, Soad Moubtakir, Chafik Terrafe, Rachida Aboufatima, and Abderrahman Chait. 2022. "Antianxiety and antidepressant effects of aqueous latex extract of Euphorbia resinifera." *Bangladesh Journal of Pharmacology* 17 (2): 32–41.

Baek, Ji Hyun, Jung-Yoon Heo, Maurizio Fava, David Mischoulon, Kwan Woo Choi, Eun Jin Na, Hana Cho, and Hong Jin Jeon. 2019. "Effect of Korean red ginseng in individuals exposed to high stress levels: A 6-week, double-blind, randomized, placebo-controlled trial." *Journal of Ginseng Research* 43 (3): 402–407.

Bahramsoltani, Roodabeh, Mohammad Hosein Farzaei, Marzieh Sarbandi Farahani, and Roja Rahimi. 2015. "Phytochemical constituents as future antidepressants: A comprehensive review." *Reviews in the Neurosciences* 26 (6): 699–719.

Baitharu, Iswar, Satya Narayan Deep, Vishal Jain, Dipti Prasad, and G Ilavazhagan. 2013. "Inhibition of glucocorticoid receptors ameliorates hypobaric hypoxia induced memory impairment in rat." *Behavioural Brain Research* 240: 76–86.

Ballenger, JC. 2007. *Panic Disorder and Agoraphobia*. New York: Elsevier.

Bathala, Lakshmana Rao, Ch Vekateswara Rao, SM Manjunath, S Vinuta, and Raghu Vemulapalli. 2012. "Efficacy of Ocimum sanctum for relieving stress: A preclinical study." *Journal of Contemporary Dental Practice* 13 (6): 782–786.

Baune, Bernhard T. 2017. "Are non-steroidal anti-inflammatory drugs clinically suitable for the treatment of symptoms in depression-associated inflammation?" *Current Topics in Behavioral Neurosciences* 31: 303–319.

Belmaker, RH. 2004. "Bipolar disorder." *New England Journal of Medicine* 351 (5): 476–486.

Ben-Azu, Benneth, Adegbuyi Oladele Aderibigbe, Abayomi Mayowa Ajayi, Aya-Ebi Okubo Eneni, Solomon Umukoro, and Ezekiel O Iwalewa. 2018. "Involvement of GABAergic, BDNF and Nox-2 mechanisms in the prevention and reversal of ketamine-induced schizophrenia-like behavior by morin in mice." *Brain Research Bulletin* 139: 292–306.

Bhattacharyya, D, TK Sur, U Jana, and PK Debnath. 2008. "Controlled programmed trial of Ocimum sanctum leaf on generalized anxiety disorders." *Nepal Medical College Journal* 10 (3): 176–179.

Borges Filho, Carlos, Cristiano Ricardo Jesse, Franciele Donato, Lucian Del Fabbro, Marcelo Gomes de Gomes, André Tiago Rossito Goes, Leandro Cattelan Souza, Renata Giacomeli, Michelle Antunes, and Cristiane Luchese. 2016. "Neurochemical factors associated with the antidepressant-like effect of flavonoid chrysin in chronically stressed mice." *European Journal of Pharmacology* 791: 284–296.

Bremner, J Douglas, DS Charney, DJ Stein, E Hollander, and BO Rothbaum. 2010. "Neural circuits in fear and anxiety." 55–71.

Bui, Bich Phuong, Phuong Linh Nguyen, Ha Thi Thu Do, and Jungsook Cho. 2022. "Anxiolytic effect of Korean red Ginseng through upregulation of serotonin and GABA transmission and BDNF expression in immobilized mice." *Journal of Ginseng Research* 46 (6): 819–829.

Calabrese, Carlo, William L. Gregory, Michael Leo, Dale Kraemer, Kerry Bone, and Barry Oken. 2008. "Effects of A standardized Bacopa monnieri extract on cognitive performance, anxiety, and depression in the elderly: A randomized, double-blind, placebo-controlled trial." *The Journal of Alternative and Complementary Medicine* 14 (6): 707–713.

Caputo, Lucia, Marina D Reguilon, José Miñarro, Vincenzo De Feo, and Marta Rodriguez-Arias. 2018. "Lavandula angustifolia essential oil and linalool counteract social aversion induced by social defeat." *Molecules* 23 (10): 2694.

Cárdenas-Rodríguez, Noemí, María Eva González-Trujano, Eva Aguirre-Hernández, Matilde Ruíz-García, Aristides Sampieri, Elvia Coballase-Urrutia, and Liliana Carmona-Aparicio. 2014. "Anticonvulsant and antioxidant effects of Tilia americana var. mexicana and flavonoids constituents in the pentylenetetrazole-induced seizures." *Oxidative Medicine and Cellular Longevity* 2014: 329172.

Carmona, Fabio, Fernando Saraiva Coneglian, Priscila Alves Batista, Davi Casale Aragon, Mateus Andrea Angelucci, Edson Zangiacomi Martinez, and Ana Maria Soares Pereira. 2019. "Aloysia polystachya (Griseb.) Moldenke (Verbenaceae) powdered leaves are effective in treating anxiety symptoms: A phase-2, randomized, placebo-controlled clinical trial." *Journal of Ethnopharmacology* 242: 112060.

Carpenter, Jessica M, Mary K Jourdan, Emily M Fountain, Zulfiqar Ali, Naohito Abe, Ikhlas A Khan, and Kenneth J Sufka. 2016. "The effects of Sceletium tortuosum (L.) NE Br. extract fraction in the chick anxiety-depression model." *Journal of Ethnopharmacology* 193: 329–332.

Casarotto, Plinio C, Felipe V Gomes, Leonardo BM Resstel, and Francisco S Guimarães. 2010. "Cannabidiol inhibitory effect on marble-burying behaviour: Involvement of CB1 receptors." *Behavioural Pharmacology* 21 (4): 353–358.

Chimakurthy, Jithendra, and Talasila EGK Murthy. 2010. "Effect of curcumin on quinpirole induced compulsive checking: An approach to determine the predictive and construct validity of the model." *North American Journal of Medical Sciences* 2 (2): 81–87.

Craske, Michelle G, Linda Street, and David H Barlow. 1989. "Instructions to focus upon or distract from internal cues during exposure treatment of agoraphobic avoidance." *Behaviour Research and Therapy* 27 (6): 663–672.

Demir, Enver Ahmet, Hasan Serdar Gergerlioglu, and Mehmet Oz. 2016. "Antidepressant-like effects of quercetin in diabetic rats are independent of hypothalamic–pituitary–adrenal axis." *Acta Neuropsychiatrica* 28 (1): 23–30.

Deravi, Niloofar. 2022. "Ginko biloba leaves extract for the treatment of anxiety, stress, depression." *ASEAN Journal of Psychiatry* 23 (4): 1–7.

Dhawan, Kamaldeep, Suresh Kumar, and Anupam Sharma. 2001. "Anxiolytic activity of aerial and underground parts of Passiflora incarnata." *Fitoterapia* 72 (8): 922–926.

Dhindsa, Ryan S, and David B Goldstein. 2016. "From genetics to physiology at last." *Nature* 530 (7589): 162–163.

Dhingra, D., and A. Bhankher. 2014. "Behavioral and biochemical evidences for antidepressant-like activity of palmatine in mice subjected to chronic unpredictable mild stress." *Pharmacological Reports* 66 (1): 1–9. https://doi.org/10.1016/j.pharep.2013.06.001.

Dhingra, Dinesh, and Rekha Valecha. 2014. "Evidence for involvement of the monoaminergic system in antidepressant-like activity of an ethanol extract of Boerhaavia diffusa and its isolated constituent, punarnavine, in mice." *Pharmaceutical Biology* 52 (6): 767–774.

Domschke, Katharina, Christa Hohoff, Lena S Mortensen, Tilmann Roehrs, Jürgen Deckert, Volker Arolt, and Bernhard T Baune. 2008. "Monoamine oxidase A variant influences antidepressant treatment response in female patients with major depression." *Progress in Neuro-Psychopharmacology and Biological Psychiatry* 32 (1): 224–228.

Dyck, Garrison JB, Zaid H Maayah, Dean T Eurich, and Jason RB Dyck. 2021. "Understanding the potential benefits of CBD for patients with schizophrenia: A narrative review." *Schizophrenia Bulletin Open* 3 (1): 1–11.

Ebrahimi, Hossein, Abbas Mardani, Mohammad Hasan Basirinezhad, Azam Hamidzadeh, and Fatemeh Eskandari. 2022. "The effects of Lavender and Chamomile essential oil inhalation aromatherapy on depression, anxiety and stress in older community-dwelling people: A randomized controlled trial." *Explore* 18 (3): 272–278.

Ellenbroek, Bart A, Xue-Xiang Zhang, and Guo-Zhang Jin. 2006. "Effects of (-) stepholidine in animal models for schizophrenia 1." *Acta Pharmacologica Sinica* 27 (9): 1111–1118.

Fajemiroye, James O, Dayane M da Silva, Danillo R de Oliveira, and Elson A Costa. 2016. "Treatment of anxiety and depression: Medicinal plants in retrospect." *Fundamental & Clinical Pharmacology* 30 (3): 198–215.

Garabadu, Debapriya, Ankit Shah, Ausaf Ahmad, Vijaya B Joshi, Bhagawati Saxena, Gautam Palit, and Sairam Krishnamurthy. 2011. "Eugenol as an anti-stress agent: Modulation of hypothalamic–pituitary–adrenal axis and brain monoaminergic systems in a rat model of stress." *Stress* 14 (2): 145–155.

Georgiadou, G, PA Tarantilis, and N Pitsikas. 2012. "Effects of the active constituents of Crocus Sativus L., crocins, in an animal model of obsessive–compulsive disorder." *Neuroscience Letters* 528 (1): 27–30.

Ge, Jin-Fang, Lei Peng, Jiang-Qun Cheng, Chun-Xiao Pan, Jie Tang, Fei-Hu Chen, and Jun Li. 2013. "Antidepressant-like effect of resveratrol: Involvement of antioxidant effect and peripheral regulation on HPA axis." *Pharmacology Biochemistry and Behavior* 114: 64–69.

Ghasemi, T, K Abnous, F Vahdati, S Mehri, BM Razavi, and H Hosseinzadeh. 2015. "Antidepressant effect of Crocus sativus aqueous extract and its effect on CREB, BDNF, and VGF transcript and protein levels in rat hippocampus." *Drug Research* 65 (7): 337–343.

Ghoneim, Mohamed M, and Michael W O'Hara. 2016. "Depression and postoperative complications: An overview." *BMC Surgery* 16 (1): 1–10.

Girdhar, Shikha, Manish M Wanjari, Sunil K Prajapati, and Amit Girdhar. 2010. "Evaluation of anti-compulsive effect of methanolic extract of Benincasa hispida Cogn. fruit in mice." *Acta Poloniae Pharmaceutica-Drug Research* 67 (4): 417–421.

Grace, Anthony A, and Felipe V Gomes. 2019. "The circuitry of dopamine system regulation and its disruption in schizophrenia: Insights into treatment and prevention." *Schizophrenia Bulletin* 45 (1): 148–157.

Grande, Iria, Gabriel Rodrigo Fries, Mauricio Kunz, and Flavio Kapczinski. 2010. "The role of BDNF as a mediator of neuroplasticity in bipolar disorder." *Psychiatry Investigation* 7 (4): 243.

Griswold, Kim S, Dianne M Loomis, and Patricia A Pastore. 2021. "Mental health and illness." *Primary Care: Clinics in Office Practice* 48 (1): 131–145.

Guzmán-Gutiérrez, Silvia Laura, Herlinda Bonilla-Jaime, Rocío Gómez-Cansino, and Ricardo Reyes-Chilpa. 2015. "Linalool and β-pinene exert their antidepressant-like activity through the monoaminergic pathway." *Life Sciences* 128: 24–29.

Yao, Ying, Hai-Ying Huang, Yuan-Xiao Yang, and Jian-You Guo. 2015. "Cinnamic aldehyde treatment alleviates chronic unexpected stress-induced depressive-like behaviors via targeting cyclooxygenase-2 in mid-aged rats." *Journal of Ethnopharmacology* 162: 97–103.

Ham, Hyeon Joo, Yong Sun Lee, Jaesuk Yun, Sang-Bae Han, Dong Ju Son, and Jin Tae Hong. 2020. "Anxiolytic-like effects of the ethanol extract of Magnolia obovata leaves through its effects on GABA-benzodiazepine receptor and neuroinflammation." *Behavioural Brain Research* 383: 112518.

Harris, Sharonda, Shakevia Johnson, Jeremy W Duncan, Chinelo Udemgba, Jeffrey H Meyer, Paul R Albert, Gwen Lomberk, Raul Urrutia, Xiao-Ming Ou, and Craig A Stockmeier. 2015. "Evidence revealing deregulation of the KLF11-MAO A pathway in association with chronic stress and depressive disorders." *Neuropsychopharmacology* 40 (6): 1373–1382.

Haybar, Habib, Ahmad Zare Javid, Mohammad Hosein Haghighizadeh, Einollah Valizadeh, Seyede Marjan Mohaghegh, and Assieh Mohammadzadeh. 2018. "The effects of Melissa officinalis supplementation on depression, anxiety, stress, and sleep disorder in patients with chronic stable angina." *Clinical Nutrition ESPEN* 26: 47–52.

Hernandez-Leon, Alberto, María E González-Trujano, and Alonso Fernández-Guasti. 2017. "The anxiolytic-like effect of rutin in rats involves GABAA receptors in the basolateral amygdala." *Behavioural Pharmacology* 28 (4): 303–312.

Heydari, Naval, Mansure Dehghani, Maasumeh Emamghoreishi, and Marzieh Akbarzadeh. 2019. "Effect of Melissa officinalis capsule on the mental health of female adolescents with premenstrual syndrome: A clinical trial study." *International Journal of Adolescent Medicine and Health* 31 (3): 1–5.

Hosseini, Mahmoud, Mohammad Hossein Boskabady, and Mohammad Reza Khazdair. 2021. "Neuroprotective effects of Coriandrum sativum and its constituent, linalool: A review." *Avicenna Journal of Phytomedicine* 11 (5): 436.

Huang, Xing, Ning Li, Yiqiong Pu, Tong Zhang, and Bing Wang. 2019. "Neuroprotective effects of ginseng phytochemicals: Recent perspectives." *Molecules* 24 (16): 2939.

Ivanova Stojcheva, Emilija, and José Carlos Quintela. 2022. "The effectiveness of rhodiola rosea L. Preparations in alleviating various aspects of life-stress symptoms and stress-induced conditions—encouraging clinical evidence." *Molecules* 27 (12): 3902.

Jamalludin, Nurul Bariah, and Nurhuda Manshoor. 2022. "Ethnobotanical importance, phytochemical constituents, and pharmacological properties of Withania somnifera." *Journal of Herbmed Pharmacology* 11 (3): 320–329.

Jensen, AG, SH Hansen, and EØ Nielsen. 2001. "Adhyperforin as a contributor to the effect of Hypericum perforatum L. in biochemical models of antidepressant activity." *Life Sciences* 68 (14): 1593–1605.

Kakuda, Takami, Eiichi Hinoi, Akihiro Abe, Ayumu Nozawa, Masato Ogura, and Yukio Yoneda. 2008. "Theanine, an ingredient of green tea, inhibits [3H] glutamine transport in neurons and astroglia in rat brain." *Journal of Neuroscience Research* 86 (8): 1846–1856.

Kalshetty, Padmaja, Urmila Aswar, Subhash Bodhankar, Arulmozhi Sinnathambi, V Mohan, and Prasad Thakurdesai. 2012. "Antidepressant effects of standardized extract of Centella asiatica L in olfactory bulbectomy model." *Biomedicine & Aging Pathology* 2 (2): 48–53.

Kayashima, T., K. Nagao, M. Umino, H. Kaikiri, S. Shibata, and K. Matsubara. 2020. "Anti-stress effects of rosemary (*Rosmarinus officinalis* L.) leaf extract on intestinal goblet cells and immobility of forced-swimming test in BALB/c mice." *Bioscience, Biotechnology, and Biochemistry* 84 (11): 2385–2389.

Karcı, Canan Kuygun, and Gonca Gül Celik. 2020. "Nutritional and herbal supplements in the treatment of obsessive compulsive disorder." *General Psychiatry* 33 (2): e100159.

Khan, Imran, Nasiara Karim, Waqar Ahmad, Abeer Abdelhalim, and Mary Chebib. 2016. "GABA-A receptor modulation and anticonvulsant, anxiolytic, and antidepressant activities of constituents from Artemisia indica Linn." *Evidence-Based Complementary and Alternative Medicine* 2016: 1–13.

Kim, Hee Jin, Pitna Kim, and Chan Young Shin. 2013. "A comprehensive review of the therapeutic and pharmacological effects of ginseng and ginsenosides in central nervous system." *Journal of Ginseng Research* 37 (1): 8–29.

Kochlamazashvili, Gaga, Christian Henneberger, Olena Bukalo, Elena Dvoretskova, Oleg Senkov, Patricia M Lievens, Ruth Westenbroek, Andreas K Engel, William A Catterall, and Dmitri A Rusakov. 2010. "The extracellular matrix molecule hyaluronic acid regulates hippocampal synaptic plasticity by modulating postsynaptic L-type Ca2+ channels." *Neuron* 67 (1): 116–128.

Koyanagi, Ai, Hans Oh, Jordan DeVylder, Jae Il Shin, Karel Kostev, Lee Smith, Louis Jacob, Guillermo F López Sánchez, Adel S Abduljabbar, and Josep Maria Haro. 2022. "Psychotic experiences among informal caregivers: Findings from 48 low-and middle-income countries." *Social Psychiatry and Psychiatric Epidemiology* 57 (9): 1771–1780.

Kumar, S, R Madaan, and A Sharma. 2008. "Estimation of apigenin, an anxiolytic constituent, in Turnera aphrodisiaca." *Indian Journal of Pharmaceutical Sciences* 70 (6): 847.

Kumbol, Victor Wumbor-Apin, Wonder Kofi Mensah Abotsi, Edmund Ekuadzi, and Eric Woode. 2018. "Albizia zygia root extract exhibits antipsychotic-like properties in murine models of schizophrenia." *Biomedicine & Pharmacotherapy* 106: 831–841.

LeDoux, Joseph E. 1994. "Emotion, memory and the brain." *Scientific American* 270 (6): 50–57.

Lee, Bun-Hee, and Yong-Ku Kim. 2010. "The roles of BDNF in the pathophysiology of major depression and in antidepressant treatment." *Psychiatry Investigation* 7 (4): 231.

Lee, Seungjoo, Dong Hyun Kim, Ji Wook Jung, Ju Hee Oh, Ho Jae Park, Chan Park, Youngbuhm Huh, Jae Hoon Cheong, Tae Hwan Oh, and Jong Hoon Ryu. 2007. "Schizandra chinensis and Scutellaria baicalensis counter stress behaviors in mice." *Phytotherapy Research: An International Journal Devoted to Pharmacological and Toxicological Evaluation of Natural Product Derivatives* 21 (12): 1187–1192.

Lee, Seungyeop, and Dong-Kwon Rhee. 2017. "Effects of ginseng on stress-related depression, anxiety, and the hypothalamic–pituitary–adrenal axis." *Journal of Ginseng Research* 41 (4): 589–594.

Li, Song, Che Wang, Wei Li, Kazuo Koike, Tamatsu Nikaido, and Min-Wei Wang. 2007. "Antidepressant-like effects of piperine and its derivative, antiepilepsirine." *Journal of Asian Natural Products Research* 9 (5): 421–430.

López, Víctor, Birgitte Nielsen, Maite Solas, Maria J Ramírez, and Anna K Jäger. 2017. "Exploring pharmacological mechanisms of lavender (Lavandula angustifolia) essential oil on central nervous system targets." *Frontiers in Pharmacology* 8: 280.

Machado-Vieira, Rodrigo, Marcio G Soeiro-De-Souza, Erica M Richards, Antonio L Teixeira, and Carlos A Zarate Jr. 2014. "Multiple levels of impaired neural plasticity and cellular resilience in bipolar disorder: Developing treatments using an integrated translational approach." *The World Journal of Biological Psychiatry* 15 (2): 84–95.

Magaji, Mohammed Garba, Aliyu Muhammad Musa, Musa Ismail Abdullahi, Jamilu Ya'u, and Isa Marte Hussaini. 2015. "Isolation of bergenin from the root bark of Securinega virosa and evaluation of its potential sleep promoting effect." *Avicenna Journal of Phytomedicine* 5 (6): 587.

Magee, William J, William W Eaton, Hans-Ulrich Wittchen, Katherine A McGonagle, and Ronald C Kessler. 1996. "Agoraphobia, simple phobia, and social phobia in the National Comorbidity Survey." *Archives of General Psychiatry* 53 (2): 159–168.

Markowitz, Jeffrey S, Myrna M Weissman, Robert Ouellette, Jennifer D Lish, and Gerald L Klerman. 1989. "Quality of life in panic disorder." *Archives of General Psychiatry* 46 (11): 984–992.

McCutcheon, Robert A, Anissa Abi-Dargham, and Oliver D Howes. 2019. "Schizophrenia, dopamine and the striatum: From biology to symptoms." *Trends in Neurosciences* 42 (3): 205–220.

Miyaoka, Tsuyoshi, Rei Wake, Motohide Furuya, Kristian Liaury, Masa Ieda, Kazunori Kawakami, Keiko Tsuchie, Michiyo Fukushima, Kotomi Ishihara, and Tomoko Araki. 2013. "Yokukansan (TJ-54) for treatment of very-late-onset schizophrenia-like psychosis: An open-label study." *Phytomedicine* 20 (7): 654–658.

Murlanova, Kateryna, Netanela Cohen, Anna Pinkus, Liudmila Vinnikova, Mikhail Pletnikov, Michael Kirby, Jonathan Gorelick, Elyashiv Drori, and Albert Pinhasov. 2022. "Antidepressant-like effects of a chlorogenic acid-and cynarine-enriched fraction from Dittrichia viscosa root extract." *Scientific Reports* 12 (1): 1–10.

Murphy, Kevin, ZJ Kubin, Jennifer N Shepherd, and R Ettinger. 2010. "Valeriana officinalis root extracts have potent anxiolytic effects in laboratory rats." *Phytomedicine* 17 (8–9): 674–678.

Musazzi, Laura, Paolo Tornese, Nathalie Sala, and Maurizio Popoli. 2018. "What acute stress protocols can tell us about PTSD and stress-related neuropsychiatric disorders." *Frontiers in Pharmacology* 9: 758.

Myburg, Theo, Anél Petzer, and Jacobus P Petzer. 2022. "The inhibition of monoamine oxidase by harmine derivatives." *Results in Chemistry* 4: 100607.

Nazemi, Hamidreza, Mahmoud Mirzaei, and Elham Jafari. 2019. "Antidepressant activity of curcumin by monoamine oxidase–A inhibition." Journal Of Advances In Chemistry *B* 1: 3–9.

Nouri, Mohammad, Fereshteh Farajdokht, Mohammadali Torbati, Fatemeh Ranjbar, Sanaz Hamedyazdan, Saeed Sadigh-Eteghad, and Mostafa Araj-Khodaei. 2019. "Antidepressant and anxiolytic effect of Echium amoenum in restraint stress model: The role of neuroinflammation in the prefrontal cortex and hippocampus." *Iranian Red Crescent Medical Journal* 21 (10): e95438.

Oliveira, Ana I, Cláudia Pinho, Bruno Sarmento, and Alberto CP Dias. 2016. "Neuroprotective activity of Hypericum perforatum and its major components." *Frontiers in Plant Science* 7: 1004.

Patil, Rashmi, and Urmila. Aswar. 2021. "Role of pterostilbene in metabolic diseases through SIRT1 pathway-a review." *Turkish Journal of Agriculture-Food Science and Technology* 9: 2581–2586.

Patriquin, Michelle A, and Sanjay J Mathew. 2017. "The neurobiological mechanisms of generalized anxiety disorder and chronic stress." *Chronic Stress* 1: 1–10.

Perera, Joachim, Joon Heng Tan, S Jeevathayaparan, Srikumar Chakravarthi, and Nagaraja Haleagrahara. 2011. "Neuroprotective effects of alpha lipoic acid on haloperidol-induced oxidative stress in the rat brain." *Cell & Bioscience* 1 (1): 1–6.

Prajapati, RP, MV Kalaria, VP Karkare, SK Parmar, and NR Sheth. 2011. "Effect of methanolic extract of Lagenaria siceraria (Molina) Standley fruits on marble-burying behavior in mice: Implications for obsessive–compulsive disorder." *Pharmacognosy Research* 3 (1): 62.

Puri, Prerna, and Nidhi Purohit. 2016. "Development of stress scale for caretakers of mentally disabled children." *The International Journal of Indian Psychology* 3 (4) 82: 103.

Qin, Yanjie, Nanxi Wang, Xinlin Zhang, Xuemei Han, Xuejia Zhai, and Yongning Lu. 2018. "IDO and TDO as a potential therapeutic target in different types of depression." *Metabolic Brain Disease* 33 (6): 1787–1800.

Rabbani, M, SE Sajjadi, A Jafarian, and G Vaseghi. 2005. "Anxiolytic effects of Salvia reuterana Boiss. on the elevated plus-maze model of anxiety in mice." *Journal of Ethnopharmacology* 101 (1–3): 100–103.

Ren, Zhaoxiang, Pengju Yan, Liushuai Zhu, Huicui Yang, Yafei Zhao, Brian P Kirby, John L Waddington, and Xuechu Zhen. 2018. "Dihydromyricetin exerts a rapid antidepressant-like effect in association with enhancement of BDNF expression and inhibition of neuroinflammation." *Psychopharmacology* 235 (1): 233–244.

Reus, Victor I, Laura J Fochtmann, A Evan Eyler, Donald M Hilty, Marcela Horvitz-Lennon, Michael D Jibson, Oscar L Lopez, Jane Mahoney, Jagoda Pasic, and Zaldy S Tan. 2016. "The American Psychiatric Association practice guideline on the use of antipsychotics to treat agitation or psychosis in patients with dementia." *American Journal of Psychiatry* 173 (5): 543–546.

Rodrigues-Amorim, Daniela, Tania Rivera-Baltanás, Joao Bessa, Nuno Sousa, Maria de Carmen Vallejo-Curto, Cynthia Rodríguez-Jamardo, María Elena de Las Heras, Roberto Díaz, Roberto Carlos Agís-Balboa, and JM Olivares. 2018. "The neurobiological hypothesis of neurotrophins in the pathophysiology of schizophrenia: A meta-analysis." *Journal of Psychiatric Research* 106: 43–53.

Salve, Jaysing, Sucheta Pate, Khokan Debnath, and Deepak Langade. 2019. "Adaptogenic and anxiolytic effects of ashwagandha root extract in healthy adults: A double-blind, randomized, placebo-controlled clinical study." *Cureus* 11 (12): e6466.

Sarris, Jerome, Gerard J Byrne, Lachlan Cribb, Georgina Oliver, Jenifer Murphy, Patricia Macdonald, Sonia Nazareth, Diana Karamacoska, Samantha Galea, and Anika Short. 2019. "L-theanine in the adjunctive treatment of generalized anxiety disorder: A double-blind, randomised, placebo-controlled trial." *Journal of Psychiatric Research* 110: 31–37.

Sathyanarayanan, Vidya, Tinku Thomas, Suzanne JL Einöther, Rajendra Dobriyal, MK Joshi, and Srinivasan Krishnamachari. 2013. "Brahmi for the better? New findings challenging cognition and anti-anxiety effects of Brahmi (Bacopa monniera) in healthy adults." *Psychopharmacology* 227 (2): 299–306.

Saxena, Ram Chandra, Rakesh Singh, Parveen Kumar, Mahendra P Singh Negi, Vinod S Saxena, Periasamy Geetharani, Joseph Joshua Allan, and Kudiganti Venkateshwarlu. 2012. "Efficacy of an extract of ocimum tenuiflorum (OciBest) in the management of general stress: A double-blind, placebo-controlled study." *Evidence-Based Complementary and Alternative Medicine* 2012: 894509.

Schweizer, Edward. 1995. "Generalized anxiety disorder: Longitudinal course and pharmacologic." *Psychiatric Clinics of North America* 18 (4): 843–857.

Seol, Geun Hee, Hyun Soo Shim, Pill-Joo Kim, Hea Kyung Moon, Ki Ho Lee, Insop Shim, Suk Hyo Suh, and Sun Seek Min. 2010. "Antidepressant-like effect of Salvia sclarea is explained by modulation of dopamine activities in rats." *Journal of Ethnopharmacology* 130 (1): 187–190.

Seo, Ho-Jun, Sheng-Min Wang, Changsu Han, Soo-Jung Lee, Ashwin A Patkar, Prakash S Masand, and Chi-Un Pae. 2015. "Curcumin as a putative antidepressant." *Expert Review of Neurotherapeutics* 15 (3): 269–280.

Shafiee, Mojtaba, Soheil Arekhi, Alireza Omranzadeh, and Amirhossein Sahebkar. 2018. "Saffron in the treatment of depression, anxiety and other mental disorders: Current evidence and potential mechanisms of action." *Journal of Affective Disorders* 227: 330–337.

Sharma, Kailash, Milind Parle, and Monu Yadav. 2016. "Evaluation of antipsychotic effect of methanolic extract of Ocimum sanctum leaves on laboratory animals." *Journal of Applied Pharmaceutical Science* 6 (5): 171–177.

Shen, Manjun, YI Yang, Ying Wu, Beibei Zhang, Haishan Wu, Lu Wang, Hui Tang, and Jindong Chen. 2019. "L-theanine ameliorate depressive-like behavior in a chronic unpredictable mild stress rat model via modulating the monoamine levels in limbic–cortical–striatal–pallidal–thalamic-circuit related brain regions." *Phytotherapy Research* 33 (2): 412–421.

Shoval, G., and A. Weizman. 2005. "The possible role of neurotrophins in the pathogenesis and therapy of schizophrenia." *European Neuropsychopharmacology* 15(3): 319–329.

Siegel, Jerome M. 2004. "The neurotransmitters of sleep." *Journal of Clinical Psychiatry* 65 (16): 4–7.

Sotille, Robert, Herpreet Singh, Anne Weisman, and Thomas Vida. 2022. "Unraveling the mysteries of mental illness with psilocybin." *Cureus* 14 (5): e25414.

Sousa, F, C Melo, A Monteiro, V Lima, S Gutierrez, B Pereira, J Barbosa-Filho, S Vasconcelos, M Fonteles, and G Viana. 2004. "Antianxiety and antidepressant effects of riparin III from Aniba riparia (Nees) Mez (Lauraceae) in mice." *Pharmacology Biochemistry and Behavior* 78 (1): 27–33.

Stefanovic, Bojana, Natasa Spasojevic, Predrag Jovanovic, Nebojsa Jasnic, Jelena Djordjevic, and Sladjana Dronjak. 2016. "Melatonin mediated antidepressant-like effect in the hippocampus of chronic stress-induced depression rats: Regulating vesicular monoamine transporter 2 and monoamine oxidase A levels." *European Neuropsychopharmacology* 26 (10): 1629–1637.

Stein, Dan J. 2002. "Obsessive-compulsive disorder." *The Lancet* 360 (9330): 397–405.
Suzuki, O, Y Katsumata, M Oya, S Bladt, and H Wagner. 1984. "Inhibition of monoamine oxidase by hypercin." *Planta Medica* 50 (3): 272–274.
Taïwe, Germain Sotoing, Elisabeth Ngo Bum, Emmanuel Talla, Amadou Dawe, Fleur Clarisse Okomolo Moto, Gwladys Temkou Ngoupaye, Neteydji Sidiki, Bernard Dabole, Paul Désiré Djomeni Dzeufiet, and Théophile Dimo. 2012. "Antipsychotic and sedative effects of the leaf extract of Crassocephalum bauchiense (Hutch.) Milne-Redh (Asteraceae) in rodents." *Journal of Ethnopharmacology* 143 (1): 213–220.
Tao, Guoxin, Yoshifumi Irie, Dian-Jun Li, and Wing Ming Keung. 2005. "Eugenol and its structural analogs inhibit monoamine oxidase A and exhibit antidepressant-like activity." *Bioorganic & Medicinal Chemistry* 13 (15): 4777–4788.
Taylor, L.H., and K.A. Kobak. 2000. "An open-label trial of St. John's Wort (*Hypericum perforatum*) in obsessive-compulsive disorder." *Journal of Clinical Psychiatry* 61 (8): 575–578.
Temml, Veronika, Susanne Kuehnl, Daniela Schuster, Stefan Schwaiger, Hermann Stuppner, and Dietmar Fuchs. 2013. "Interaction of Carthamus tinctorius lignan arctigenin with the binding site of tryptophan-degrading enzyme indoleamine 2, 3-dioxygenase." *FEBS Open Bio* 3: 450–452.
Turner, David T, Edel McGlanaghy, Pim Cuijpers, Mark Van Der Gaag, Eirini Karyotaki, and Angus MacBeth. 2018. "A meta-analysis of social skills training and related interventions for psychosis." *Schizophrenia Bulletin* 44 (3): 475–491.
Unno, Keiko, Naoki Tanida, Naoto Ishii, Hiroyuki Yamamoto, Kazuaki Iguchi, Minoru Hoshino, Atsushi Takeda, Hayato Ozawa, Tsutomu Ohkubo, and Lekh Raj Juneja. 2013. "Anti-stress effect of theanine on students during pharmacy practice: Positive correlation among salivary α-amylase activity, trait anxiety and subjective stress." *Pharmacology Biochemistry and Behavior* 111: 128–135.
Vasha, Zannatun Nayem, Bidyut Sharma, Israt Jahan Esha, Jabir Al Nahian, and Johora Akter Polin. 2023. "Depression detection in social media comments data using machine learning algorithms." *Bulletin of Electrical Engineering and Informatics* 12 (2): 987–996.
Vijeepallam, Kamini, Vijayapandi Pandy, Thubasni Kunasegaran, Dharmani D Murugan, and Murali Naidu. 2016. "Mitragyna speciosa leaf extract exhibits antipsychotic-like effect with the potential to alleviate positive and negative symptoms of psychosis in mice." *Frontiers in Pharmacology* 7: 464.
Viola, H, C Wasowski, M Levi De Stein, C Wolfman, R Silveira, F Dajas, J Medina, and A Paladini. 1995. "Apigenin, a component of Matricaria recutita flowers, is a central benzodiazepine receptors-ligand with anxiolytic effects." *Planta Medica* 61 (3): 213–216.
Wang, Bo, Yong-Jie Lian, Xin Dong, Wei Peng, Lin-Lin Liu, Wen-Jun Su, Hong Gong, Ting Zhang, Chun-Lei Jiang, and Jia-Si Li. 2018. "Glycyrrhizic acid ameliorates the kynurenine pathway in association with its antidepressant effect." *Behavioural Brain Research* 353: 250–257.
Wang, Qi, Jun Mei, and Jing Xie. 2022. "The effects of lemon balm (Melissa officinalis L.) Essential oil on the stress response, anti-oxidative ability, and kidney metabolism of sea bass during live transport." *Animals* 12 (3): 339.
Xiang, X., X. Wang, Y. Wu, J. Hu, Y. Li, S. Jin, and X. Wu. 2022. "Activation of GPR55 attenuates cognitive impairment, oxidative stress, neuroinflammation, and synaptic dysfunction in a streptozotocin-induced Alzheimer's mouse model." *Pharmacology Biochemistry & Behavior* 214: 173340.
Yamada, Takashi, Takehiko Terashima, Keiko Wada, Sakiko Ueda, Mitsuyo Ito, Tsutomu Okubo, Lekh Raj Juneja, and Hidehiko Yokogoshi. 2007. "Theanine, r-glutamylethylamide, increases neurotransmission concentrations and neurotrophin mRNA levels in the brain during lactation." *Life Sciences* 81 (16): 1247–1255.

Yang, Shui-Jin, Hai-Yang Yu, Dan-Yu Kang, Zhan-Qiang Ma, Rong Qu, Qiang Fu, and Shi-Ping Ma. 2014. "Antidepressant-like effects of salidroside on olfactory bulbectomy-induced pro-inflammatory cytokine production and hyperactivity of HPA axis in rats." *Pharmacology Biochemistry and Behavior* 124: 451–457.

Yang, Longfei, Yinghao Zhao, Yicun Wang, Lei Liu, Xingyi Zhang, Bingjin Li, and Ranji Cui. 2015. "The effects of psychological stress on depression." *Current Neuropharmacology* 13 (4): 494–504.

Zakaria, Fauziahanim, Muhammad Tayyab Akhtar, Wan Ibrahim Wan Norhamidah, Abu Bakar Noraini, Azira Muhamad, Shamarina Shohaimi, Hafandi Ahmad, Intan Safinar Ismail, Nor Hadiani Ismail, and Khozirah Shaari. 2023. "Centella asiatica (L.) Urb. Extract ameliorates branched-chain amino acid (BCAA) metabolism in acute reserpine-induced stress zebrafish model via 1H nuclear magnetic resonance (NMR)-based metabolomics approach." *Comparative Biochemistry and Physiology Part C: Toxicology & PhArmacology* 264: 109501.

Zeng, Min-Jie, Li-Ping Zhou, Ya-Qin Li, Fan Zhao, Zhi-Yuan Shang, Xue-Yang Deng, Zhan-Qiang Ma, Qiang Fu, Shi-Ping Ma, and Rong Qu. 2018. "Baicalin exerts neuroprotective effects via inhibiting activation of GSK3β/NF-κB/NLRP3 signal pathway in a rat model of depression." *International Immunopharmacology* 64: 175–182.

Zhang, Meng-Di, Xue Tao, Rui-Le Pan, Li-Sha Wang, Chen-Chen Li, Yun-Feng Zhou, Yong-Hong Liao, Shan-Guang Chen, Qi Chang, and Xin-Min Liu. 2020. "Antidepressant-like effects of cajaninstilbene acid and its related mechanisms in mice." *Fitoterapia* 141: 104450.

Zheng, Xingxing, Ying Cheng, Yiwei Chen, Yisong Yue, Yingchun Li, Sizhe Xia, Yang Li, Huanhuan Deng, Junli Zhang, and Yanjun Cao. 2019. "Ferulic acid improves depressive-like behavior in prenatally-stressed offspring rats via anti-inflammatory activity and HPA axis." *International Journal of Molecular Sciences* 20 (3): 493.

6 The Role of Epigenetic Modifications by Phytoconstituents in Neurological Ailments

Aarthi Narasimhan, Sri Raagavee Sivakumar, Karthikeyan Subbiahanadar Chelladurai, Kamarajan Rajagopalan, and Jackson Durairaj Selvan Christyraj

6.1 NEUROLOGICAL DISORDER MECHANISM

Neurological disorders are termed as the dysfunction condition that affects the brain, spinal cord, and the nerves that run all over the human body. The determined cause of neurological disorder differs but involves genetic disorders, infections, congenital abnormalities, lifestyle, and some other causes including malnutrition, brain injury, spinal cord injury, or nerve injury. Certain neurological disorders manifest before birth and are considered congenital. Some other disorders are caused by additional conditions like tumor, degeneration, trauma, infections, and other structural defects. The primary cause of neurological disabilities is due to the damage occurring in the nervous system.

There are many commonly known neurological disorders such as Alzheimer's disease (AD), autism, Parkinson's disease (PD), multiple sclerosis (MS), meningitis, muscular dystrophy, tumors, neuromuscular disorders, AD, learning disabilities, cerebral palsy, acute spinal cord injury, amyotrophic lateral sclerosis (ALS), ataxia Bell's palsy, brain tumors, cerebral aneurysm, epilepsy and seizures, Guillain-Barré syndrome, headache, head injury, hydrocephalus lumbar disk disease (herniated disk), and encephalitis.

Genetic disorders arise when a harmful change to the gene called mutation takes place. DNA, or deoxyribonucleic acid, the building block of genes, contains the instructions for cells functioning and the characteristics of every individual. Each biological parent contributes half to the children's genes, and children may also inherit a gene mutation from one or both of the biological parents. Mutations, in the DNA, can cause genetic changes which could make children more likely to have a genetic and neurological disorder.

Genetic diseases include:

Chromosomal: The chromosomal structures that hold the genes and DNA within each cell are affected by this type. People with these conditions lack chromosome material or have duplicated chromosomes.

 DOI: 10.1201/9781003389781-6

Multifactorial complex: A combination of gene mutations and other factors is the cause of these disorders. They include smoking or drinking, chemical exposure, certain medications, and diet.

Monogenic (single gene): A single gene mutation responsible for certain disorders.

There are more than 600 neurological diseases identified according to the United States National Library of Medicine. A variety of symptoms can be caused by structural, biochemical, or electrical abnormalities in the brain, spinal cord, or other nerves including loss of feeling or tingling, weakness or loss of muscle strength, loss of sight or double vision, memory loss, impaired mental ability, lack of coordination, rigidity of the muscles, tremors and seizures, back pain that radiates to the feet, toes, or other parts of the body, muscle wasting and slurred speech.

6.1.1 Autism

A group of developmental disabilities characterized by stereotyped behaviors, limited interest, and abnormal social interaction and communication is termed to be autism spectrum disorder (ASD) (Won et al. 2013). Kanner was the first to describe autism based on the observation of 11 children who appeared with "severe loneliness from the very beginning of life, not responding to anything that comes to them from the outside world" as the earliest case. He suggested that childhood schizophrenia could be caused by the behavioral concoction of autism, obsession, stereotypy, and echolalia (Kanner 1943). Autism was termed "infantile autism" in 1980, "autism disease" in 1987, and, more recently, "autism" or the umbrella term "ASD" from Kanner's diagnosis of "childhood or early-onset schizophrenia." It is commonly recognized that ASD is a condition with significant hereditary components on the basis of several research studies that have been conducted to elucidate the pathogenic mechanisms underlying ASD. The concordance rates for autism in monozygotic twins can outstretch up to 90%, and in dizygotic twins, it can reach up to 10% (Abrahams and Geschwind 2008).

6.1.2 SCD

The autosomal recessive disease, with a multisystem condition that causes progressive organ failure, painful ischemic vaso-occlusion episodes, and persistent hemolytic anemia. With millions of people suffering globally, sickle cell disease (SCD) is considered to be the most prevalent monogenetic disease. The term "sickle cell disease" refers to a group of ailments brought on by the inheritance of hemoglobin S (HbS) (Houwing et al. 2019). The sixth codon of the β-globin gene conversion was influencing the essential amino acid of valine production instead of glutamic acid causing SC (Meher et al. 2019). An aggregate of hemoglobin molecules and unsaturated oxygen molecules coupled to form a specific linkage drives the formulation of a long chain consisting of hemoglobin molecules. Abnormal production of hemoglobin has turned the physical behavior of erythrocytes that intimate the actual pathologic indications of SCD. Repeated alignment leads to the cause of birefringent toxoids (Harris 1950). In the 1950 signs of SCA, including differences in heart and kidney, as well as pulmonary involvement and stress. Generally, sickled cells are rigid and fixed and have long tapering on both ends. Capillary blockades are generated by these sickle cells. Sickle cell anemia also influences cardiac, disease, thrombosis, and infarctions. Interestingly, an SCD patient has cardiac hypertrophy without any form of endocardial abnormalities or myocardial damage (Zimmerman and Barnett 1944).

6.1.3 Parkinson's Disease

The most obvious progress has been made in understanding the proximate causes, which stated that PD is a neurological disorder characterized by the degeneration of dopaminergic neurons in the midbrain's substantia nigra. Rather than the normal synthesis and pumping movement of neurotransmitter dopamine (DA) in regulating the brain regions, up to 70% of these dopaminergic neurons in substantia nigra will be killed during the development of PD (Zeng et al. 2018). It is a neurodegenerative disorder that gets worse over time and mostly affects old age people, in comparison with younger age. It is considered to be the second most neurodegenerative disease (Sherer et al. 2012).

6.1.4 Alzheimer's Disease

Progressive mental deterioration that can occur in middle or old age as a result of generalized brain degeneration. It is the most common cause of early senility and so it is also termed as senile dementia. The disease often results in the following behaviors – impaired memory, thinking, and confusion. The locus coeruleus, the brain stem nuclei (such as the raphe nucleus), reticular formation, amygdala, substantia nigra, striatum, hypothalamus, thalamus, and claustrum, as well as some areas of the cerebral cortex, are the cell types that are impacted by AD (Kocahan and Doğan 2017). Depending on the expression of neurotransmitters, neuromodulators, and neuropeptides, different neuronal types are influenced in different regions. The degenerative process causes loss of neurons and brain shrinkage (Duyckaerts et al. 2009).

6.1.5 Multiple Sclerosis

The demyelinating inflammatory condition of the central nervous system (CNS), known as MS, is an autoimmune disease. The main pathology of MS is immune destruction of myelin basic protein (MBP) throughout the nervous system, especially brain and spinal cord (Lutton et al. 2004). Myelinated CNS tracts are the targets in people with MS, but the immune trigger is unknown. Breakdown of myelin sheath, blood-brain barrier, axonal damage, gliosis, and sclerotic plaque formation occurs in the region of inflammation as a result of abnormal fibrosis and hardening of the neuronal myelin sheath (World Health Organization 2006).

6.2 EPIGENETIC MECHANISM OF NEUROLOGICAL DISORDERS

Regions of a genome, gene products, and cellular pathways that are linked to the genetic disease can be regulated by epigenetic processes. The entire genome, which includes the loci implicated in the pathogenesis of any neurological disease with a hereditary component, is modulated by epigenetic mechanisms. DNA methylation, histone modifications, nucleosome and higher-order chromatin remodeling, ncRNAs, and RNA editing are some of the main epigenetic mechanisms. Neurodevelopmental problems are also linked to epigenetic mutations in various ways. For instance, syndromic and non-syndromic forms of intellectual and developmental disability are brought on by mutations in a number of genes encoding proteins that both directly and indirectly control chromatin structure (Qureshi and Mehler 2013).

The MECP2 gene, which codes for a multifunctional protein belonging to the methyl-CpG-binding domain family, is one of the most important examples for Rett syndrome, an autistic spectrum illness, is primarily brought on by MECP2 mutations, which include missense, nonsense, frameshift, and major deletion variants. Newborn encephalopathy, intellectual syndromic and developmental disabilities, and Angelman-like syndrome are also associated with MECP2 mutations (Van Bokhoven 2011). Lysine methylation is known to be dysregulated in medulloblastoma and other cancers. Lysine methylation is also involved in controlling cell identity, DNA repair, cell cycle, stress responses, and transcription (Robinson et al. 2012). The emergence, progression, and risk of neurological disease can be altered by genetic variation in the genes encoding the epigenetic factors. For instance, brain volume measures on neuroimaging, indicators of illness severity, are linked with single-nucleotide polymorphisms (SNPs) in 3HDAC gene loci – rs2522129 (SIRT4), rs2675231 (HDAC11), and rs2389963 (HDAC9) (Inkster et al. 2013). Another notable example is the International Stroke Genetics Consortium's huge genome-wide association research, which showed that an SNP in an intron of the HDAC9 gene on chromosome 7p21.1 (rs11984041) is linked to risk of large vessel ischemic stroke (Bellenguez et al. 2012). In the majority of population-based studies, the extended haplotypes of HLA-DRB1*15 (HLA-DRB11501-DRB50101-DQA10102-DQB10602) have been replicated and demonstrated to increase the risk of MS (Lincoln et al. 2005). Nearly 17 autosomal dominant and recessive gene mutations have been identified for the cause of variants in familial PD. Parkin, ubiquitin carboxyl-terminal hydrolase L1 (UCH-L1), phosphatase and tensin homolog inducible kinase 1 (PINK1), leucine-rich repeat kinase 2 (LRRK2), and glucocerebrosidase (GBA) are some examples of mutation inducing genes (Houlden and Singleton 2012). The epigenetic mechanism of muscular dystrophy includes DNA methylation, histone modification, and regulatory non-coding RNA expression play a promising role in muscular regeneration. The Regenerative capacity is directly linked to the presence of satellite cells (Souza et al. 2015). Muscular dystrophy progression occurs due to the failure of satellite cells to divide asymmetrically and to maintain damage repair cycle. The lack of Dystrophin decreases the serine-threonine protein kinase 2 (MARK2) which leads to the abnormal mitotic division.

6.3 PHYTOCHEMICALS FOR NEUROLOGICAL DISORDER

Neurological disorders include neurotraumatic disorders as well as neurodegenerative illnesses like AD, PD, Huntington's disease (HD), MS, and lot more. Neurodegenerative diseases occur due to genetic and environmental factors. These neurodegenerative disorders frequently exhibit neuroinflammation, deposition of certain aggregated proteins, and oxidative stress as pathogenic characteristics. Symptoms identified and diagnosed at the early stages of disorders can be useful for the proper treatment and prevention to further progression and high risk of neurodegenerative illness.

6.3.1 Epigallocatechin-3-Gallate

The epigallocatechin-3-gallate (EGCG) polyphenol is the most prevalent one to be found in tea leaves. The chemical mechanism through which EGCG prevents the uptake of DA and exerts neuroprotection is what prevents neuronal cell death. Tyrosine hydroxylase (TH)-positive cells are severely lost in PD, and one study found that taking EGCG and tea at the same time may stop the loss of these cells in the substantia nigra. By interfering with catechol-O-methyltransferase (COMT), an enzyme that prevents the metabolism of DA, it performs this effect. Moreover, EGCG is believed to control the proteolytic cleavage of α-Synuclein, turning it into a less toxic form. Synuclein aggregates are observed in ND. In protein kinase C (PKC) intracellular signaling pathway in SH-SY5Y cells, EGCG demonstrates neuroprotection, and it is shown that PKC inhibitors participate in the method of neuroprotection by EGCG (Koh et al. 2003). The activity of EGCG is concentration-dependent and suppresses the neurotoxin 6-hydroxydopamine (6-OHDA) by phosphorylating PKC. While EGCG at high doses tends to have anti-proliferative effects and inhibit angiogenesis, EGCG at low concentrations protects neurons. It inhibits oxidation at low doses, but at larger quantities, it has pro-oxidant characteristics (Castellano-González et al. 2016).

6.3.2 Berberine

Berberine has the ability to cease the activity of different enzymes which is implicated in the development of AD. It has also been demonstrated that the ability of berberine to protect the neuronal cells from neurotoxicity. Also, it has been discovered that berberine is effective in treating cobalt chloride-induced hypoxia. It functions as a scavenger of reactive oxygen species (ROS) produced as a result of the hypoxic conditions, which in turn suppresses numerous agents that promote apoptosis and provide neuroprotection (Hsu et al. 2013). By scavenging free radicals, berberine activates the PI3K/Akt/Nrf2 pathway and has neuroprotective benefits. Moreover, it has been noted that berberine has anti-apoptotic effects via upregulating Bcl-2, downregulating caspase 1 and 3, and caspase 3 expression. When given for prolonged periods of time, berberine is to blame for the degeneration of dopaminergic neurons in the substantia nigra. Berberine may prevent ischemic stroke by scavenging radical species, or, to put it another way, by lowering oxidative stress (Asai et al. 2007). Because it can solubilize β-amyloid plaques in mouse models, berberine is regarded to have a promising function as a treatment for AD. By reducing NeuN, laminin, MMP-9, and gelatinase activity, berberine is expected to lessen the effects of an ischemic stroke (Hong et al. 2012).

6.3.3 Curcumin

Curcumin, considered to be one of the key components of turmeric, which has a prominent position among Indian spices, is curcumin. Due to its therapeutic benefits, curcumin is used to treat conditions like diabetes, biliary diseases, coughing, and hepatic illnesses. Curcumin has been linked to a number of mechanisms, one of which is the ability to attach to amyloid plaques by inhibiting NF-, thus lowering the pathogenesis of AD (Nam et al. 2014). Moreover, studies have demonstrated that curcumin promotes neuronal regeneration by enhancing brain-derived neurotrophic factor (BDNF) levels in a PD model through activating Trk/PI3K signaling pathways (Hoppe et al. 2013). Inflammation makes the situation worse in ND cases, and curcumin tends to lower IL-6 and TNF-α expression. It has been discovered that AD impairs the operation of brain macrophages, preventing them from phagocytosing properly.

Phytochemicals such as polyphenols, curcumin, and sulforaphane are known to exert a positive impact on the muscular function at the molecular level. Curcumin is a phytochemical component of spice turmeric, which exerts anti-inflammatory effects by inducing the activation of heat shock proteins (Dunsmore et al. 2001), oxygen free radical formation (Chattopadhyay et al. 2006), inhibiting the p38 kinase activity (Jin and Li 2007) and preventing the production and release of cytokines (Jobin et al. 1999). The effects of curcumin in preventing muscular dystrophy occur significantly by inhibiting the activity of NF-KB which is responsible for the loss of muscle mass (Penner et al. 2001).

6.3.4 Resveratrol

Resveratrol is a phenolic substance found in tea, wine, peanuts, and grapes. Further research revealed that resveratrol raises IL-10 levels, which restrict TNF-α and NF-κB levels to promote anti-inflammatory properties. After resveratrol injection, extracellular signal-regulated kinase/1/2 (ERK1/2)/CREB signaling pathways were activated, which helps neurons survive by boosting the levels of BDNF and GDNF release (Anastacio et al. 2014). In the course of ND, glial cells release inflammatory cytokines as well as a number of neurotoxic chemicals, including nitric oxide and superoxide. Resveratrol provides neuroprotection by preventing glial cell activation (Ma et al. 2014). It has been discovered that resveratrol plays a part in plaque destabilization and plaque prevention. In rotenone-induced PD cell models, resveratrol has neuroprotective benefits by activating the AMPK-SIRT1-autophagy pathway (Wu et al. 2011).

6.3.5 Apigenin

Apigenin is one of the phytochemical components obtained from *Matricaria recutita* L. Apigenin is a CNS active molecule and interacts with other neurotransmission systems by acting as a partial agonist of gamma amino butyric acid (GABA) receptors. This indicates that apigenin exerts an anti-epileptic mechanism by binding with benzodiazepine receptors and by exhibiting reduced GABA-activated chloride currents (Avallone et al. 2000).

6.3.6 Limonoids

Limonoids are rare and highly oxygenated molecules found in plants. Due to its high limonoid content, *Melia toosendan*, a member of the Meliaceae family, contains limonoids in abundance and the plant extract stimulates neuronal development which is similar to that of NGF (nerve growth factor), which does so by activating ERK and protein kinase A (PKA). It has been discovered that limonoids increase NGF levels in a PC12 rat cell line, promoting neuronal differentiation and proliferation (Roy and Saraf 2006). Limonoids activate ERKs and PKA to enhance neuronal differentiation. PKA is activated by an increase in cAMP, although limonoids can also activate PKA without an increase in cAMP (Yoon et al. 2015).

6.4 EPIGENETIC MODIFICATIONS OF THE PHYTOCHEMICAL

Nowadays, epigenetics is understood to be the study of gene function modifications that are heritable during mitosis and/or meiosis but do not involve a change in DNA sequence. Dietary phytochemicals play a significant role in the regulation of normal physiological processes. Natural substances called phytochemicals have been demonstrated to have protection against oxidative stress and neuroinflammation. The immune system is stimulated by phytochemicals, platelet aggregation is reduced, and hormone metabolism is controlled (Farooqui 2012). These phytochemicals activate stress response pathways, where the cells will encounter a defense mechanism, in addition to regulation of gene expression and enzyme metabolism for normal functioning of the body. phytochemicals participate in a downward signal transduction cascade and serve as a ligand, binding to specific receptors on cell membranes or nuclei thereby producing antioxidative activity (Si and Liu 2007).

The use of naturally occurring some of the dietary phytochemicals on a daily basis such as polyphenols, genistein, curcumin, sulforaphane, phenyl isothiocyanate, lycopene, resveratrol, quercetin, indole-3-carbinol, ellagitannin, and organosulfur compounds are used to treat several disorders including cancer. The phytochemicals overlap the number of mechanisms of action including the induction of detoxification of enzymes, antioxidant effects, alteration of hormonal metabolism, etc (Wang et al. 2012). The dynamic nature of DNA methylation, active mechanisms for DNA demethylation, diverse activities of 5-methylcytosine and its oxidized derivatives, the incorporation of histone variations into chromatin, nucleosome occupancy, and dynamics are all included in the current understanding of epigenetics. These epigenetic alterations appear to control crucial gene networks controlling physiological processes linked to the health benefits of certain diets, offering a justifiable and straightforward method to prevent or perhaps treat these disorders. The effectiveness of food and exercise in treating cancer, as well as cardiovascular disease, diabetes, obesity, and rheumatoid arthritis and in some other neurological disorders like ADs (Müller et al. 2001).

Histones can undergo a wide range of changes to their amino acid residues. Acetylation, methylation, phosphorylation, ubiquitination, and sumoylation are some examples of histone post-translational changes. Nucleosomes are the basic building blocks of chromatin found in cell nuclei (Sassone-Corsi 2013). Each nucleosome is made up of two copies of each of the four core histones and 147 DNA base pairs tightly wound around an octamer of histone proteins (H2A, H2B, H3, and H4). The DNA between the nucleosomal core particles is where the linker histone H1 interacts, and its job is to stabilize higher-order chromatin structures.

Moreover, each histone protein has an N-terminal tail with numerous possible modification sites and a core globular domain (Wang et al. 2013). Lysine, arginine, serine, and threonine amino acids are the main residues that serve as the substrates for these changes (Rothbart and Strahl 2014). According to the position of the change, these modifications have been linked to the activation or repression of gene transcription, strongly indicating the presence of a histone code. According to this theory, particular histone modifications cause proteins connected to the chromatin to interact, resulting in a distinct regulatory response for gene expression (Strahl and Allis 2000).

Histone methylation is carried and there have been studies linking specific processes to various types of changes occurring in the genes. For instance, chromatin aperture has been linked to H3K4, H3K36, and H3K79. Yet, several distinct roles have also been linked to the methylation of these residues. On the other hand, promoter regions have been linked to H3K4 trimethylation. H3K36 dimethylation has been linked to the transcription's elongation of the RNA POL II (Lin et al. 2007). Histone acetylation might be explained by the chemistry of the modification, which involves adding an acetyl group (COCH3) to an amino-terminal residue and lowering the positive charge of histones as a result. This interaction with DNA causes a small reduction in the compaction of chromatin (Shahbazian and Grunstein 2007).

The covalent joining of cytosine methyl groups, which are primarily found in the context of dinucleotide 50-CpG-3, results in DNA methylation (Klose and Bird 2006). When DNA duplicates, a process keeps these patterns from changing because DNA methylation creates patterns that are set throughout embryonic development. A group of proteins known as DNA methyltransferases carry out the DNA methylation mechanism (DNMTs). These proteins are divided into two groups: One for de novo methylation and the other for methylation maintenance. Depending on the DNA substrate, both enzymes have different properties. For instance, DNA methyl transferase 1 maintains DNA methylation (DNMT1). During DNA replication, these proteins add methyl groups to already-existing methyl patterns on a new strand of DNA (Fraga et al. 2005).

REFERENCES

Abrahams, Brett S., and Daniel H. Geschwind. "Advances in autism genetics: On the threshold of a new neurobiology." Nature reviews genetics 9, no. 5 (2008): 341–355.

Anastacio, Janine R., Carlos A. Netto, Cibele C. Castro, Eduardo F. Sanches, Daniele C. Ferreira, Cristie Noschang, Rachel Krolow, Carla Dalmaz, and Aline Pagnussat. "Resveratrol treatment has neuroprotective effects and prevents cognitive impairment after chronic cerebral hypoperfusion." Neurological research 36, no. 7 (2014): 627–633.

Asai, Masashi, Nobuhisa Iwata, Ayumu Yoshikawa, Yoshimi Aizaki, Shoichi Ishiura, Takaomi C. Saido, and Kei Maruyama. "Berberine alters the processing of Alzheimer's amyloid precursor protein to decrease Aβ secretion." Biochemical and biophysical research communications 352, no. 2 (2007): 498–502.

Avallone, Rossella, Paola Zanoli, Giulia Puia, Matthias Kleinschnitz, Peter Schreier, and Mario Baraldi. "Pharmacological profile of apigenin, a flavonoid isolated from Matricaria chamomilla." Biochemical pharmacology 59, no. 11 (2000): 1387–1394.

Castellano-González, Gloria, Nicolas Pichaud, J. William O. Ballard, Alban Bessede, Helder Marcal, and Gilles J. Guillemin. "Epigallocatechin-3-gallate induces oxidative phosphorylation by activating cytochrome c oxidase in human cultured neurons and astrocytes." Oncotarget 7, no. 7 (2016): 7426.

Céline Bellenguez, Steve Bevan, Andreas Gschwendtner, Chris CA Spencer, Annette I. Burgess et al.; International Stroke Genetics Consortium (ISGC), Wellcome Trust Case Control Consortium 2 (WTCCC2). "Genome-wide association study identifies a variant in HDAC9 associated with large vessel ischemic stroke." Nature genetics 44, no. 3 (2012): 328–333.

Chattopadhyay, Ishita, Uday Bandyopadhyay, Kaushik Biswas, Pallab Maity, and Ranajit K. Banerjee. "Indomethacin inactivates gastric peroxidase to induce reactive-oxygen-mediated gastric mucosal injury and curcumin protects it by preventing peroxidase inactivation and scavenging reactive oxygen." Free radical biology and medicine 40, no. 8 (2006): 1397–1408.

Dunsmore, Katherine E., Philip G. Chen, and Hector R. Wong. "Curcumin, a medicinal herbal compound capable of inducing the heat shock response." Critical care medicine 29, no. 11 (2001): 2199–2204.

Duyckaerts, C., B. Delatour, and M. C. Potier. "Classification and basic pathology of Alzheimer disease." Acta neuropathologica 118, no. 1 (2009): 5–36.

Farooqui, Akhlaq A.. "Effect of lifestyle, aging, and phytochemicals on the onset of neurological disorders." Phytochemicals, signal transduction, and neurological disorders (2012): 1–29.

Fraga, Mario F., Esteban Ballestar, Maria F. Paz, Santiago Ropero, Fernando Setien, Maria L. Ballestar, and Damia Heine-Suñer et al. "Epigenetic differences arise during the lifetime of monozygotic twins." Proceedings of the national academy of sciences 102, no. 30 (2005): 10604–10609.

Harris, John W. "Studies on the destruction of red blood cells. VIII. Molecular orientation in sickle cell hemoglobin solutions." Proceedings of the society for experimental biology and medicine 75, no. 1 (1950): 197–201.

Hong, Jeong-Seok, Yeun-Kyung Chu, Hyung Lee, Byung-Hoon Ahn, Jae-Hyung Park, Mi-Jung Kim, and Sunghye Lee et al. "Effects of berberine on hippocampal neuronal damage and matrix metalloproteinase-9 activity following transient global cerebral ischemia." Journal of neuroscience research 90, no. 2 (2012): 489–497.

Hoppe, Juliana B., Karine Coradini, Rudimar L. Frozza, Claudia M. Oliveira, André B. Meneghetti, Andressa Bernardi, Elisa Simões Pires, Ruy CR Beck, and Christianne G. Salbego. "Free and nanoencapsulated curcumin suppress β-amyloid-induced cognitive impairments in rats: Involvement of BDNF and Akt/GSK-3β signaling pathway." Neurobiology of learning and memory 106 (2013): 134–144.

Houlden, Henry, and Andrew B. Singleton. "The genetics and neuropathology of Parkinson's disease." Acta neuropathologica 124 (2012): 325–338.

Houwing, M. E., P. J. De Pagter, E. J. Van Beers, B. J. Biemond, E. Rettenbacher, A. W. Rijneveld, and E. M. Schols et al. "Sickle cell disease: Clinical presentation and management of a global health challenge." Blood reviews 37 (2019): 100580.

Hsu, Ya-Yun, Yu-Ting Tseng, and Yi-Ching Lo. "Berberine, a natural antidiabetes drug, attenuates glucose neurotoxicity and promotes Nrf2-related neurite outgrowth." Toxicology and applied pharmacology 272, no. 3 (2013): 787–796.

Inkster, Becky, Eva MM Strijbis, Maria Vounou, Ludwig Kappos, Ernst-Wilhelm Radue, Paul M. Matthews, and Bernard MJ Uitdehaag et al. "Histone deacetylase gene variants predict brain volume changes in multiple sclerosis." Neurobiology of aging 34, no. 1 (2013): 238–247.

Jin, Bingwen, and Yi-Ping Li. "Curcumin prevents lipopolysaccharide-induced atrogin-1/MAFbx upregulation and muscle mass loss." Journal of cellular biochemistry 100, no. 4 (2007): 960–969.

Jobin, Christian, Cynthia A. Bradham, Maria Pia Russo, Booker Juma, Acharan S. Narula, David A. Brenner, and R. Balfour Sartor. "Curcumin blocks cytokine-mediated NF-κB activation and proinflammatory gene expression by inhibiting inhibitory factor i-κB kinase activity." The journal of immunology 163, no. 6 (1999): 3474–3483.

Kanner, Leo. "Autistic disturbances of affective contact." Nervous child 2, no. 3 (1943): 217–250.

Klose, Robert J., and Adrian P. Bird. "Genomic DNA methylation: The mark and its mediators." Trends in biochemical sciences 31, no. 2 (2006): 89–97.

Kocahan, Sayad, and Zumrut Doğan. "Mechanisms of Alzheimer's disease pathogenesis and prevention: The brain, neural pathology, N-methyl-D-aspartate receptors, tau protein and other risk factors." Clinical psychopharmacology and neuroscience 15, no. 1 (2017): 1.

Koh, Seong-Ho, Seung H. Kim, Hyugsung Kwon, Younjoo Park, Ki Sok Kim, Chi Won Song, and Juhan Kim et al. "Epigallocatechin gallate protects nerve growth factor differentiated PC12 cells from oxidative-radical-stress-induced apoptosis through its effect on phosphoinositide 3-kinase/Akt and glycogen synthase kinase-3." Molecular brain research 118, no. 1–2 (2003): 72–81.

Li, Bing, Michael Carey, and Jerry L. Workman. "The role of chromatin during transcription." Cell 128, no. 4 (2007): 707–719.

Lincoln, Matthew R., Alexandre Montpetit, M. Zameel Cader, Janna Saarela, David A. Dyment, Milvi Tiislar, and Vincent Ferretti et al. "A predominant role for the HLA class II region in the association of the MHC region with multiple sclerosis." Nature genetics 37, no. 10 (2005): 1108–1112.

Lutton, J. D., R. Winston, and T. C. Rodman. "Multiple sclerosis: Etiological mechanisms and future directions." Experimental biology and medicine 229, no. 1 (2004): 12–20.

Ma, Teng, Meng-Shan Tan, Jin-Tai Yu, and Lan Tan. "Resveratrol as a therapeutic agent for Alzheimer's disease." BioMed research international 2014 (2014).

Meher, Satyabrata, Siris Patel, Kishalaya Das, Snehadhini Dehury, Bimal Prasad Jit, Mahendra M. Maske, Padmalaya Das, Bisnu Prasad Dash, and Pradeep Kumar Mohanty. "Association of plasma homocysteine level with vaso-occlusive crisis in sickle cell anemia patients of Odisha, India." Annals of hematology 98 (2019): 2257–2265.

Müller, Horst, F. Wilhelmi de Toledo, and K-L. Resch. "Fasting followed by vegetarian diet in patients with rheumatoid arthritis: A systematic review." Scandinavian journal of rheumatology 30, no. 1 (2001): 1–10.

Nam, Sung Min, Jung Hoon Choi, Dae Young Yoo, Woosuk Kim, Hyo Young Jung, Jong Whi Kim, and Miyoung Yoo et al. "Effects of curcumin (Curcuma longa) on learning and spatial memory as well as cell proliferation and neuroblast differentiation in adult and aged mice by upregulating brain-derived neurotrophic factor and CREB signaling." Journal of medicinal food 17, no. 6 (2014): 641–649.

Penner, C. Gail, Gyu Gang, Curtis Wray, Josef E. Fischer, and Per-Olof Hasselgren. "The transcription factors NF-κB and AP-1 are differentially regulated in skeletal muscle during sepsis." Biochemical and biophysical research communications 281, no. 5 (2001): 1331–1336.

Qureshi, Irfan A., and Mark F. Mehler. "Understanding neurological disease mechanisms in the era of epigenetics." JAMA neurology 70, no. 6 (2013): 703–710.

Robinson, Giles, Matthew Parker, Tanya A. Kranenburg, Charles Lu, Xiang Chen, Li Ding, and Timothy N. Phoenix et al. "Novel mutations target distinct subgroups of medulloblastoma." Nature 488, no. 7409 (2012): 43–48.

Rothbart, Scott B., and Brian D. Strahl. "Interpreting the language of histone and DNA modifications." Biochimica et biophysica acta (BBA)-Gene regulatory mechanisms 1839, no. 8 (2014): 627–643.

Roy, Amit, and Shailendra Saraf. "Limonoids: Overview of significant bioactive triterpenes distributed in plants Kingdom." Biological and pharmaceutical bulletin 29, no. 2 (2006): 191–201.

Sassone-Corsi, Paolo. "When metabolism and epigenetics converge." Science 339, no. 6116 (2013): 148–150.

Shahbazian, M. D., and M. Grunstein. "Functions of site-specific histone acetylation and deacetylation." Annual review of biochemistry 76 (2007): 75–100.

Sherer, Todd B., Sohini Chowdhury, Katherine Peabody, and Deborah W. Brooks. "Overcoming obstacles in Parkinson's disease." Movement disorders 27, no. 13 (2012): 1606–1611.

Si, Hongwei, and Dongmin Liu. "Phytochemical genistein in the regulation of vascular function: New insights." Current medicinal chemistry 14, no. 24 (2007): 2581–2589.

Souza, Gustavo Torres de, Rafaella de Souza Salomão Zanette, Danielle Luciana Aurora Soares do Amaral, Francisco Carlos da Guia, Claudinéia Pereira Maranduba, Camila Maurmann de Souza, and Ernesto da Silveira Goulart Guimarães et al. "Satellite cells: Regenerative mechanisms and applicability in muscular dystrophy." Stem cells international 2015 (2015).

Strahl, Brian D., and C. David Allis. "The language of covalent histone modifications." Nature 403, no. 6765 (2000): 41–45.

Van Bokhoven, Hans. "Genetic and epigenetic networks in intellectual disabilities." Annual review of genetics 45 (2011): 81–104.

Wang, Junjun, Zhenlong Wu, Defa Li, Ning Li, Scott V. Dindot, M. Carey Satterfield, Fuller W. Bazer, and Guoyao Wu. "Nutrition, epigenetics, and metabolic syndrome." Antioxidants & redox signaling 17, no. 2 (2012): 282–301.

Wang, Jun, Jin-Tai Yu, Meng-Shan Tan, Teng Jiang, and Lan Tan. "Epigenetic mechanisms in Alzheimer's disease: Implications for pathogenesis and therapy." Ageing research reviews 12, no. 4 (2013): 1024–1041.

Won, Hyejung, Won Mah, and Eunjoon Kim. "Autism spectrum disorder causes, mechanisms, and treatments: Focus on neuronal synapses." Frontiers in molecular neuroscience 6 (2013): 19.

World Health Organization. Neurological disorders: public health challenges. World Health Organization. (2006).

Wu, Yuncheng, Xinqun Li, Julie Xiaohong Zhu, Wenjie Xie, Weidong Le, Zhen Fan, Joseph Jankovic, and Tianhong Pan. "Resveratrol-activated AMPK/SIRT1/autophagy in cellular models of Parkinson's disease." Neurosignals 19, no. 3 (2011): 163–174.

Yoon, Yeo Cho, Sung-Hee Kim, Min Jung Kim, Hye Jeong Yang, Mee-Ra Rhyu, and Jae-Ho Park. "Limonin, a component of dictamni radicis cortex, inhibits eugenol-induced calcium and cAMP levels and PKA/CREB signaling pathway in non-neuronal 3T3-L1 cells." Molecules 20, no. 12 (2015): 22128–22136.

Zeng, Xian-Si, Wen-Shuo Geng, Jin-Jing Jia, Lei Chen, and Peng-Peng Zhang. "Cellular and molecular basis of neurodegeneration in Parkinson disease." Frontiers in aging neuroscience 10 (2018): 109.

Zimmerman, S. L., and Roy Barnett. "Sickle cell anemia simulating coronary occlusion." Annals of internal medicine 21, no. 6 (1944): 1045–1049.

7 Applications of Phyto-Nanotechnology in Neurological Disorders

Manoj Kumar Srinivasan, Nivedha Jayaseelan, Kanimozhi Kaliyamoorthi, Kamalesh Balakumar Venkatesan, Saravanan Alamelu, and Sathish-Kumar Kamaraj

7.1 INTRODUCTION

Herbal medicines and other natural therapies have been used to treat illness for centuries. Chemicals such as phenolic acids, flavonoids, polyphenols, and alkaloids are found in herbs. Citric acid, vitamin C, and other phytonutrients found in plants function synergistically to treat a given disease or illness (Akhtar et al, 2013). Herbal therapeutic compounds need to have their active principle modified in a scientific way so that they may be used for continuous and targeted release. With this method, patients are more likely to take their medication as directed, and more frequent dosing is not required. Nanotechnology has been shown in numerous studies to solve the bioavailability and toxicity problems that plague conventional dosing methods (Mitchell et al, 2021). From $5.2 billion in 2021, the global nanotechnology sector is expected to grow to $23.6 billion by 2026, a compound annual growth rate of 35.5%. The report comes from Business Wire (2021) and originates in Dublin.

Research into phytonanoformulations has the potential to increase the availability of novel products with fewer side effects than conventional and synthetic herbal formulations. Many different methods of administering herbal medicines were tested to see if they could effectively preserve the physicochemical and biological properties of natural substances, despite their wide variety of molecular structures. In order to further expand a rigorous assessment of plant-derived pharmaceuticals in terms of safety and quality, the World Health Organization (WHO) has supplied the necessary technical assistance.

About 65% of Indians are curious about trying out traditional medicines. Herbal remedies are gaining popularity even in the sophisticated world. Traditional medicines are always in high demand in developed countries. Just behind China is India as a major supplier of herbs. The Indian Ayurveda, Yoga and Naturopathy, Unani, Siddha and Homeopathy (AYUSH) system has a $5 billion domestic market and 8,000 herbal remedies. More than $500 million in annual revenue is made from the practice of Ayurveda, Siddha, and Unani. By 2023, the market for pharmaceuticals derived from plants is expected to generate revenues of $111.1 billion worldwide. In India, as in many

DOI: 10.1201/9781003389781-7

Western countries, the phytopharmaceutical industry is only getting off the ground. Opportunities for phytonanoformulations are promising in light of the potential for future growth in demand for phytopharmaceuticals as a result of advancements in nanotechnology and the benefits of nanoformulations over conventional commodities.

Pharmaceutical companies will be able to advance their study of phytonano-pharmaceuticals with the use of innovative drug delivery systems (Business World, 2021). Nanoparticles in the 1–100 nm range are synthesized for use in a wide variety of nanotherapeutics and diagnostic agents through the integration of scientific and technological skills. These are some of the ways in which nanoparticles excel over more conventional methods: (Pandit et al, 2022). Enhanced solubility increases bio-availability, which in turn increases targeted drug distribution by prolonging drug retention in the body. Maintaining a happy medium between a medicine's thera-peutic efficacy and its side effects requires a service provider with extraordinary capability and dependability.

Neurodegenerative disorders (NDs) are characterized by various abnormalities in the structure and function of neurons and other nervous system cells. These chronic, deteriorating injuries can lead to mental and physical decline over time. Other forms of dementia and Alzheimer's disease (AD), multiple sclerosis (MS), Parkinson's dis-ease (PD) and PD-related disorders, Huntington's disease (HD), and amyotrophic lateral sclerosis (ALS) are some of the most frequent neurodegenerative diseases (NDs). Predominant risk factors for NDs are thought to include genetic predisposition, advancing age, sedentary lifestyle, poor diet, chemicals, some viruses, and exposure to certain environmental contaminants (Hodjat et al, 2017; Przedborski et al, 2003). People are living longer than ever before, yet with that improvement in longevity comes an increase in degenerative brain diseases. According to a study by the WHO, the prevalence of death caused by dementia and other age-related NDs more than doubled between 2000 and 2016. The WHO (2006) predicts that in the near future, the healthcare system will face enormous emotional, social, and financial strain due to the prevalence of mental and emotional health difficulties. Existing treatments for NDs are not without serious side effects; new, safer approaches are needed (Durães et al, 2018). Natural substances cannot reach the brain and nervous system because of the blood-brain barrier's (BBB's) poor diffusion properties (Dwivedi et al, 2019). As a result, nanotechnology, and more especially phyto-nanomedicine or pharma-ceutical nanotechnology, enables superior drug delivery systems for the treatment of NDs (Figure 7.1) by enhancing molecular monitoring, control, construction, and diagnostics (Maravajhala et al, 2012; Ochekpe et al, 2009). One method that shows promise for getting over these restrictions and increasing bioavailability is the use of nanoformulations of natural substrates (Ratheesh et al, 2017). The current methods of treating neurodegenerative diseases with pharmaceuticals are compared in this chapter with the potential future applications of phytomedicine and nanotechnology.

7.2 NEURODEGENERATIVE DISEASE

The slow and steady loss of neuronal cells is a hallmark of neurodegenerative dis-eases. When things deteriorate here, neurons always die. This causes problems with walking (ataxia) or cognitive function (dementia), and balance due to a disruption in

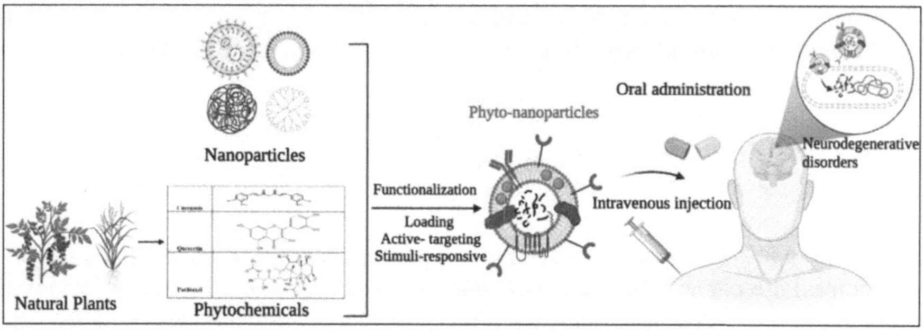

FIGURE 7.1 Nanoparticles mediated drug delivery of phytochemicals for neurodegenerative disorders.

neuromuscular control (Emard et al, 1995; Higuma et al, 2017). In recent years, gait analysis has received a lot of attention from medical researchers. Research on gait issues has mostly concentrated on neurodegenerative diseases like ALS, PD, MS, AD, HD, and others that cause dementia. Figure 7.2 shows the different types of neurodegenerative diseases and their neurons.

The condition is the most common kind of dementia. Common early stage symptoms of AD include difficulty remembering recent events and general disorientation. Cognitive abilities like language and visual-spatial awareness also change over time. The aforementioned degradation is caused by deficits in cognition and gait

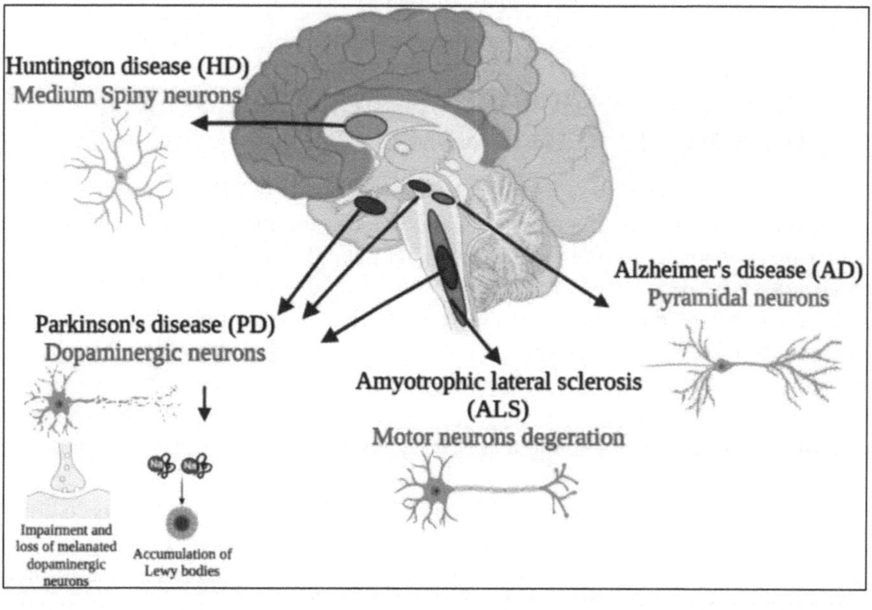

FIGURE 7.2 Neurodegenerative diseases and the types of neurons affected (Bhattacharya et al, 2022).

(Valkanova et al, 2018; Valkanova and Ebmeier, 2017). Stuttering speech, apraxia, and hyperkinesia are all symptoms of AD, along with unsteady walking and torso motion. Gait abnormalities are common in the early stages of AD and include a higher stride-to-stride variability, decreased cadence (longer stride time/gait cycle), shorter stride length, and slower gait. After AD, it is the second most prevalent form of neurodegeneration. Bradykinesia is a sign of PD, the most prevalent hypokinetic movement condition. In the latter stages of the condition, tremors, body rigidity (hypertonia), loss of postural reflexes, flexed posture, and freezing are characteristic symptoms. Patients with PD often demonstrate several gait abnormalities, including stride shortening, slowness, increased axial rigidity, diminished rhythmicity, and increased festination (Vallabhajosula et al, 2013).

MS is a disease of the central nervous system (CNS) that causes slow degeneration in young people. Symptoms include lethargy, weak muscles, trouble moving the arms or legs, clumsiness, and an overall unbalanced feeling. Gait is commonly significantly affected by motor weakness, spasticity, ataxia, and sensory impairment in the early stages of neurological diseases (Kelleher et al, 2010). The degeneration of motor neurons (motoneurons) in the CNS is a hallmark of ALS. Loss of walking ability is one of the first symptoms of advanced ALS. Instability in the timing between steps and changes in the fluctuation dynamics are two common symptoms of ALS (Hausdorff et al, 2000).

The dysfunctional neural process that causes HD's dementia, tremors, and depression. Patients' clumsiness and unsteadiness deteriorate as the disease progresses. HD patients walked more slowly, took shorter steps and strides, stayed still for a longer stance phase, and swung their arms less often (Pyo et al, 2017). In Table 7.1, a summary of relationships among gait characteristics and the above-listed diseases is given.

7.3 TREATMENT OF NEURODEGENERATIVE DISEASES: THE ROLE OF PHYTO-NANOMEDICINE

The rapid global expansion of neurodegenerative disorders (ND) during the past few years can be attributed to the rising global population of adults aged 60 and up. The main cause for concern is not the increasing number of cases, but the absence of viable therapies. Evidence suggests that NDs were first identified in the early 20th century, although this information has only recently been made public. Sadly, there are currently no effective therapies for these illnesses. At times, diseases can be halted or their progression slowed by using integrative medicine. Researchers believe that as the state of modern science and integrative medicine improves, the prevalence of scientific approaches to ND treatment will increase (Cavalu et al, 2020; Velmurugan et al, 2018).

Multiple synthetic pharmaceuticals have demonstrated positive results in clinical trials for the treatment of chronic diseases like Alzheimer's, Parkinson's, and autism. Many synthetic medications are unfit for therapeutic use because of their numerous negative side effects. Scientists are gradually shifting away from the use of these synthetic medications due to their negative side effects and toward the use of phytochemicals, which have fewer side effects. Phytochemicals have the

TABLE 7.1
Gait Characteristics in the Most Common Neurodegenerative Diseases (Cicirelli et al, 2021).

Neurodegenerative Diseases	Symptoms	Gait Characteristics
Alzheimer's disease (AD)	Apraxia, hyperkinesia, abnormalities in walking and trunk movements.	Increased support time Lower cadence Decreased walking speed Greater stride-to-stride variability Decreased stride length
Parkinson's disease (PD)	Hypertonia, hypokinetic movement, bradykinesia, flexed posture, tremor, loss of postural reflexes, freezing, and festination.	Decreased walking speed Reduced swing time Reduced stride length Higher double support time Increased cadence
Multiple sclerosis (MS)	Spasticity, motor weakness, sensory disturbance, and ataxia.	Increased double support time Decreased walking speed Shorter step length Reduced cadence
Amyotrophic lateral sclerosis (ALS)	Altered gait rhythm, perturbations in the fluctuation dynamics, weakness in feet, legs, or ankles.	Increased stride time Decreased walking speed Increased stride time variability
Huntington's disease (HD)	Emotional problems, uncontrolled movements, loss of thinking, and psychiatric disorders.	Increased stance/swing phase Decreased single support time Decreased step/stride length Decreased walking speed

potential to be employed as medications because of their antioxidant, anti-amyloid, anti-inflammatory, and anticholinesterase capabilities (Hajialyani et al, 2019; Xu et al, 2018).

The existence of BBB is the most significant factor limiting ND therapeutic options. As a result of a lot of hard work, numerous nano methods have been used to resolve this issue (Bhaskar et al, 2010). Many drug delivery carriers have taken an interest in polymeric nanoparticles (PNPs) due to their large drug-carrying capacity, long half-life in circulation, and resistance to drug degradation. This provides ligands with a wide variety of surface-handling strategies for BBB crossing (Roney et al, 2005). There is speculation that NP-based drug delivery systems facilitate drug entry across the BBB and enhance medication uptake in the brain. Nanoparticles, when used in medicine, are touted for their biodegradability and reduced toxicity to the body's peripheral organs (Caruso et al, 2011). It is possible for both non-toxic and toxic nanomaterials to cross the BBB and reach the cells below. The BBB can be traversed by nanomaterials via paracellular pathways. In contrast, the essential structure of the BBB is preserved and the BBB is unaffected by drug administration when non-invasive approaches are used (Xie et al, 2019). Nanocarriers allow medications to be delivered to the brain without the need for injection (Poovaiah et al, 2018). The ideal nanocarriers may be manufactured without altering the drug's characteristics, as is currently believed.

Nanomaterials can alter autophagy, inflammatory events (i.e., overexpression and/or suppression of pro-inflammatory cytokines and chemokines), the BBB, and neuronal tissue regeneration, and they can even prevent neural apoptosis (Kim et al, 2017; Re et al, 2012; Zheng et al, 2016). Because of their strong affinity for A, nanostructures mitigate the disease's negative effects on Alzheimer's patients. However, nano-based techniques enhance dopamine transport and release in the brains of persons with PD (Re et al, 2012).

Important advantages, such as enhanced biodegradability and biocompatibility, enhanced drug pharmacokinetics and therapeutic effectiveness, and decreased drug-related adverse effects, may result from treating NDs with NPs (Ratheesh et al, 2017). Evidence suggests that the BBB can be crossed by the herbal compound ginsenoside-NPs, which then protects nerve cells (Aalinkeel et al, 2018). The cytotoxic effects of A42 on human neuroblastoma SH-SYYY cells were attenuated in vitro by poly lactic-co-glycolic acid (PLGA)-functionalized quaternary ammonium compounds (PLGA@QC)-NPs, which were also found to be cytotoxic-free. AD rats treated with PLGA@QC)-NPs showed enhanced performance on the Morris Water Maze and the New Object Recognition tests (Sun et al, 2019). Choline-loaded NPs, lectin-laden NPs conjugated with Solanum tuberosum lectin, and bacoside (a loaded PLGA-NP) have all been shown to be effective therapy for AD (Li et al, 2011; Zhang et al, 2019).

Herbal extracts fortified with NPs provided many advantages to NDs, including improved biocompatibility and biodegradability of pharmaceuticals, enhanced therapeutic efficacy, elimination of pharmacokinetic constraints, decreased side effects, controlled release, and precise localization of medication effects. Therapy efficacy could be increased with the use of a variety of nanoparticle materials that reduce reactive oxygen species (ROS), have powerful antioxidant properties, and prevent A from sticking together. Despite the fact that nanocarriers can cross the BBB more easily and are therefore useful weapons for delivering targeted chemicals to the brain, a number of challenges remain. When blood channels are manipulated, clots develop, or blood cells are destroyed, platelets can clump together and cause problems (Niu et al, 2019; Ramanathan et al, 2018). It's potentially harmful if NPs aren't distributed uniformly throughout the brain. The work of NPs like cerium oxide, iron, silica, and gold in the body is obscured by their inorganic components.

Neurotoxicity can result from a buildup of these substances in the brain due to their ability to alter mitochondrial activity, neuronal inflammation, autophagy, and apoptosis (Niu et al, 2019). Current treatments for ND appear to be inadequate, thus scientists are looking toward plant-based therapeutics based on nanotechnology. Nano theragnostic is one technique for the treatment of ND that is gaining popularity among scientists worldwide. It's been found that nanoparticles can be useful in both medical diagnosis and treatment. Because it is so direct, Tripathy et al (2018) say, this treatment has a lot of weight in the medical field. It is also possible to modify the method for individual patients and account for variations in disease severity, which expands the method's potential application (Bar-Zeev et al, 2017). Using light that can be adjusted to activate nanoparticles, chemical engineers have developed a new nanotheranostic device, greatly expanding the scope of the science (Bhattacharya

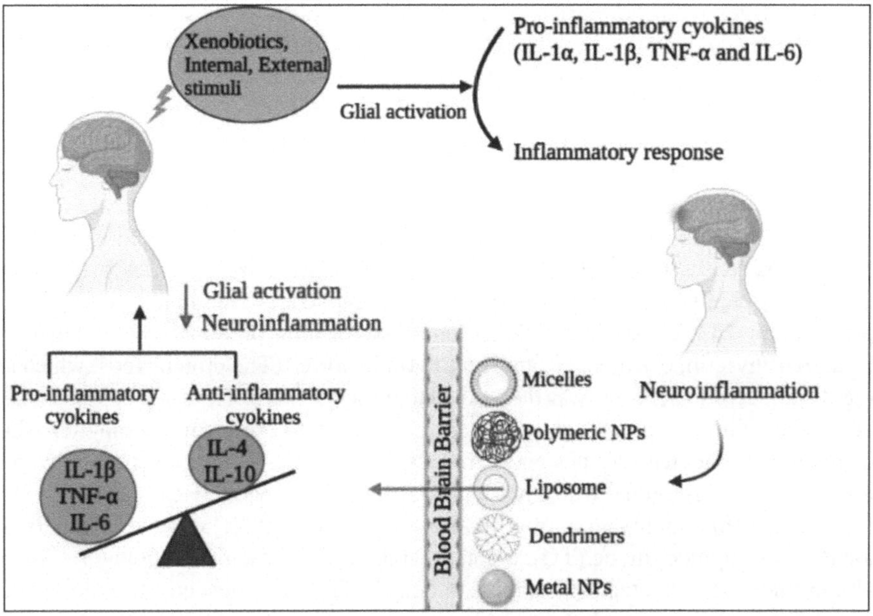

FIGURE 7.3 Schematic representation of nanotechnology-based therapy for CNS disorders.

et al, 2022). In Figure 7.3, a schematic illustration of nanotechnology-based therapy for CNS disorders.

7.4 RECENT TRENDS OF PHYTO-NEURO MEDICINE

Green chemistry and nanotechnology have recently made impressive strides, and these developments have the potential to greatly improve the theragnostic capacities of the biological sciences (Khalil et al, 2017; Ovais et al, 2018). While this offers great therapeutic promise, it has not yet been applied to the development of therapies for NDs like PD and AD. The use of chemically synthesized substances, while convenient, has a number of drawbacks, including their toxicity and high cost (Gaude et al, 2017; Khalil et al, 2017; Ovais et al, 2017). Some of the molecules used in the chemical synthesis of NPs have been shown to stick to the NP surface, which makes them inappropriate for use in biological systems, as reported in a single study. As a result, there has been a shift in focus to the manufacture of materials via environmentally friendly chemistry and processes (Emmanuel et al, 2017). Medicinal plants or pure phytochemicals with therapeutic qualities have traditionally been used in the green chemistry-based technique because they offer chelation and stability to nanoparticles (Ovais et al, 2017, 2018).

Biogenic gold nanoparticles derived from *Terminalia arjuna* have recently been shown to shield neurons, as demonstrated by research from the labs of Suganthy et al (2018). The study's findings demonstrated that biogenic gold nanoparticles are not only beneficial to the neurological system but also harmless to live organisms.

Low quantities of these particles inhibited acetylcholine esterase activity, slowed a fibrillation, and reduced the stability of mature fibrils. Trehalose improved the ability of the gold NPs to prevent protein aggregation and disrupt mature fibrils. One possible application of this is in photothermal therapy (Tripathy et al, 2018). For gold nanoparticles to be more effective at protecting neurons, it is desirable to combine them with anti-amyloidogenic drugs. A second strategy employs *Bacopa monnieri*-derived biogenic platinum nanoparticles as a neuroprotective agent. The use of phytochemicals to improve the performance of NPs has yielded significant benefits. Recently, polyphenols were added to selenium nanoparticles by coating them with epigallocatechin gallate (EGCG), a polyphenol found in tea. Neurons are protected by EGCG, and the action of amyloid-forming proteins such as amyloid beta, transthyretin, α-synuclein, and huntingtin is slowed. The protein Tet-1, which is highly attracted to neurons, was then added to these particles. The same approach has been taken with curcumin and its derivatives, leading to promising therapy for AD. Curcumin or other natural substances that assist in curing AD can be delivered by the use of benzothiazolinone, which has a high affinity for amyloid (Reddy et al, 2018).

To target this quality in animal models, Yusuf et al (2021) created PLGA NPs loaded with thymoquinone (TQ), a potent antioxidant and anti-inflammatory. Male albino mice were treated with streptozotocin (SZT), and subsequently, they were given TQ-loaded PLGA NPs coated with polysorbate 80, which simulates AD oxidative stress by inhibiting superoxide dismutase (SOD) activity (P-80-TQN). These nanoparticles, with an average particle size of 226 nm and a zeta potential of 45.6 mV, were prepared through the single-emulsion solvent evaporation process. Initial TQ release occurred in 2 hours, followed by a prolonged, gradual release (stabilized dipole-dipole interactions taking place between TQ and PLGA components). P-80-TQ NPs were able to cross the BBB because they were aided in entering cells via LDL receptors (mediated by the polysorbate coating). These systems significantly altered SOD activity at the site between seven and twenty-eight days after their arrival (increase). Simultaneously, people were given the "Despair test" to assess how being in the company of animals influenced their mental health.

The spirocyclic alkaloid rhynchophylline (RIN), which is found in the Uncaria genus, has multiple pharmacological effects, including the preservation of nerve cells. RIN inhibits solubility and excessive stimulation of hippocampal neurons, which occurs in AD. The first study on how to make and test an injectable RIN treatment for AD that targets the brain was presented by Xu et al (2020). To improve the therapeutic efficacy and selectivity of RIN, methoxy polyethylene glycol NPs coated with Tween 80 were prepared via nanoprecipitation. T80-coated RIN-loaded PLGA NPs are safe to use because they do not cause hemolysis. Using bEnd.3 cells to establish an in vitro BBB model, researchers looked into how these nanosystems crossed the BBB and found that they transported more efficiently than either free RIN or RIN-coated NPs. Treatment of the brain with T80 has been shown to be beneficial in healthy C57BL/6 mice. T80 RIN NPs increased the number of viable PC12 cells that had been injured by A25-35. These studies showed that enclosing RIN in PLGA nanoparticles did not compromise its neuroprotective properties.

In comparison to currently available treatments for neurodegenerative disorders, phyto-nanomedicines have a brighter future. Nanomedicine derived from phytochemicals has the potential to cure neurodegenerative illnesses in the future due to its many beneficial qualities. More study is required to understand the entire scope of neuroprotective benefits of these substances, how they function, and whether or not combination therapy could be more effective as a neuroprotectant (Stone et al, 2020).

Phyto-nanomedicines provide unquestionable optimism for the future development of anti-ND therapies due to their lower risk of adverse effects and enhanced effectiveness against their targets. There are, however, some restrictions to bear in mind. Kumar et al (2020) discovered there was a limitation to the effectiveness of nano theragnostic in treating NDs. They discovered that because each person's brain, appearance, and genome are unique, there is no single approach that can be applied to treat a given problem across the board. Also, researcher demonstrated that the NPs were being absorbed too quickly, proving that this technique was ineffective. Incorrect injection technique may result in NPs being distributed inadvertently to different tissues or organs, rather than the intended target. It is also well-known that there is no reliable means to monitor the progress of the treatment. This makes it challenging for the healthcare providers to assess the efficacy of their treatment. The treatment and approach are not only ineffective but also prohibitively costly. Nanoformulations, as depicted in Figure 7.4, are utilized to boost the efficacy of natural compounds. Table 7.2 summarizes the phyto-nanoformulations used in neurological disorders.

7.4.1 Protein-Based Nanoparticles

Biopolymer-based nanoparticles, especially protein nanoparticles, are extensively used as medical and functional instruments due to their non-toxicity and ease of biodegradation (Jacob et al, 2018). Proteins have numerous applications in both biology and industry, making them ideal building blocks for a diverse range of nanoparticles. Cells can take in protein nanoparticles because of their small size. Biodegradability, stability, and the ability to modify the surface of protein nanoparticles are just a few of the benefits they offer as a drug delivery system. Their particle size is simple to control, and they pose no toxicity issues, such as immunogenicity. The stability, efficacy, and half-life of the medicine could be enhanced by preventing its breakdown by enzymes and elimination by the kidneys. Due to their lack of immunogenicity, many treatments for cancer can benefit from the usage of protein nanoparticles, tumor treatments, and vaccines. Protein nanoparticles can be slowly liberated from their biodegradable polymer capsules. When designing nanoparticles, controlling their size and surface area is crucial for achieving desirable pharmacological activity through the controlled release of active chemicals to target sites.

The protein content of silk is extremely high; in fact, fibroin can make up as much as 85% of the protein content (Numata and Kaplan, 2010). Na_2CO_3 is used for "degumming," or removing the sericin layer, from silk. Indeed, this is one of the most common ways to obtain fibroin (Chopra et al, 2021). Fibrin nanoparticles are widely used because they are stable, biodegradable, immune-reaction free, non-toxic to living tissue, and versatile. The zeta potential of fibroin nanoparticles is negative. When a positively charged polymer such as poly(ethyleneimine) (PEI) or chitosan is added to the surface, crosslinking results. Many variables influence fibrin

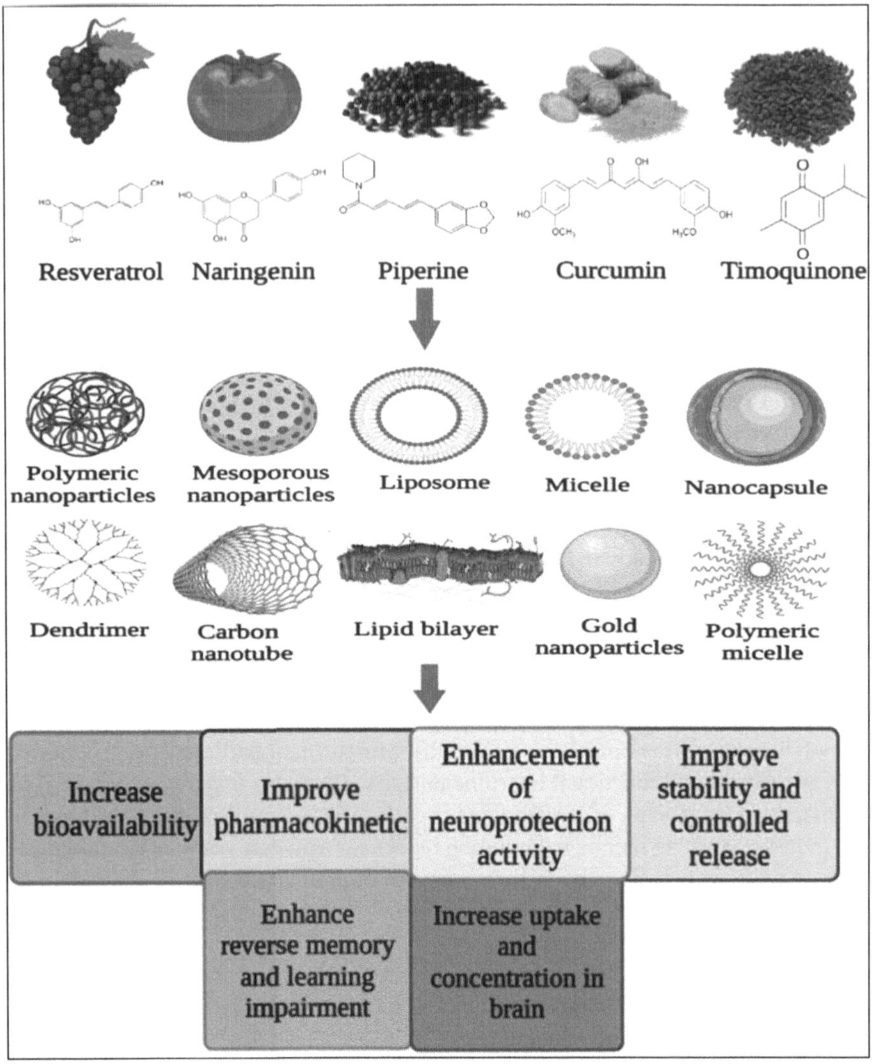

FIGURE 7.4 Nanoformulations used to improve the effectiveness of natural compounds.

nanoparticles's release profile, formation stability, particle size distribution, and drug encapsulation. Factors include the fibroin's molecular weight (MW), its crystallinity, the nature of the drug to be encapsulated, and the conditions of the manufacturing facility. Scientists have worked hard to devise methods of dispersing and utilizing fibroin nanoparticles in the administration of medicines because of the problems they can solve for low-molecular-weight drugs. Improved medication solubility, stability, and reduced drug breakdown and toxicity have all been shown in small-molecule pharmaceuticals encapsulated in fibroin nanoparticles. This improves their receptivity to medicinal treatment.

TABLE 7.2
Phyto-nanoformulations in Neurodegenerative Disorders (Moradi et al, 2020).

Type of Nanoformulation	Benefits
Curcumin	
Lipid-polyethylene glycol polylactide nanoparticles	Decrease the amyloid-beta aggregation
Cur-loaded lipid core nanocapsules	Alzheimer's disease
Entrapped PEG-PLA	Significant increase in memory cue
Drug-loaded lipid-based nanoformulation	Parkinson's disease
Selenium-conjugated PLGA nanospheres	Alzheimer's disease
Solid lipid NPs	Increase 3-nitro1-propionic acid-enhanced Huntington's disease in rats
PLGA-associated Cur NPs coupled with Tet-1 peptide	Alzheimer's disease
Lactoferrin NPs	Protect SKN-SH neuroblastoma cell line of dopaminergic cells
Plain liposomes and anti-transferrin antibody-tagged liposomes	Increase the brain permeation of drug in Alzheimer's disease patients
Quercetin	
PLGA nanocapsule	Higher brain intake and enhanced bioavailability
Nanocrystals	Parkinson's disease
Nano lipidic carriers	Elevated drug concentration in plasma and target-specific delivery
Solid lipid NPs	Retention of memory in animal models
Ferulic Acid	
Solid lipid NPs	Induced phosphoinositide 3-kinases (PI3Ks) pathway in ischemic neural injuries mode. Repressed Aβ-promoted cell death, reduced ROS (reactive oxygen species) production, and inhibited the apoptosis pathway
Resveratrol	
Chitosan-over-layered PLGA NPs	Reduced concentration of inflammatory cytokines, enhance the neuroprotective IL-10 concentration
SLNPs conjugated with apolipoprotein E	Bioavailability and concentration get enhanced thereby permeation of the drug in the brain improved
Nanoemulsion loaded with vitamin E	Positive effects in Parkinson's disease
Lipid-core NPs	Aβ-provoked neuroinflammation was regulated
PS80-layered poly(lactide) NPs	Decrease neuronal damage properties
Gallic Acid	
GA-entrapped chitosan NPs (GANP)	Scopolamine-intensified amnesia in vivo
Piperine	
Intranasal chitosan NPs	Showed more efficacy in the AD model
Tween-modified monoolein cubosomes	Higher potency over conventional drugs and ability to re-establish the perception function

The 585 amino acids found in HSA's (human somatostatin) structure are responsible for its 66 kDa MW (Langer et al, 2003). Blood is where HSA concentrations are highest. All three of the primary HSA subunits (A, B, and C) may be detected in the whole protein. The HSA's major binding sites are located in subunits IIA and IIIA and are referred to as Sudlow's sites I and II. (Jahanban-Esfahlan et al, 2016). HSA's ability to carry other substances increases their solubility in blood means that even hydrophobic molecules may be transported with relative ease. The HSA can direct the delivery of specific substances to specific organs and tissues. For up to 10 hours, the temperature can be kept at 60°C, and the pH can change from 4 to 9. Because they are organic solvents, they are also quite stable (Kratz, 2008). The fact that they are entirely natural means that they pose no threat to human health and do not weaken the immune system in any way. It's easy to use and dissolves quickly. These features account for the widespread use of HSAs and bovine in studies of drug delivery and protein interactions. Due to HSA's strong affinity for these compounds, a matrix of HSA nanoparticles may be able to effectively incorporate a wide variety of medications.

7.4.2 Polymeric Nanoparticles

PNPs are biodegradable and biocompatible polymer carriers that encase a medication within a colloidal particle between 1 and 1000 nm in size (Chopra et al, 2021; Langer et al, 2003). Polyacrylate, polylactide, and polylactide polyglycolide copolymers are three of the most widely used polymers. There has been a lot of research done on lactide glycolide copolymer. Chitosan, alginate, and albumin are just a few examples of the many different types of natural polymers that can be used. Researchers have been studying curcumin (also known as "Indian solid gold") for years to establish whether or not it is useful against cancer and other ailments. Animal studies have shown that curcumin can reduce the accumulation of amyloid beta and tau phosphorylation.

In addition, it promotes the growth of neural stem cells and new neurons in the hippocampus (Li et al, 2019). Curcumin's nanoparticle form is thought to improve its targeting of the neurons that are damaged in AD due to the compound's poor water solubility (AD). Curcumin nanoparticles coated in a PLGA that specifically target Tet-1 were tested for their ability to reduce amyloid and increase antioxidant activity in patients with AD (Mathew et al, 2012). Nanoparticles were created with a zeta potential of 230–220 mV and an average size of 150–200 nm by evaporating the solvent. These nanoparticles can produce light and are totally soluble in water. Cell viability assays have shown that these nanoparticles are safe for human cells to be exposed to. In vitro uptake fluorescence measurements showed that GI-1 glioma cells absorbed nanoparticles targeted with Tet-1 peptide at a much higher rate than they did non-targeted nanoparticles.

7.4.3 Liposomes-Based Drug Delivery Systems

Methods for carrying liposomes across the BBB remain in their infancy. Cell internalization is induced by absorption via electrostatic interactions due to negative charges on the BBB and cationic liposomal drug delivery methods (Joshi et al, 2015).

The BBB can be crossed by equipping liposomes with glucose and GSH as nutrition. Because receptor-mediated transcytosis can link specific ligands to a diverse set of BBB receptors, it has therapeutic potential. Nanoliposomes covalently conjugated with curcumin and the human immunodeficiency virus Trans-Activator of Transcription (TAT) peptide enhance peptide binding and BBB bridging (Noble et al, 2014; Sancini et al, 2013). As for how TAT is taken up, endocytosis and micropinocytosis were proposed to be the most common ways. Thiolmaleimide is used to covalently attach TAT to nanoliposomes. According to high-performance liquid chromatography–mass spectrometry (MS)/MS, the range of TAT-CurcNL sizes is between 196.5 and 3.2 nm. Absorption of nanoliposomes by human brain capillary endothelial cells (hCMEC and D3) was shown to rise by a factor of three after TAT functionalization was demonstrated by MS, confocal microscopy, and a radioactivity assay utilizing [3H]-sphingomyelin (Erdoğar et al, 2018).

Furthermore, Mourtas employed the thin-film hydration method to create multifunctional nanoliposomes loaded with curcumin-lipid derivatives. The commercially available functionalized lipid DSPE-PEG2000SH may be traced back to the first reaction of DSPE-PEG2000 and 4-methoxytrityl-thiol, which resulted in DSPEPEG2000-S-Mmt. The DSPE-PEG2000-S-Mmt synthesis was carried out with the aid of DIPEA (diisopropylethylamine), which facilitated a rapid reaction with no byproducts (Mourtas et al, 2014). For its production, a polyethylene glycol (PEG) spacer was inserted between curcumin and lipids. DSPEPEG2000-SH was reacted with curcumin to form the DPSPEPEG2000-CURC derivative when its thiol protection was removed. Successful incorporation of nanoliposome membranes into this unique synthetic material was achieved. It has been shown that the addition of DIPEA to the second phenol-protons of the curcumin moiety allows for the accurate identification of A deposits in the postmortem tissues of AD patients. Ginkgo biloba, one of the most popular herbal supplements, has roots in ancient Chinese medicine. The enzymes SOD, catalase, glutathione peroxidase, and glutathione reductase all see an increase in activity when exposed to antioxidants in the CNS. Improved memory and learning abilities have been linked to Ginkgo biloba's ability to boost antioxidant activity in the brain's hippocampus (Naik et al, 2006). The niosome is a surfactant-based non-ionic bilayer vesicle. While liposomes can only carry hydrophilic medications, niosomes can carry both hydrophilic and hydrophobic pharmaceuticals (McCall and Sirianni, 2013; Uchegbu and Vyas, 1998). Drugs used to treat conditions of the CNS may be able to cross the BBB within niosomes. There is a difference in niosome size between freeze-dried powder (661 nm) and spray-dried powder (680 nm). The zeta potential of spray-dried niosomes was greater than that of freeze-dried niosomes. Because of the strong electrostatic repulsion between particles, spray-dried noisome powders with a high zeta potential are more stable (Huang and Zhang, 2018; Jin et al, 2013).

7.4.4 Green Synthesized Nanoparticles

An alternative to nanotechnology that is less harmful to the environment is green chemistry or green technology. Several studies have shown that metal nanoparticles are superior to microbes for creating nanoparticles for use in herbal extracts.

The organic molecules may decompose into biodegradable materials because of the presence of polyphenols. This ayurvedic herb, known as shankhapushpi, has been shown to have hypotensive, immunomodulatory, and anticonvulsant effects. It has been shown that iron oxide nanoparticles benefit from the antioxidant activity of the ayurvedic medication *Convolvulus pluricaulis*. Researchers are currently trying to piece together the molecular process behind these microscopic particles (Poka et al, 2017). In traditional Chinese medicine, the *Pulicaria undulata* plant is used as an insect repellant, an anti-inflammatory, and an anti-epileptic. AgNPs from *Pulicaria undulata* interact better with proteins as their concentration in the brain rises, limiting fibril formation by decreasing protein conformation and self-association (Dehvari and Ghahghaei, 2018).

7.4.5 Carbon Nanotubes (CNTs) and Nanofibers

Several inorganic nano-drug delivery methods have emerged as therapeutic agents for a variety of illnesses, including NDs. These include CNTs, calcium phosphate nanoparticles, superparamagnetic iron oxide nanoparticles, layered double hydroxides, and mesoporous silica nanoparticles. Availability at target sites, retention effect, stability, and drug accumulation can all be improved using inorganic nanocarbon systems, which can also withstand prolonged systemic circulation. Also, these nanostructures can control the rate at which drugs are released, as well as help with drug monitoring and imaging. Furthermore, CNTs are an excellent choice for nanopharmacology because of their adaptability to many stimuli (including chemicals, temperature, pressure, pH, and electric and magnetic fields) (Naz et al, 2019). The utilization of carbon-based nanostructures such as CNTs is one of the most important neurotech methods. CNTs, or carbon nanotubes, are a type of carbon allotrope having a spherical nanostructure. Research on CNTs to improve their electrical stimulation is widespread. PD is just one of several mental and neurological conditions that can be helped by deep brain stimulation. There can be issues with the use of these stimulating electrodes if the immune system has an adverse reaction to their presence. Producing nanofibers poses less of a threat to the environment than doing so with CNTs. Unexpectedly, nanofibers are being used in the creation and production of brain prostheses. It's possible that electrospun nanofibers will outperform other nano techniques in a number of settings (Ganesan et al, 2015; Modi et al, 2009). CNTs can be utilized as scaffolds on their own or in conjunction with other biodegradable biomaterials to aid in neuro-engineering applications such as regeneration, neuroprotection, interface, stimulation, and neuronal differentiation. This is due to their electrical properties, structural characteristics, and appropriate biological effects on cell growth (Xiang et al, 2020).

7.5 CURRENT PROS AND FUTURE SCOPE

The BBB presents a significant challenge for medications designed to treat illnesses of the CNS. Drug delivery systems with a phyto-nano emphasis have seen a lot of progress in recent years. BBB is a major obstacle to the delivery of therapeutic medications to damaged brain tissue since it is part of the body's intrinsic defense mechanism. In order to develop efficient, innovative targeted-drug delivery systems, a thorough understanding of the functions of various brain cell types in the etiology of neuro-disorders is required. Although many nanoformulations have demonstrated encouraging results in preclinical and clinical studies, translating them from the lab to the bedside has proven challenging due to worries about toxicity and other side effects, as well as issues like aggregation and rapid clearance caused by their nano-size. Nanoformulations with a well-defined mechanism of action and pharmacokinetics that are intended for delivery to the brain should undergo extensive toxicological testing, both in combination with and independently of pharmaceuticals.

7.6 CONCLUSIONS

Nanotechnology-based medicine has proven to be important in medical science, resulting in novel therapy options for neurological diseases such as cancer. However, there are some potential difficulties, such as a lack of advanced equipment for accurate and scalable nanomaterial production, the difficulty of evaluating its efficacy and safety, and other material limits that must be addressed first. The following elements should make future nanoparticle manufacturing easier: (1) High efficacy and low cost; (2) toxicity-free, biocompatibility, and no impact on the pathological system, including thrombosis and inflammation; (3) the ability to penetrate multiple biological barriers and adequate targeting; (4) blood stability and resistance to RES clearance; and (5) loaded molecules with potential therapeutic efficacy on disease. To summarize, treating neurological problems and cancer is a difficult task. As a result, there is a need for multifunctional therapeutics, which might be reduced by using nanomaterials.

REFERENCES

Aalinkeel, R., H. L. Kutscher, A. Singh, K. Cwiklinski, N. Khechen, S. A. Schwartz, and S. D Mahajan. 2018. "Neuroprotective effects of a biodegradable poly (lactic-co-glycolic acid)-ginsenoside Rg3 nanoformulation: A potential nanotherapy for Alzheimer's disease?" *Journal of Drug Targeting 26* (2): 182–193.

Akhtar, M. S., J. Panwar, and Y. S Yun. 2013. "Biogenic synthesis of metallic nanoparticles by plant extracts." *ACS Sustainable Chemistry & Engineering 1* (6): 591–602.

Ali, T., M. J. Kim, S. U. Rehman, A. Ahmad, and M. O Kim. 2017. "Anthocyanin-loaded PEG-gold nanoparticles enhanced the neuroprotection of anthocyanins in an aβ 1–42 mouse model of Alzheimer's disease." *Molecular Neurobiology 54*: 6490–6506.

Bar-Zeev, M., Y. D. Livney, and Y. G Assaraf. 2017. "Targeted nanomedicine for cancer therapeutics: Towards precision medicine overcoming drug resistance." *Drug Resistance Updates 31*: 15–30.

Bhaskar, S., F. Tian, T. Stoeger, W. Kreyling, J. M. de la Fuente, V. Grazú, and D Razansky. 2010. "Multifunctional nanocarriers for diagnostics, drug delivery and targeted treatment across blood-brain barrier: Perspectives on tracking and neuroimaging." *Particle and Fibre Toxicology 7* (1): 1–25.

Bhattacharya, T., G. A. B. E. Soares, H. Chopra, M. M. Rahman, Z. Hasan, S. S. Swain, and S Cavalu. 2022. "Applications of phyto-nanotechnology for the treatment of neurodegenerative disorders." *Materials 15* (3): 804.

Business World. 2022. New segment in the opening-phytopharmaceuticals. https://thepharmatimes.in/new-segment-is-opening-phytopharmaceuticals/.

Caruso, G., M. Caffo, C. Alafaci, G. Raudino, D. Cafarella, S. Lucerna, and F Tomasello. 2011. "Could nanoparticle systems have a role in the treatment of cerebral gliomas?" *Nanomedicine: Nanotechnology, Biology and Medicine 7* (6): 744–752.

Cavalu, S., I. V. Antoniac, A. Mohan, F. Bodog, C. Doicin, I. Mates, and A Semenescu. 2020. "Nanoparticles and nanostructured surface fabrication for innovative cranial and maxillofacial surgery." *Materials 13* (23): 5391.

Chopra, H., S. Kumar, and I Singh. 2021. "Biopolymer-based scaffolds for tissue engineering applications." *Current Drug Targets 22* (3): 282–295.

Chopra, H., S. Kumar, and I Singh. 2022. "Strategies and therapies for wound healing: A review." *Current Drug Targets 23* (1): 87–98.

Cicirelli, G., D. Impedovo, V. Dentamaro, R. Marani, G. Pirlo, and T. R. D'Orazio. 2021. "Human gait analysis in neurodegenerative diseases: A review." *IEEE Journal of Biomedical and Health Informatics 26* (1): 229–242.

Dehvari, M., and A Ghahghaei. 2018. "The effect of green synthesis silver nanoparticles (AgNPs) from Pulicaria undulata on the amyloid formation in α-lactalbumin and the chaperon action of α-casein." *International Journal of Biological Macromolecules 108*: 1128–1139.

Business Wire. 2021. The "Global Nanotechnology Market 2021–2026, Dublin-(Business Wire). Accessed on November 29, 2021.

Durães, F., M. Pinto, and E Sousa. 2018. "Old drugs as new treatments for neurodegenerative diseases." *Pharmaceuticals 11* (2): 44.

Dwivedi, N., J. Shah, V. Mishra, M. Tambuwala, and P Kesharwani. 2019. "Nanoneuromedicine for management of neurodegenerative disorder." *Journal of Drug Delivery Science and Technology 49*: 477–490.

Emard, J. F., J. P. Thouez, and D Gauvreau. 1995. "Neurodegenerative diseases and risk factors: A literature review." *Social Science & Medicine 40* (6): 847–858.

Emmanuel, R., M. Saravanan, M. Ovais, S. Padmavathy, Z. K. Shinwari, and P Prakash. 2017. "Antimicrobial efficacy of drug blended biosynthesized colloidal gold nanoparticles from Justicia glauca against oral pathogens: A nanoantibiotic approach." *Microbial Pathogenesis 113*: 295–302.

Erdoğar, N., S. Akkın, and E. Bilensoy. 2018. "Nanocapsules for drug delivery: An updated review of the last decade." *Recent Patents on Drug Delivery & Formulation 12* (4): 252–266.

Ganesan, P., H. M. Ko, I. S. Kim, and D. K Choi. 2015. "Recent trends in the development of nanophytobioactive compounds and delivery systems for their possible role in reducing oxidative stress in Parkinson's disease models." *International Journal of Nanomedicine 10*: 6757.

Gaude, T. T., G. A. B. e Soares, R. N. S. Priolkar, B. Biradar, and S Mamledesai. 2017. "Synthesis of 4-hydroxy-1-(phenyl/methyl)-3-[3-(substituted amino)-2-nitropropanoyl] quinolin-2 (1H)-ones as antimicrobial and antitubercular agents." *Indian Journal of Heterocyclic Chemistry 27* (03): 223–228.

Hajialyani, M., M. Hosein Farzaei, J. Echeverría, S. M. Nabavi, E. Uriarte, and E Sobarzo-Sánchez. 2019. "Hesperidin as a neuroprotective agent: A review of animal and clinical evidence." *Molecules 24* (3): 648.

Hausdorff, J. M., Lertratanakul, A., Cudkowicz, M. E., Peterson, A. L., Kaliton, D., and Goldberger, A. L. 2000. "Dynamic markers of altered gait rhythm in amyotrophic lateral sclerosis." *Journal of Applied Physiology 88* (6) :2045–2053.

Higuma, M., N. Sanjo, H. Mitoma, M. Yoneyama, and T Yokota. 2017. "Whole-day gait monitoring in patients with Alzheimer's disease: A relationship between attention and gait cycle." *Journal of Alzheimer's Disease Reports 1* (1): 1–8.

Hodjat, M., S. Rahmani, F. Khan, K. Niaz, M. Navaei–Nigjeh, S. Mohammadi Nejad, and M Abdollahi. 2017. "Environmental toxicants, incidence of degenerative diseases, and therapies from the epigenetic point of view." *Archives of Toxicology 91*: 2577–2597.

Huang, W., and C Zhang. 2018. "Tuning the size of poly (lactic-co-glycolic acid)(PLGA) nanoparticles fabricated by nanoprecipitation." *Biotechnology Journal 13* (1): 1700203.

Jacob, J., J. T. Haponiuk, S. Thomas, and S Gopi. 2018. "Biopolymer based nanomaterials in drug delivery systems: A review." *Materials Today Chemistry 9*: 43–55.

Jahanban-Esfahlan, A., S. Dastmalchi, and S Davaran. 2016. "A simple improved desolvation method for the rapid preparation of albumin nanoparticles." *International Journal of Biological Macromolecules 91*: 703–709.

Jin, Y., J. Wen, S. Garg, D. Liu, Y. Zhou, L. Teng, and W Zhang. (2013). Development of a novel niosomal system for oral delivery of Ginkgo biloba extract. *International Journal of Nanomedicine* 8, 421–430.

Joshi, S., R. P. Singh-Moon, J. A. Ellis, D. B. Chaudhuri, M. Wang, R. Reif, and R. M Straubinger. 2015. "Cerebral hypoperfusion-assisted intraarterial deposition of liposomes in normal and glioma-bearing rats." *Neurosurgery 76* (1): 92.

Kelleher, K. J., W. Spence, S. Solomonidis, and D Apatsidis. 2010. "The characterisation of gait patterns of people with multiple sclerosis." *Disability and Rehabilitation 32* (15): 1242–1250.

Khalil, A. T., M. Ovais, I. Ullah, M. Ali, Z. K. Shinwari, S. Khamlich, and M Maaza. 2017. "Sageretia thea (Osbeck.) mediated synthesis of zinc oxide nanoparticles and its biological applications." *Nanomedicine 12* (15): 1767–1789.

Kim, M. J., S. U. Rehman, F. U. Amin, and M. O Kim. 2017. "Enhanced neuroprotection of anthocyanin-loaded PEG-gold nanoparticles against Aβ1-42-induced neuroinflammation and neurodegeneration via the NF-KB/JNK/GSK3β signaling pathway." *Nanomedicine: Nanotechnology, Biology and Medicine 13* (8): 2533–2544.

Kratz, F. 2008. "Albumin as a drug carrier: Design of prodrugs, drug conjugates and nanoparticles." *Journal of Controlled Release 132* (3): 171–183.

Kumar, A., R. K. Chaudhary, R. Singh, S. P. Singh, S. Y. Wang, Z. Y. Hoe, and X Dai. 2020. "Nanotheranostic applications for detection and targeting neurodegenerative diseases." *Frontiers in Neuroscience 14*: 305.

Langer, K., S. Balthasar, V. Vogel, N. Dinauer, H. Von Briesen, and D Schubert. 2003. "Optimization of the preparation process for human serum albumin (HSA) nanoparticles." *International Journal of Pharmaceutics 257* (1–2): 169–180.

Lewis, P. A., and J. E Spillane. 2018. *The Molecular and Clinical Pathology of Neurodegenerative Disease*. Academic Press.

Li, J., Y. Han, M. Li, and C Nie. 2019. "Curcumin promotes proliferation of adult neural stem cells and the birth of neurons in Alzheimer's disease mice via notch signaling pathway." *Cellular Reprogramming 21* (3): 152–161.

Li, J., L. Zhou, D. Ye, S. Huang, K. Shao, R. Huang, and C Jiang. 2011. "Choline-derivate-modified nanoparticles for brain-targeting gene delivery." *Advanced Materials 23* (39): 4516–4520.

Maravajhala, V., S. Papishetty, and S Bandlapalli. 2012. "Nanotechnology in development of drug delivery system." *International Journal of Pharmaceutical Sciences and Research 3* (1): 84.

Mathew, A., T. Fukuda, Y. Nagaoka, T. Hasumura, H. Morimoto, Y. Yoshida, and D. S Kumar. 2012. "Curcumin loaded-PLGA nanoparticles conjugated with Tet-1 peptide for potential use in Alzheimer's disease." *PLoS One 7* (3): e32616.

McCall, R. L., and R. W Sirianni. (2013). PLGA nanoparticles formed by single-or double-emulsion with vitamin E-TPGS. *JoVE (Journal of Visualized Experiments) 27* (82), e51015.

Mitchell, Michael J., Margaret M. Billingsley, Rebecca M. Haley, Marissa E. Wechsler, Nicholas A. Peppas, and Robert Langer. 2021. "Engineering precision nanoparticles for drug delivery." *Nature Reviews Drug Discovery 20* (2): 101–124.

Modi, G., V. Pillay, Y. E. Choonara, V. M. Ndesendo, L. C. du Toit, and D Naidoo. 2009. "Nanotechnological applications for the treatment of neurodegenerative disorders." *Progress in Neurobiology 88* (4): 272–285.

Moradi, S. Z., S. Momtaz, Z. Bayrami, M. H. Farzaei, and M Abdollahi. 2020. "Nanoformulations of herbal extracts in treatment of neurodegenerative disorders." *Frontiers in Bioengineering and Biotechnology 8*: 238.

More, M. P., S. R. Pardeshi, C. V. Pardeshi, G. A. Sonawane, M. N. Shinde, P. K. Deshmukh, and A. D Kulkarni. 2021. "Recent advances in phytochemical-based nano-formulation for drug-resistant cancer." *Medicine in Drug Discovery 10*: 100082.

Mourtas, S., A. N. Lazar, E. Markoutsa, C. Duyckaerts, and S. G Antimisiaris. 2014. "Multifunctional nanoliposomes with curcumin–lipid derivative and brain targeting functionality with potential applications for Alzheimer disease." *European Journal of Medicinal Chemistry 80*: 175–183.

Naik, S. R., V. W. Pilgaonkar, and V. S Panda. 2006. "Evaluation of antioxidant activity of Ginkgo biloba phytosomes in rat brain." *Phytotherapy Research: An International Journal Devoted to Pharmacological and Toxicological Evaluation of Natural Product Derivatives 20* (11): 1013–1016.

Naz, S., M. Shamoon, R. Wang, L. Zhang, J. Zhou, and J Chen. 2019. "Advances in therapeutic implications of inorganic drug delivery nano-platforms for cancer." *International Journal of Molecular Sciences 20* (4): 965.

Niu, X., J. Chen, and J Gao. 2019. "Nanocarriers as a powerful vehicle to overcome blood-brain barrier in treating neurodegenerative diseases: Focus on recent advances." *Asian Journal of Pharmaceutical Sciences 14* (5): 480–496.

Noble, G. T., J. F. Stefanick, J. D. Ashley, T. Kiziltepe, and B Bilgicer. 2014. "Ligand-targeted liposome design: Challenges and fundamental considerations." *Trends in Biotechnology 32* (1): 32–45.

Numata, K., and D. L Kaplan. 2010. "Silk-based delivery systems of bioactive molecules." *Advanced Drug Delivery Reviews 62* (15): 1497–1508.

Ochekpe, N. A., P. O. Olorunfemi, and N. C Ngwuluka. (2009). Nanotechnology and drug delivery part 1: Background and applications. *Tropical Journal of Pharmaceutical Research, 8* (3): 265–274.

Ovais, M., A. Nadhman, A. T. Khalil, A. Raza, F. Khuda, M. F. Sohail, and Z. K Shinwari. 2017. "Biosynthesized colloidal silver and gold nanoparticles as emerging leishmanicidal agents: An insight." *Nanomedicine 12* (24): 2807–2819.

Ovais, M., N. Zia, I. Ahmad, A. T. Khalil, A. Raza, M. Ayaz, and Z. K Shinwari. 2018. "Phyto-therapeutic and nanomedicinal approaches to cure Alzheimer's disease: Present status and future opportunities." *Frontiers in Aging Neuroscience 10*: 284.

Pandit, C., A. Roy, S. Ghotekar, A. Khusro, M. N. Islam, T. B. Emran, and D. A Bradley. 2022. "Biological agents for synthesis of nanoparticles and their applications." *Journal of King Saud University-Science 34* (3): 101869.

Poka, L. P., G. K. Mohan, K. V. Rao, and K Shanker. 2017. "Neuroprotective effect of green synthesized iron oxide nanoparticles using aqueous extract of Convolvulus pluricaulis plant in the management of Alzheimer's disease." *Phytopathology 9*: 703–709.

Poovaiah, N., Z. Davoudi, H. Peng, B. Schlichtmann, S. Mallapragada, B. Narasimhan, and Q Wang. 2018. "Treatment of neurodegenerative disorders through the blood–brain barrier using nanocarriers." *Nanoscale 10* (36): 16962–16983.

Przedborski, S., M. Vila, and V Jackson-Lewis. 2003. "Series introduction: Neurodegeneration: What is it and where are we?" *The Journal of Clinical Investigation 111* (1): 3–10.

Pyo, S. J., H. Kim, I. S. Kim, Y. M. Park, M. J. Kim, H. M. Lee, and S. B Koh. 2017. "Quantitative gait analysis in patients with Huntington's disease." *Journal of Movement Disorders 10* (3): 140.

Ramanathan, S., G. Archunan, M. Sivakumar, S. T. Selvan, A. L. Fred, S. Kumar, and P Padmanabhan. 2018. "Theranostic applications of nanoparticles in neurodegenerative disorders." *International Journal of Nanomedicine 13*: 5561.

Ratheesh, G., L. Tian, J. R. Venugopal, H. Ezhilarasu, A. Sadiq, T. P. Fan, and S Ramakrishna. 2017. "Role of medicinal plants in neurodegenerative diseases." *Biomanufacturing Reviews 2*: 1–16.

Reddy, P. H., M. Manczak, X. Yin, M. C. Grady, A. Mitchell, S. Tonk, and A. P Reddy. 2018. "Protective effects of Indian spice curcumin against amyloid-β in Alzheimer's disease." *Journal of Alzheimer's Disease 61* (3): 843–866.

Re, F., M. Gregori, and M Masserini. 2012. "Nanotechnology for neurodegenerative disorders." *Maturitas 73* (1): 45–51.

Roney, C., P. Kulkarni, V. Arora, P. Antich, F. Bonte, A. Wu, and T. M Aminabhavi. 2005. "Targeted nanoparticles for drug delivery through the blood–brain barrier for Alzheimer's disease." *Journal of Controlled Release 108* (2–3): 193–214.

Sancini, G., Gregori, M., Salvati, E., Cambianica, I., Re, F., et al. 2013. "Functionalization with TAT-peptide enhances blood-brain barrier crossing in vitro of nanoliposomes carrying a curcumin-derivative to bind amyloid-b peptide." *Journal of Nanomedicine and Nanotechnology 4*: 171. doi: 10.4172/2157-7439.1000171.

Stone, N. L., A. J. Murphy, T. J. England, and S. E O'Sullivan. 2020. "A systematic review of minor phytocannabinoids with promising neuroprotective potential." *British Journal of Pharmacology 177* (19): 4330–4352.

Suganthy, N., V. Sri Ramkumar, A. Pugazhendhi, G. Benelli, and G Archunan. 2018. "Biogenic synthesis of gold nanoparticles from Terminalia arjuna bark extract: Assessment of safety aspects and neuroprotective potential via antioxidant, anticholinesterase, and antiamyloidogenic effects." *Environmental Science and Pollution Research 25*: 10418–10433.

Sun, J., C. Wei, Y. Liu, W. Xie, M. Xu, H. Zhou, and J Liu. 2019. "Progressive release of mesoporous nano-selenium delivery system for the multi-channel synergistic treatment of Alzheimer's disease." *Biomaterials 197*: 417–431.

Tripathy, N., R. Ahmad, and G Khang. (2018). Inorganic Nanotheranostics: Strategy Development and Applications. In *Drug Delivery Nanosystems for Biomedical Applications*, edited by M. Saquib Hasnain, A. K. Nayak, and T. M. Aminabhavi (pp. 377–419). Elsevier.

Uchegbu, I. F., and S. P Vyas. 1998. "Non-ionic surfactant based vesicles (niosomes) in drug delivery." *International Journal of Pharmaceutics 172* (1–2): 33–70.

Valkanova, V., and K. P Ebmeier. 2017. "What can gait tell us about dementia? Review of epidemiological and neuropsychological evidence." *Gait & Posture 53*: 215–223.

Valkanova, V., P. Esser, N. Demnitz, C. E. Sexton, E. Zsoldos, A. Mahmood, and K. P Ebmeier. 2018. "Association between gait and cognition in an elderly population based sample." *Gait & Posture 65*: 240–245.

Vallabhajosula, S., T. A. Buckley, M. D. Tillman, and C. J Hass. 2013. "Age and Parkinson's disease related kinematic alterations during multi-directional gait initiation." *Gait & Posture 37* (2): 280–286.

Velmurugan, B. K., B. Rathinasamy, B. P. Lohanathan, V. Thiyagarajan, and C. F Weng. 2018. "Neuroprotective role of phytochemicals." *Molecules 23* (10): 2485.

Xiang, C., Y. Zhang, W. Guo, and X. J Liang. 2020. "Biomimetic carbon nanotubes for neurological disease therapeutics as inherent medication." *Acta Pharmaceutica Sinica B 10* (2): 239–248.

Xie, J., Z. Shen, Y. Anraku, K. Kataoka, and X Chen. 2019. "Nanomaterial-based blood-brain-barrier (BBB) crossing strategies." *Biomaterials 224*: 119491.

Xu, R., J. Wang, J. Xu, X. Song, H. Huang, Y. Feng, and C Fu. (2020). Rhynchophylline loaded-mPEG-PLGA nanoparticles coated with tween-80 for preliminary study in Alzheimer's disease. *International Journal of Nanomedicine* 15, 1149–1160.

Xu, J., E. A. Wold, Y. Ding, Q. Shen, and J Zhou. 2018. "Therapeutic potential of oridonin and its analogs: From anticancer and antiinflammation to neuroprotection." *Molecules 23* (2): 474.

Yusuf, M., M. Khan, M. M. Alrobaian, S. A. Alghamdi, M. H. Warsi, S. Sultana, and R. A Khan. 2021. "Brain targeted polysorbate-80 coated PLGA thymoquinone nanoparticles for the treatment of Alzheimer's disease, with biomechanistic insights." *Journal of Drug Delivery Science and Technology 61*: 102214.

Zhang, M., X. Zang, M. Wang, Z. Li, M. Qiao, H. Hu, and D Chen. 2019. "Exosome-based nanocarriers as bio-inspired and versatile vehicles for drug delivery: Recent advances and challenges." *Journal of Materials Chemistry B 7* (15): 2421–2433.

Zheng, W., M. Wei, S. Li, and W Le. 2016. "Nanomaterial-modulated autophagy: Underlying mechanisms and functional consequences." *Nanomedicine 11* (11): 1417–1430.

8 Application of Neurophytomedicine for Protective Intervention and Remediation of Multiple Diseases

Sumanta Bhattacharya

8.1 INTRODUCTION

The paradigm shift in climate patterns due to the rapidly increasing rate of environmental degradation causes an increase in the chronicity of diseases. It also enhances the frequency of serious health issues like cancer and other severe genetic problems. Conventional allopathic medicine focuses solely on treating the diseases' underlying causes. However, the byproducts formed by the chemical reactions between the chemicals present in the medicines and the body chemicals cause serious side effects in the human body, which in turn cause the initiation of other health issues in the body. Furthermore, allopathic medicines only react with the chemicals that cause diseases, whereas traditional phytomedicines cause significant changes in current lifestyle and food habits. The primary and secondary metabolites present in different plant species act as the major components of phytomedicines. The prevalence of neurological diseases is increasing in human society due to an increase in social, political, and economic stresses in our daily lives. The modern lifestyle based on rapid urbanization and nuclearization of family structures results in the prevalence of serious neurological health issues. Because of the disruption to our daily lives and economic activities, the recent outbreak of the COVID-19 pandemic has had a serious impact on our mental health. Under this circumstance, the study of neurophytomedicines gains importance in the field of medical science.

The advancement of science and technology facilitates the development of neuropharmacology. The efficiency and interactivity of neurophytomedicines can be improved by using modern technologies such as nanotechnology, biotechnology, genetic engineering, and so on. The production rate of bioactive components in plants can be improved through advanced biotechnological studies. Because of the progress made in molecular biotechnology, proteins and nucleic acids can be studied in vitro, where they perform more effectively than in vivo, or inside the human body. Immune stimulators produced from plants, such as tomatine, polysaccharides, saponine, etc., have been shown to be more effective in the treatment of neurological

DOI: 10.1201/9781003389781-8

illnesses. Recombinant DNA technology, tissue culture, transgenesis, etc., are all examples of cutting-edge in situ and in vitro biotechnological processes that allow for the modification of plant metabolism, which in turn results in the production of secondary metabolites with enhanced in vivo activity in the human body. An efficient immune response can be induced against acute neurological illnesses through the synthetic synthesis of plant-based consumable vaccines using cutting-edge nanobiotechnology and genetic engineering. The advancement of neurophytomedicines leads to the achievement of sustainability in the treatment of neurological diseases and plays an important role in the achievement of sustainable development goals and good health and well-being. Aside from that, the growing demand for medicinal plants encourages plant cultivation, resulting in growth in the agricultural sector. The application of nanotechnology and biotechnology improves the productivity rate of medicinal plants, leading to the sustainable growth of the sector and an increase in the income of the marginal farmers attached to the cultivation process. Moreover, a significant portion of the tribal population of the Indian subcontinent is attached to the plantation and cultivation of a large section of rare plants having potential therapeutic characteristics. The promotion of these plants in the medical sector boosts the growth and development of these populations.

8.2 APPLICATION OF PHYTOMEDICINE IN HUMAN MEDICAL CARE

As per the report of the Directorate General of Commercial Intelligence and Statistics, Government of India, the trade of herbal medicines has been increasing since the last few years, indicating the increasing importance of herbal medicines replacing conventional allopathic medicines (Figure 8.1). India is also an important

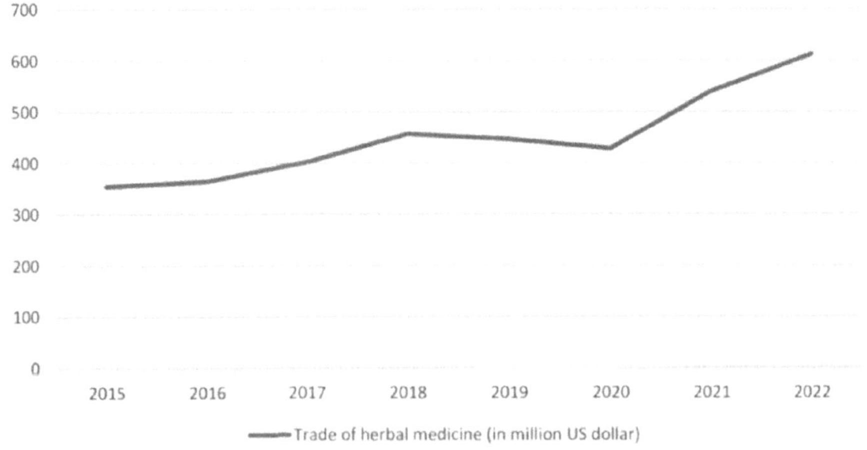

FIGURE 8.1 A line graph plotting the amount of the trade of herbal medicine in India in different years. Significant increase in the trade is observed from 2020 to 2022.

exporter of herbal medicines due to its rich reserves of medicinal plants. Allopathic medicines act based on the repair of the main cause of the disease, ignoring the impacts on other physiological mechanisms in the human body.

The secondary metabolites that plant species produce have immense potential to heal health-related issues in animals. Phytomedicines are manufactured from the secondary metabolites extracted from plants, which facilitate the overall maintenance of balance in the human body, whereas allopathic medicines are focused on the remediation of the cause of diseases only. The antioxidant and anti-inflammatory properties of plant secondary metabolites aid in the treatment of chronic health issues such as neurological and other health issues. Secondary metabolites of various types, such as terpenoids, flavonoids, alkaloids, and so on, effectively intervene in the human body mechanism and aid in maintaining the balance between the energy of movement, digestion, lubrication, and structure. The byproducts formed through the interaction between chemicals present in allopathic medicines and the human body create additional health issues, whereas the primary and secondary metabolites present in plant species modify the physiological mechanisms of the body. Secondary metabolites are now the primary ingredients in the production of various modern pharmaceutical products. These pharmaceutical products based on secondary metabolites are more effective in treating modern health issues. Climate change causes a slew of new, sometimes incurable, and undiagnosable diseases. Basically, the major use of phytomedicine is for health promotion and therapy for chronic, as opposed to life-threatening, conditions. However, phytomedical remedies are safe to use for cancer and new infectious diseases. Phytomedicines are regarded as Reese resources of biochemical, biomedical, and bioactive ingredients that can be used in drug development, regardless of whether the drugs are pharmacopoeial, non-pharmacopoeial, or synthetic. Numerous medicinal plants play an important role in the development of human health and sustainability around the globe. As we know, the products from botanicals or plants are used to treat diseases or to maintain human health. Consider that phytomedicine takes a longer time to work compared to synthetic or chemical pharmaceutical drugs. Quality control and management play an important role in the collection of herbal medicine from various resources. Adult Russian, substitution, and lack of skilled personnel are the main reasons for the unavailability of genuine herbal drugs. With the use of advanced quality control techniques and suitable standards maintenance, there is a need to supervise the quality of medicinal herbal products.

The active ingredients present in plant products induce the immune system of the human body by accelerating the activity of lymphocytes, enhancing phagocytosis, and increasing the rate of interferon production. The presence of a large number of vitamins in plant extracts corrects the human body's vitamin deficiency. The wide range of phytosterols, saponins, flavonoids, triterpenes, and carotenoids present in the plant species exhibit potential anti-carcinogenic activities, facilitating the use of herbal products in the treatment of cancer. Nowadays, the chemotherapeutic agents prepared from the plant extracts exhibit more efficiency in killing the tumor cells than the conventional chemical-based agents without exhibiting any severe side effects. [1] The inclusion of herbal products in the daily diet improves the immunity of the human body, leading to the prevention of diseases, including infections caused by highly contagious viruses like COVID-19. [2] Plant species also play an important role in the sustainable treatment of gynecological issues like the management of premenstrual tension, menopause symptoms, and period pains. The daily consumption of plant products maintains the metabolic, emotional, and nutrient balance of the human body, lowering the risks of increasing blood pressure and diabetes and also maintaining the amount of cholesterol, triglycerides, and other components in the human body, mitigating the risks of severe chronic health issues. To treat illnesses of the nervous system, scientists have developed a wide range of nanomedicine approaches. Due to the fact that the human nervous system is so delicate, neurophytomedicines pose less of a risk to the patient. Polymeric nanoparticles, dendrimers, micelles, liposomes, etc., are examples of nanosized drug carriers used to deliver bioactive phytochemicals to their intended sites of action.

The extraction of bioactive components from plant species is an ancient practice in preparing medicinal products. The integration of advanced technology in the extraction processes and analysis of the biomolecules present diversify the application of phytomedicines in the health care industry with greater efficiency. The emergence of super infectious viruses leads us to innovate new kinds of therapeutic viruses. The development of phytomedicines becomes a new hope in this regard due to their capability of improving overall body function instead of treating a particular disease. [3] The functional groups present in different types of secondary metabolites present in plant species effectively react with the proteins and affect the receptor, enzymatic, and transcription mechanisms of the body. Moreover, secondary metabolites are also capable of reacting with the purine and pyrimidine bases of DNA, resulting in a significant alteration of the genomic sequence. In this way, herbal medicines remediate genetic diseases. The biotechnological modification of herbal drugs with mutagens plays a significant role in the treatment of cancer-related diseases (Figure 8.2). [4]

8.3 THERAPEUTIC APPROACH TO NEUROLOGICAL DISEASES THROUGH NEUROPHYTOMEDICINE

The disruption in the mechanism of the transformation of signals from the senses to the central nervous system and from the central nervous system to body muscles causes the prevalence of neurodevelopmental diseases. The phytoconstituents present in the plant extracts form stable chemical bonds with the receptors and signals to facilitate the steady flow of signals from the nervous system to different parts of the human body. The interaction between the bioactive constituent and different biomolecules responsible for neurological activities becomes the key to understanding the bioactivity of phytoconstituents in the treatment of different neurological diseases. Phytochemicals facilitate the maintenance of overall chemical balance in the brain instead of the target-based approach of allopathic medicines. Flavonoids, phenols, alkaloids, fatty acids, terpenes, saponins, etc., are some important secondary metabolites used for the treatment of several neurodegenerative, neuropsychiatric, and cognitive disorders. [5] Phytochemicals control the function of receptors for the major inhibitory neurotransmitters. Several bioactive constituents, like polyphenols, activate specific transcription factors that facilitate the synthesis of RNAs. These RNAs activate specific signal transduction pathways by the translation of favorable protein molecules, which results in the resistance of neurons to various stress factors,

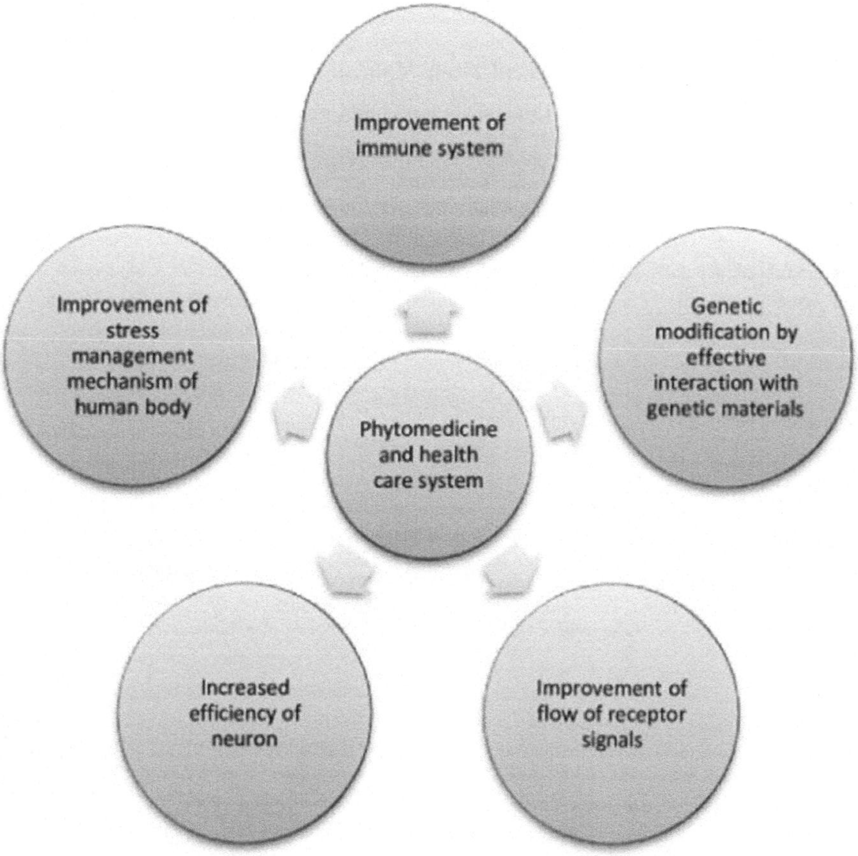

FIGURE 8.2 Phytomedicine and health care system.

leading to the prevention of neuropsychiatric and neurodegenerative diseases. [6] The neuroprotective activities of phytochemicals make the phytomedicines more effective toward the overall improvement of the nervous system, mitigating the risk of neurological diseases (Table 8.1) (Figure 8.3).

The aromatic amino acid pathway and the mevalonic acid pathway are the two main metabolic mechanisms that transform the polysaccharides created by photosynthesis in plants into bioactive components. These bioactive substances don't contribute significantly to the process of plant growth because they are byproducts of the metabolic process. However, these bioactive substances help the plant bodies defend themselves from diseases and herbivorous predators. Additionally, a number of bioactive elements help plants recover from nitrogenous wastes and survive in climatically stressful situations like drought and flooding. [7] Typically, bioactive substances fall into one of the groups indicated in Table 8.1.

With the exception of amides and amino acids, alkaloids are heterocyclic nitrogen-containing chemicals that are generated from plant extracts. It has alkaline properties and is mostly produced from amino acids. Decarboxylation of amino acids

TABLE 8.1
List of Some Phytochemicals and Their Neurological Effects.

Bioactive Constituent	Name of the Plant	Neurological Effect
Berberine	*Berberis aquifolium*, *Berberis vulgaris*, *Berberis aristata*, *Hydrastis canadensis*, *Tinospora cordifolia*	Cholinesterase inhibition, MAO inhibition, hypocholesterolemic effect, mitigation of endoplasmic reticulum stress leads to the prevention of Alzheimer's disease
Resveratrol (one type of polyphenolic extract)	Japanese knotweed, grapes, apples, blueberries, plums, peanut	Antitumor, anti-inflammatory, hypoglycemic, and neuroprotective effects, induction of passive avoidance task mechanism
Curcumin	Turmeric, *Curcuma longa*, ginger	Treatment of Alzheimer's diseases, Parkinson's disease
Quercetin (a flavonol type flavonoid)	Onion, grapes, cherries, broccoli, citrus trees	Neuroprotective action, anti-oxidant effects
Myricitine	Grapes, jambolana fruits, berries	Anti-oxidant, anti-inflammatory characteristics
Baicalein	*Scutellaria baicalensis*	Neuroinflammation blockage, homeostasis regulation of neurotransmitters, induction of apoptosis in brain tumor cells
Glycyrrhizin	*Glycyrrhiza glabra*	Treatment of inflammatory illness, suppression of locomotor impairment, inhibition of the degeneration of neurons
Chicoric acid	*Echinacea purpurea*	Prevention of motor dysfunction, inhibition of glial hyperactivation, neuroprotective activities
Mecasin	Brassica napus	Anti-inflammatory properties
Morin	*Psidium guajava*, *Maclura pomifera*, *Maclura tinctoria*	Activity against amyloids
4-Hydroxyisoleucine (bioactive amino acid)	*Trigonella foenum*, *Trigonella graecum*	Neuroprotective properties by chemical alteration of neurons leading to behavioral changes
Embelin	*Embelia ribes* Burm	Repairing of striatal neuronal damage, mitigation of oxidative and neuroinflammatory stress
Baicalein	*Scutellaria baicalensis*, *Oroxylum indicum*	Neuroprotective effect, advancement of psychological and cognitive alteration
Cannabidiol	*Cannabis sativa*	Anti-epileptic and anticonvulsant activities
Apigenin	Nuts, citrus, tea, chamomile, thyme, celery, etc.	Anti-inflammatory, antioxidant, and neurological effects

yields amines, which interact with amine oxides to produce aldehydes, which are used to make alkaloids. Aldehydes and amine groups condense in a Mannich-type reaction to produce the distinctive heterocyclic rings of alkaloids. It demonstrates a wide spectrum of phytochemical properties, such as anti-malarial, anti-asthmatic, anti-cancer, cholinomimetic, analgesic, antibacterial, anti-hyperglycemic, vasodilatory, and anti-arrhythmic activity, among others. [8, 9]

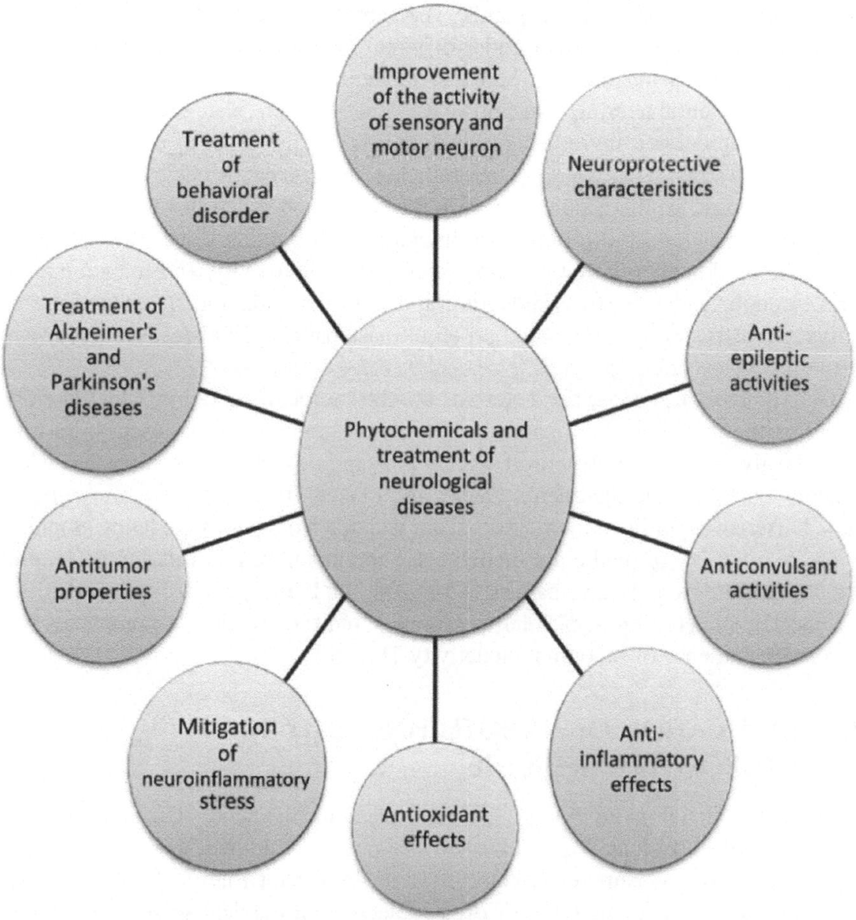

FIGURE 8.3 Phytochemicals and treatment of neurological diseases.

The polyphenolic substances known as flavonoids are what give flowers their color and scent. Due to the highly conjugated heterocyclic structures present, it functions as a pigment. Due to their anti-oxidative, anti-mutagenic, anti-carcinogenic, and anti-inflammatory properties, they are frequently used as therapeutic medicines. They can also control how human cells' enzymatic functions are carried out. Cancer, Alzheimer's disease, atherosclerosis, and other major disorders are all treated with flavonoids' anti-oxidant abilities. The low cardiovascular mortality rate of flavonoids, which has been demonstrated in recent studies on their chemical characteristics, makes a wide range of applications for them as a preventative measure for cardiac disorders possible. Flavonoids are essential for the growth and development of plants because their colorful compounds make pollination easier. [10, 11]

Anthocyanins are water-soluble plant pigments that are primarily responsible for giving fruits and vegetables their colors. These are oxygen-containing heterocyclic compounds. The colors of the plants' vegetables and fruits were caused by the oxygen atoms' resonance effect under different pH and temperature conditions. A wide range of pharmacological benefits, including antioxidant, anti-cancer, anti-inflammatory, antibacterial, and anti-obesity actions, are also exhibited by these pigments. The colored ingredients reduce the risk of the onset of chronic diseases and oxidative stress by scavenging free radicals, such as reactive oxygen and reactive nitrogen species. Anthocyanins decrease the likelihood of cardiovascular illnesses by enhancing blood lipid profiles and biomarkers, inhibiting the growth of malignant cells by downregulating cyclooxygenase enzyme activity, and inducing apoptosis through oxidative stress reduction and lipid peroxidation. By preventing the activation of the mitogen-activated protein kinase pathway, it also slows the growth of malignancies. [12, 13]

Glycosides of triterpenes and steroids are the source of saponins, which are produced by the mevalonic acid pathway. The term "saponin" is also used to describe steroidal glycosides. The structural diversity of saponins' chemical properties, which depends on their amphipathicity, causes the bioactivity of various compounds to vary. Different saponin components that are isolated from plants are being employed to create a variety of medicinal medicines. Saponins have anti-inflammatory, anti-fungal, anti-microbial, anti-cancer, anti-viral, and anti-parasitic phytochemical effects. The penetration of plasma membranes and the creation of complexes with sterols are evidence of saponins' bioactivity. [14, 15]

8.4 APPLICATION OF NANOTECHNOLOGY IN NEUROPHYTOMEDICINE

Nanophytomedicine is an emerging branch of medical science that deals with the application of nanotechnology in herbal medicine, also known as phytotherapy or phytopharmacology. Nanotechnology plays an important role in the effective targeted and non-targeted delivery of drug molecules in medical science. The nano-adsorbants lead to the effective adsorption of drug molecules, which exhibit more therapeutic activities due to an increase in bioavailability. The lower in vivo efficacy of phytomedicines is due to their poor permeability, low systematic availability, instability, extensive first-pass metabolism, and low aqueous solubility. The interaction of herbal drugs with nanoparticles increases their bioavailability in the human body and increases the possibility of drug molecules being delivered to the target. Nanoparticles are widely used as carriers of drug molecules for their increased chemical activity and ability to cross the tissue barrier. Specially engineered nanomaterials are prepared that are able to carry drug molecules, proteins, plasmids, antibodies, oligonucleotides, fluorophores, ligands, polymers, radioisotopes, tissue-engineered products, etc., to the human body and affect desired cells. Some nanomaterials, such as dendrimers, liposomes, polymeric nanoparticles, polymeric micelles, carbon nanotubes, mesoporous silicon, quantum dots, nanocrystals, nanospheres, phytosomes, etc., participate in drug delivery mechanisms. The aqueous solubility and permeability of the blood-brain barrier of the phytomedicines can be improved by reducing the size of the drug molecules to the nanoscale. The flavonoids and lignins present in the *Cuscuta chinensis* plant exhibit improved bioavailability and adsorption capability in the human body after being reduced to nanosize by the nanosuspension method. The rate of diffusion of the suspended nanomedicines is higher than that of phytomedicines into the gastrointestinal medium. The drug delivery mechanism of herbal drugs can be increased by increasing their solubility, enhancing their stability, reducing their toxicity, improving their macrophage distribution mechanism, and enhancing their pharmacological activities. Nanotechnology ensures the targeted delivery of phytomedicines in a cell- or tissue-specific way through the development of effective drug carriers. The increase in solubility and permeability of the phytomedicines leads to the transcytosis of neurophytomedicine across tight endothelial and epithelial cells. Nanotechnology also facilitates the therapeutic modality of combination therapy by enabling the co-delivery of two or more phytomedicines in the human body. The integration of phytomedicines with the nanopolymers increases the systemic drug concentration, leading to an improvement in anticancer efficacy. [16]

Nanobioremediation is the application of nanotechnology to prevent environmental degradation, where the nanoparticles, having more chemical reactivity due to having more surface area per unit mass, react with the pollutants to form less harmful products or adsorb the pollutants to make them immobile, preventing the pollutants from contaminating the environment. Nanomaterials developed in different forms like nanotubes, nanowires, films, quantum dots, colloids, etc. are placed in the contaminated groundwater and soil in the form of fertilizers, pesticides, fungicides, composts, etc. that also intensify the activity of plants in the uptake of the toxicants from the soil and groundwater, facilitating the process of purification. Nanoparticles prevent the contamination of toxicants in soil, groundwater, and surface water by reducing or oxidizing the contaminants, immobilizing the contaminants by combining with them, forming a wall between polluted and pure water to prevent the spread of pollutants, etc. [17] Nanomaterials are instructed to be more economical, efficient, and eco-friendly than the prevailing materials in each resource conservation and protection setting. [18]

The pharmacokinetic mechanism of phytomedicines can be studied by the integration of nanosensors with the phytomedicines. Smart pills are being developed by incorporating nano-based electronic devices into pharmaceutical pills that perform advanced imaging, sensing, and drug delivery. It enables medical practitioners to track the delivery pathway of drugs in the human body and study the mechanism of drug activity. "Atmo Gas Capsule" is an advanced smart pill that examines the activity of gaseous substances in the human body. Smart sensor capsules are prepared with advanced nanotechnology that is used along with the vaccines to monitor the activity of the vaccines in the human body. It is also able to be consumed orally instead of by injection. This is a groundbreaking innovation in the field of research in medical science. Nanopatch vaccines use nanoparticles to deposit the vaccines on immune cells present in the skin and lower the risk of infection. It is an easier vaccination process and eliminates the need for vaccine refrigeration. [19]

Nanoflares are specially engineered nanomaterials in the field of cancer treatment. It effectively detects the presence of cancer cells in the bloodstream. It is able to bind with the genetic target in cancer cells and generate light when the target genetic sequence is found.

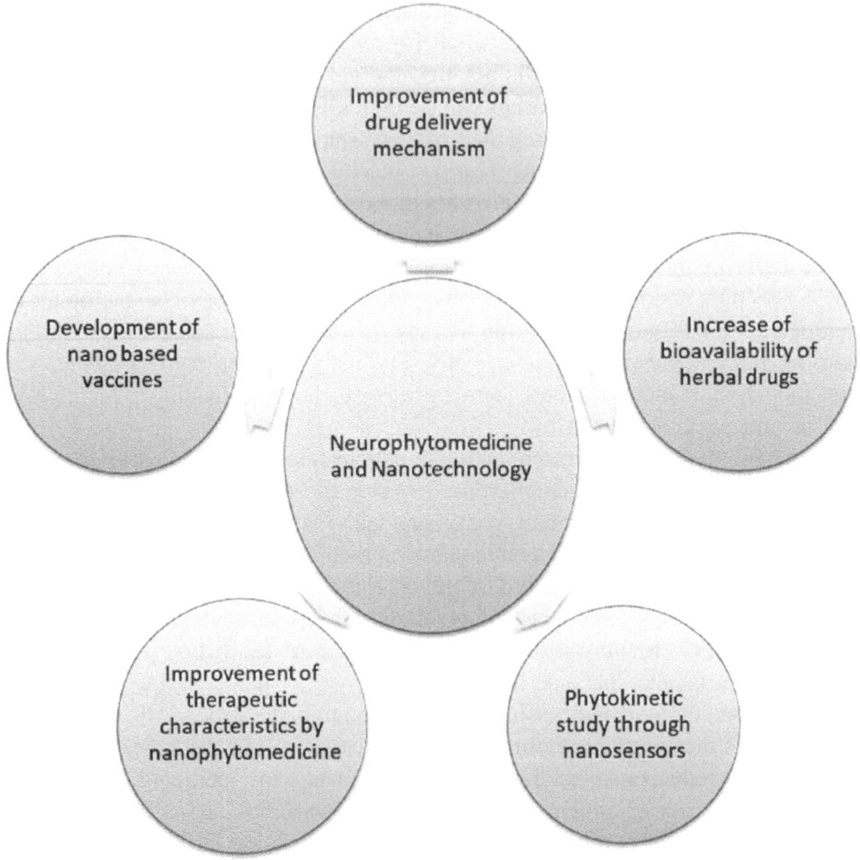

FIGURE 8.4 Neurophytomedicine and nanotechnology.

When nanobots, advanced robots created with nanotechnology and robotics, are inserted into the human body, they act as miniature surgeons. It effectively repairs the intracellular structures, resulting in the healing of particular diseases within the human body. The DNA-based nanobots are also prepared for the eradication of genetic diseases by modifying the target genome sequence. Nanomaterials also act as effective carriers of genetic material in the body (Figure 8.4). [20]

8.5 APPLICATION OF BIOTECHNOLOGY FOR BETTER USE OF NEUROPHYTOMEDICINE

Biopharmaceutical technology focuses on the synthesis of pharmaceutical products from biological resources instead of synthetic chemicals, which increases the efficiency of medicinal products by reducing the side effects. Biotechnology plays an important role in the treatment of acute diseases like cancer and other genetic diseases through targeted and non-targeted therapy, immunotherapy, hormonal therapy, targeted drug delivery, gene therapy, the manufacture of vaccines, etc. The plant extract can become a potential component of chemotherapy that reduces the possibility of effects on healthy cells during the time of chemotherapy. Bio-based nanomaterials like cantilevers (a biosensor-based nanomechanical system) and dendrimers (a nanopolymeric dendritic structure that carries drug molecules in the core) are potential carriers of drug molecules that ensure the targeted delivery of drugs with higher efficiency. Nowadays, vaccines are developed based on advanced biotechnology like genomics, proteomics, transcriptomics, metabolomics, etc., which is likely to reduce the possibility of genetic diseases in the near future. Moreover, gene therapy is proven to be more effective in healing the carcinogenic effects in the human body. The advancement of biopolymer technology is one of the latest additions to the pharmaceutical industry. Biotechnology has recently improved to the point where advanced genetic engineering can help agriculture expand sustainably and environmentally. Through modern recombinant DNA technology and transgenesis, the phenotypic traits of crops can be adjusted, resulting in the cultivation of advanced, healthy herbal crops. Plant genome modification can also be used to grow pest-resistant crops. The contaminants can be successfully remediated from the soil and water by the application of nanoparticles with fertilizers, which leads to the improvement of soil health followed by the boosting of agricultural productivity. The development of key medicinal crops through the 4th generation technology revolution will make it easier to achieve sustainable growth in the medical sciences as well as the agricultural sector (Figure 8.5).

Neurohormesis characteristics of several phytochemicals facilitate the preparation of optimum doses for the effective treatment of several neurodegenerative diseases like Alzheimer's, Parkinson's diseases, epilepsy, etc. The hormesis effects of the apparently toxic phytochemicals increase the immune response and mitigate the risk of neurological diseases. Several phytochemicals, such as resveratrol, catechin, sulforaphane, hypericin, allicin, and others, induce adaptive stress response signaling pathways, resulting in improved resistance to injury and disease. The hormesis characteristics of the phytochemicals increase the hemodynamic characteristics of the body, leading to the effective management of mental and physical stresses caused by aging. The neuron cells are more susceptible to damage with age due to the rapid rate of DNA damage. Neurohormesis enables the repair of damaged DNA, resulting in a decrease in the rate of cell death through radical scavenging and increased antioxidant activities of the phytochemicals. Mitohormesis is the process of exhibiting hormesis activities by interacting with the mitochondria, which play a central role in bioenergy production and nutrient metabolism. Berberine, an alkaloid extracted from *Coptidis rhizoma* and *Hydrastis canadensis*, directly interacts with mitochondria through the electron transport chain and reduces the oxidative stresses generated in the neuronal cells. The consumption of epicatechin, an important bioactive component belonging to the flavonol group, leads to the improvement of the cognitive function of the human brain by modifying neuronal spine concentration, hippocampal angiogenesis, and memory function. [21]

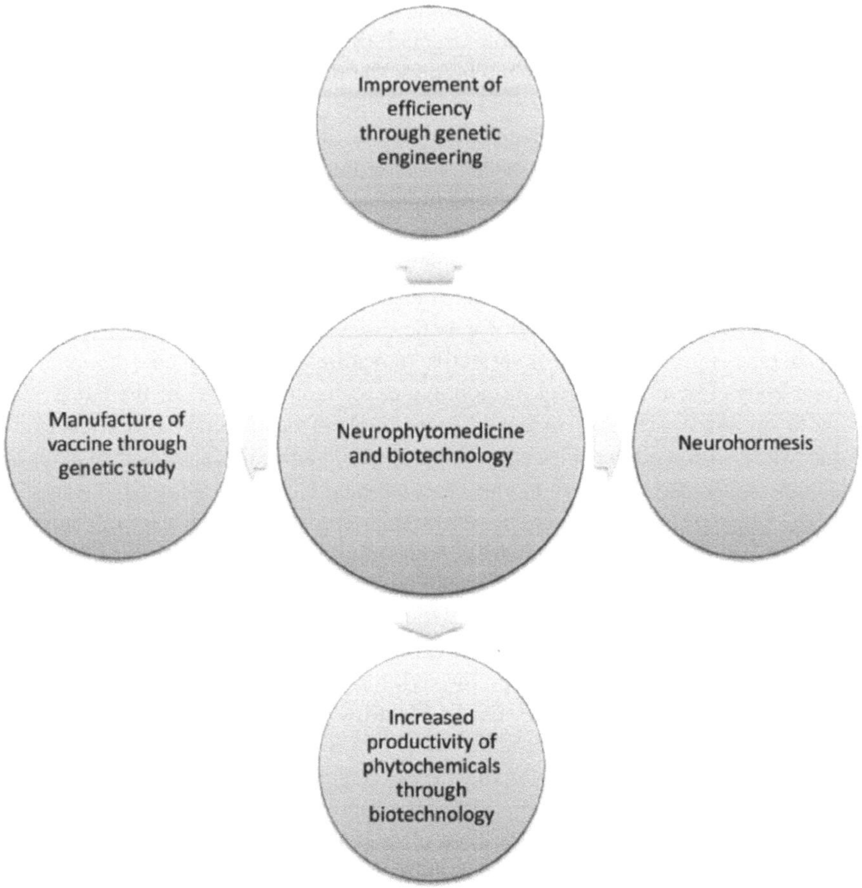

FIGURE 8.5 Neurophytomedicine and biotechnology.

8.6 FUTURE PROSPECTIVE OF PHYTOMEDICINE IN DISEASE CONTROL

The ongoing environmental issues related to climate change lead to serious health problems in humans. The conventional allopathic medication process becomes incapable of dealing with these ongoing challenges as it is unable to cure the root causes of the diseases. Under this scenario, traditional medicinal practices like Ayurveda are gaining importance. The importance of phytomedicine is increasing day by day in order to achieve sustainability in the health care sector. Under these circumstances, the advancement of phytomedicine with the help of modern technology becomes a necessary step in order to improve the health of humans. The application of modern technologies like nanotechnology, biotechnology, and genetic engineering improves the quality and efficiency of phytomedicines. The genomic, proteomic, and transcriptomic study of medicinal plants leads to the advancement of phytomedicines. Biotechnology will increase the productivity and efficiency of medicinal plants, resulting in advancements in both the health and agricultural sectors. The application of nanotechnology enables sustainable and eco-friendly growth of medicinal plants by increasing productivity through the application of nanofertilizers and nanoremediators. The in vivo efficacy of neurophytomedicines will be increased by the application of nanocarriers that will lead to effective targeted and non-targeted drug delivery. The production of bioactive secondary metabolites will be increased in the plant body by the application of specially engineered genes through nanocarriers. The phytokinetic study of neurophytomedicines can be improved by using nanosensors, which will allow for a more detailed in vivo study of neurophytomedicine mechanism. The neuroinformatic study reveals the detailed biological mechanisms of the nervous system, whereas the bioinformatic study of the plant extracts leads us to study the bioactive mechanisms of phytoconstituents. It results in the development of more effective neurophytomedicines for the sustainable treatment of neurological diseases (Figure 8.6).

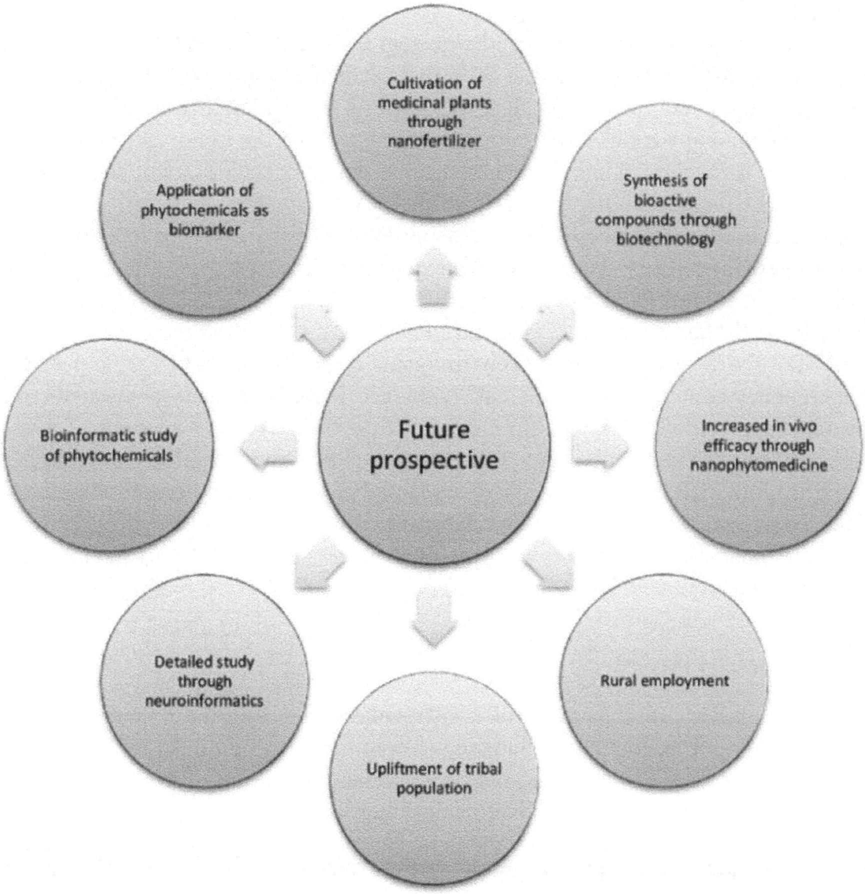

FIGURE 8.6 Future prospective of neurophytomedicine.

The detailed study of neuroinformatics facilitates the preparation of databases for the biological functions of the nervous system. The neuroinformatic study of all sensory, motor, and interneurons activities must be improved in order to develop effective phytotherapeutic techniques for the treatment of nervous system dysfunctions as well as psychological disorders.

Moreover, the phytochemicals have the potential to act as potential biomarkers by effectively interacting with the DNA, proteins, and other biomolecules. It will help to study the pharmacokinetic mechanism of the medicinal products within the nervous system, leading to the development of research in the fields of neurology, pharmacology, and herbal medicine.

Modern biotechnological innovations enable the development of new herbal plants that produce more medicinal secondary metabolites. It will facilitate the development of the field of herbal medicine and also improve the pharmacokinetic efficiency of the phytochemicals in the nervous system.

The development of nanophytomedicine improves the in vivo efficacy of phytomedicines by improving phytochemical bioavailability and delivery mechanisms in the body.

The promotion of neurophytomedicine also facilitates the development of tribal populations, as a large proportion of herbal plants belonging to tribal areas have potential therapeutic properties. It also increases the availability of herbal drugs. Furthermore, many tribes have unique knowledge about the use of phytomedicines that can be incorporated into formal medical sciences for the benefit of both sides of society. The use of nanotechnology and biotechnology increases the in vivo efficacy of neurophytomedicines by increasing the solubility and bioavailability factors of herbal products. The phytochemicals have already shown efficiency in the treatment of serious pandemics like COVID-19. The mental stress and trauma generated due to the change in social behavior resulting from the recent outbreak of COVID-19 viruses can be effectively mitigated by the use of phytomedicines. The health policy should focus on increasing the acceptability of neurophytomedicines by replacing traditional allopathic medicines for the effective treatment of sensitive nervous system diseases as well as behavioral disorders. Neurophytomedicines require additional in vitro and in vivo research before they can be used effectively. The promotion of traditional Ayurvedic plant products in the treatment of chronic diseases enhances the scope of rural employment and income for small and marginal farmers associated with it.

8.7 CONCLUSION

The allopathic approaches to medicinal treatment based on the remediation of a target disease are becoming an outdated idea in today's scenario as the prevalence of chronic diseases is increasing due to the changing nature of the environment. The improvement of overall body function and the boosting of the immune system become the basis of modern medicinal treatment, just like in ancient times. Phytomedicines are gaining importance in the effective treatment of chronic diseases due to their potential to improve the efficiency of removing the root cause of disease by keeping the balance among the physiological, mental, and emotional health of human beings. The nervous system is one of the most important and sensitive physiological systems, as it controls other body functions. The risk of allopathic neurological disease dosage increases due to the risk of side effects causing major damage to healthy neurons. On the other hand, herbal medicines focus on the repair of the overall nervous system, reducing the risk of side effects. The advancement of modern technologies like nanotechnology and biotechnology improves the efficiency of phytomedicines. The rapid development of the field of bioinformatics leads to the emergence of neuroinformatics, which enables the detailed study of the mechanisms of the nervous system. In this way, the field of herbal medicine is developing through the application of advanced techniques. The adoption of ayurvedic medicine instead of conventional allopathic medicines promotes the healthy growth of the human population under the current challenges of global warming and climate change. The increasing use of phytomedicines also facilitates the egalitarian economic development of society by expanding the scope of the primary sector. The improvement of tribal populations is also another component of the promotion of Ayurveda. The emergence of neurophytomedicine plays a significant role in the achievement of sustainable development in the health sector. It also promotes economic growth and the development of the economy.

REFERENCES

1. Khodadadi, Samaneh; Role of Herbal Medicine in Boosting Immune System; Immunopathologia Persa, Volume 1, Issue 1; January 7, 2015
2. J. Isbill, Jonathan; Kandiah, Jayanthi; Kruzliakova, Natalie; Opportunities for Health Promotion: Highlighting Herbs and Spices to Improve Immune Support and Well-Being; Integrative Medicine, Volume 19, Issue 5; October, 2020
3. Srivastava, A.; Srivastava, P.; Pandey, A.; Khanna, V. K.; Pant, A. B.; Phytomedicine: A Potential Alternative Medicine in Controlling Neurological Disorders; New Look to Phytomedicine, Advancements in Herbal Products as Novel Drug Leads, pp. 625–655; 2019; DOI: 10.1016/B978-0-12-814619-4.00025-2
4. Wink, Michael; Modes of Action of Herbal Medicines and Plant Secondary Metabolites; MDPI; Medicines, Volume 2; November 30, 2015; p. 251; September 8, 286; DOI: 10.3390/medicines2030251
5. Puri, Vivek; Kanojia, Neha; Sharma, Ameya; Huanbutta, Kampanart; Dheer, Divya; Sangnim, Tanikan; Natural Product-Based Pharmacological Studies for Neurological Disorders; Frontier Pharmacology, Neuropharmacology; November 7, 2022; DOI: 10.3389/fphar.2022.1011740
6. Phani Kumar, G.; Khanum, Farhath; Neuroprotective Potential of Phytochemicals; Pharmacogonsy Reviews, Volume 6, Issue 12; 2012; DOI: 10.4103/0973-7847.99898
7. Chikezie, Paul C.; Ibegbulem, Chiedozie O.; Mbagwu, Ferdinand N.; Bioactive Principles from Medicinal Plants; Research Journal of Phytochemistry, Volume 9, Issue 3; November 30, 2015; pp. 88–115; DOI: 10.3923/rjphyto.2015.88.115
8. Eddouks, Mohamed; Chattopadhyay, Debprasad; Feo, Vincenzo De; Cho, William C; Medicinal Plants in the Prevention and Treatment of Chronic Diseases; Evidence-Based Complementary and Alternative Medicine; Hindawi Publishing Corporation, Volume 25, 2012; DOI: 10.1155/2012/458274
9. Dogra, Kuldip S.; Chauhan, S.; Jalal, Jeewan S.; Assessment of Indian Medicinal Plants for the Treatment of Asthma; Journal of Medicinal Plant Research, Volume 9, Issue 32; November 30, 2015; p. 851; August 13, 862; DOI: 10.5897/JMPR2015.5890

10. Ahvazi, Maryam; Khalighi-Sigaroodi, Farahnaz; Charkhchiyan, Mohammad M.; Mojab, Faraz; Mozaffarian, Vali A.; Zakeri, Hamideh; Introduction of Medicinal Plant Species With Most Traditional Usage in Almut Region; Iranian Journal of Pharmaceutical Research, Volume 11, Issue 1; pp. 185–194; 2012
11. Parasuraman, Subramani; Thing, Gan S.; Dhanaraj, Sokkalingam A.; Polyherbal Formulation: Concept of Ayurveda; Pharmacognosy Reviews, Volume 8, Issue 16; November 30, 2014; pp. 73–80; DOI: 10.4103/0973-7847.134229
12. Katiyar, Chandrakant; Gupta, Arun; Kanjilal, Satyajyoti; Katiyar, Shefali; Drug Discovery from Plant Sources: An Integrated Approach; AYU, Volume 33, Issue 1; November 30, 2012; pp. 10–19; DOI: 10.4103/0974-8520.100295
13. Pan, Si-Yuan; Zhou, Shu-Feng; Gao, Si-Hua; Yu, Zhi-Ling; Zhang, Shuo-Feng; Tang, Min-Ke; Sun, Jian-Ning; Ma, Dik-Lung; Han, Yi-Fan; Fong, Wang-Fun; Ko, Kam-Ming; New Perspectives on How to Discover Drugs from Herbal Medicines: CAM's Outstanding Contribution on Modern Therapeutics; Evidence-Based Complementary and Alternative Medicine, Hindawi, 2013; DOI: 10.1155/2013/627375
14. Djordjevic, Sofia M.; From Medical Plant Raw Material to Herbal Remedies, intechopen; 2017; DOI: 10.5772/66618
15. Bisht, Lata; Verma, Ram; Allopolyherbal Formulation and Their Strategies; Journal of Phytochemistry & Biochemistry, Volume 1, Issue 1; November 30, 2017
16. Gunasekaran, Thirumurugan; Haile, Tedesse; Nigusse, Tedele; Dhanaraju, Magharla D.; Nanotechnology: an Effective Tool for Enhancing Bioavailability and Bioactivity of Phytomedicine; Asian Pacific Journal of Tropical Medicine, Volume 4; November 30, 2014; pp. S1–S7; DOI: 10.12980/APJTB.4.2014C980
17. Nzenguang, V. A.; McCutcheon, S. C.; Phytoremediation of Perchlorate. In: McCutcheon SC, Schnoor JL, editors. Phytoremediation: Transformation and Control of Contaminants. New Jersey: John Wiley and Sons, Inc.; pp. 863–885, 2003
18. Mandal, Asit; Purakayastha, Tapan; Ramana, Sivakoti; Neenu, Sathyaseelan; Bhaduri, Debarati; Chakraborty, Koushik; Manna, Madhab; Rao, Annangi S.; Status on Phytoremediation of Heavy Metals in India- a Review; International Journal of Stress Management, Volume 5, Issue 4; November 30, 2014; p. 553; December, 560; DOI: 10.5958/0976-4038.2014.00609.5
19. Idrees, Muhammad; Role of Nanotechnology in Medical Sciences: A Review; International Journal of Innovative Drug Discovery, Volume 5, Issue 1; pp. 14–24; January, 2015
20. Anjum, Sumaira; Ishaque, Sara; Fatima, Hijab; Farooq, Wajiha; Hano, Christophe; Abbasi, Bilal H.; Anjum, Iram; Emerging Applications of Nanotechnology in Healthcare Systems: Grand Challenges and Perspectives; MDPI, Pharmaceuticals, 14, 707; July 21, 2021; DOI: https://doi.org/10.3390/ph14080707
21. Sahebnasagh, Adeldeh; Eghbali, Samira; Saghafi, Fatemeh; Sureda, Antoni; Avan, Razieh; Neurohormetic Phytochemicals in the Pathogenesis of Neurodegenerative Diseases; Immun Ageing, Volume 19, Issue 36; August 11, 2022; DOI: 10.1186/s12979-022-00292-x

9 Role of Phytopharmaceuticals as Adjuncts in Treatment of Glioblastoma Multiforme

Ameya Kothekar, Rajesh Pradhan, and Rajeev Taliyan

9.1 INTRODUCTION

Glioma is an aggressive type of solid tumour associated with the central nervous system (CNS) which develops from the glial cells or supporting cells. According to the latest, World Health Organization (WHO) classification gliomas are further divided into classes/grades I, II, III, and IV. This classification is done based on pathological examination of the malignancy degree of the tumour utilizing molecular information (Louis et al, 2016). This classification is important in the clinical scenario as it helps in determining the type and mode of treatment. Grade I tumours are slow-growing neoplasms that can be treated surgically. Grade II tumours are tissue invasive and have a high chance of recurrence despite their low capacity for proliferation. Grade IV gliomas are the most treacherous and it is the final stage where the tumour becomes malignant with a significant risk of death (Louis et al, 2014).

Glioblastoma/glioblastoma multiforme (GBM), with a WHO grade 4, is the most advanced malignant brain cancer. It is the second most common brain tumour after meningioma. Intracranial radiation, ionic radiation, and rare genetic disorders like type 1 and 2 neurofibromatosis are the major factors responsible for GBMs (Salvati et al, 2003). However, other factors like environmental stresses, smoking, drinking, and organochlorides from pesticides have shown no conclusive proof to cause GBM. Poor prognosis and a median patient survival time of 12–15 months after diagnosis are the major concerns in GBM (Wen and Kesari, 2008). In total, 95% of the tumours arise in the supratentorial region and less than 5% originate in the cerebral hemispheres and brainstem (Nakada et al, 2011). Though the occurrence of this cancer is lower than other cancers like colon, lung, and breast cancer, the average years of life lost is about 20 years, thus representing a significant problem in oncology. Moreover, the prognosis of GBM has hardly changed since the early 1970s. Even after decades of study, glioblastoma is still one of the most dangerous and feared cancers. Therefore, the molecular mechanisms must be well studied which can help in better managing and designing effective therapeutic regimens with significant results. The first line of treatment is still surgery by chemotherapy and

DOI: 10.1201/9781003389781-9

radiation. Despite its moderate effect and debatable efficacy, chemotherapy has been a routine therapeutic option for glioblastoma. The role of chemotherapy is important in treating glioblastoma. GBM is renowned for the development of chemoresistance. This is a key impediment during the development of an effective therapeutic regime. Temozolomide (TMZ), an oral alkylating agent, is the first line of drug for GBM. Methylguanine-DNA methyl transferase (MGMT), a TMZ metabolite, induces intrinsic resistance to TMZ and is the reason for complications associated with TMZ. After, many studies aimed to explore more effective and efficient chemotherapeutic targets, bioactive has emerged as excellent adjuncts to traditional chemotherapy. Plants have a vast array of phytochemicals with the capacity to target invasive brain tumours and alter oncogenic pathways to aid in causing apoptosis. These phytopharmaceuticals also inhibit the activity of various oncogenic proteins that are overexpressed in cancer and are responsible for invasion, metastasis, chemotherapy resistance, and angiogenesis.

In this chapter, the scope of the bioactive as anti-cancer agents against GBM has been discussed. This chapter also includes their mechanism of action, associated limitations, and regulatory considerations.

9.2 PATHOGENESIS OF GLIOBLASTOMA MULTIFORME

Understanding the pathophysiology is essential for creating prospective chemotherapeutic drugs as well as for discovering disease biomarkers (refer Figure 9.1). Based on the presence or absence of lesions, GBM can be divided into two categories: Primary and secondary. Primary GBM arises without any pre-existing tumours or neoplasia (Agnihotri et al, 2013). Secondary tumours develop from precursor lesions like astrocytoma or from other lower-grade gliomas within 5–10 years of diagnosis (Cloughesy

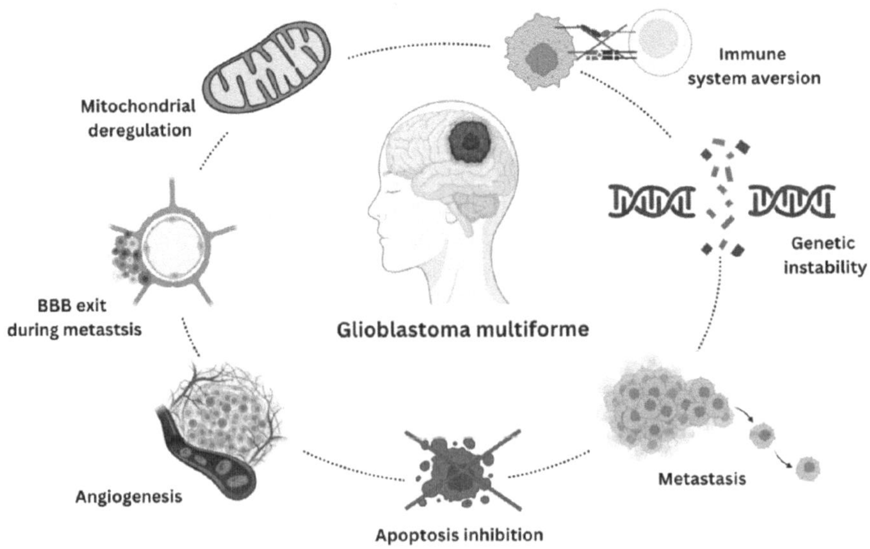

FIGURE 9.1 Schematic drawing depicting characteristics of glioblastoma multiforme (GBM).

et al, 2014). Based on the transcriptional characteristics, previously there were four classes of glioblastoma classical, neural, pro-neural, and mesenchymal. However, comprehensive longitudinal analysis of the GBM tumour transcriptome points towards the existence of three distinct forms of GBM, that is, classical, pro-neural, and mesenchymal. Tumour-specific biomarkers play an important role in the identification of therapeutic targets. MGMT, IDH (isocitrate dehydrogenase) 1, and IDH2 are some of the most common biomarkers of glioblastoma (Marumoto and Saya, 2012). A representative diagram for the pathogenesis of GBM has been depicted in Figure 9.1.

9.2.1 Unique Histopathology of Glioblastoma Multiforme

GBM is characterized by small cells with polymorphism, anaplasia, and significant anisokaryosis. These cells are polygonal or spindle-shaped containing acidophilic cytoplasm and indistinct cellular borders. Their nuclei are oval or elongated, with coarsely clumped hyperchromatic chromatin and numerous discrete nucleoli in the centre or peri-centre. Conventional histologic characteristics of glioblastoma include cellular polymorphism, nuclear atypia, a high mitotic index, and pseudo-palisading cells. These pseudopalisading cells are exclusive to malignant glioma and serve as a protective barrier. Therefore, these cells are used as prognostic markers for advanced glioma. In glioblastoma, dense migratory zones of cells surround the necrotic tissue and are termed pseudopalisades or "false palisades." Necrotic foci are among the most distinctive characteristics of GBM and have two distinct forms. One type is within the tumour's central region and is a result of insufficient blood supply. The other form consists of tiny, irregularly shaped necrotic foci surrounded by radially oriented pseudopalisading zones. Due to vascular collapse and necrosis, the pseudopalisading cells co-localize to fulfil their oxygen demand. This vicious cycle is the primary malignant pathway in cases of glioma. The presence of hypoxia also causes the activation of perivascular glioma cells. This leads to the generation of pro-angiogenic molecules that drive the formation of a new vasculature that increases and hastens the progression of GBM. Excessive vascular endothelial growth factor (VEGF) expression also causes endothelial cell over-proliferation, resulting in leaky and damaged blood vasculature. This inhibits the delivery of oxygen, medicines, and immune cells in the tumour microenvironment.

9.2.2 Oncogenic Pathways

Many critical cellular functions, such as growth, differentiation, metabolism, and survival, are regulated by signal transduction pathways. In human cancer, many of these signalling pathways are disrupted. The biology and clinical behaviour of GBM have tremendously benefited from research into these signalling pathways (Figure 9.2). A more comprehensive understanding of signal transduction will aid in the development of innovative therapies for this deadly disease.

In patients with GBM, the expression of 12 genes were found to be strongly associated with the characteristic mortality rate. These 12 genes are E2F2 (cell cycle signalling pathway), CTBP2 (Notch signalling), MAFF (Nrf2 signalling), SLC2A3 (Nrf2 signalling), ECSIT (PI3K signalling), HSP90B1 (PI3K signalling),

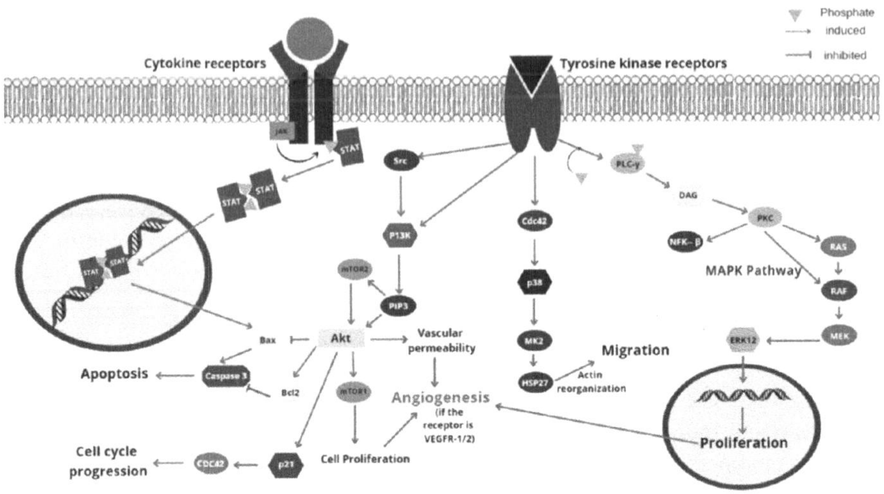

FIGURE 9.2 Oncogenic signalling pathways involved in GBM.

TNFRSF1A (PI3K signalling), PAK1 (RTK signalling), ID4 (TGF- signalling), DDB2 (p53 signalling), MDM2 (Han et al, 2022; Sanchez-Vega et al, 2018). Also, various growth factor receptors corresponding to different oncogenic signalling molecules like EGFR, PDGFR, and VEGFR are overexpressed in cases of GBM (Nakada et al, 2007). Mesenchymal-epithelial transition factor (c-Met) is a type of receptor tyrosine kinase that is needed for epithelial-to-mesenchymal transition (EMT) during developmental stages. c-Met shows abnormal expression in malignant gliomas like GBM which can be attributed to its aggressive malignancy. These growth factor receptors are involved in RTK type of paracrine/autocrine signalling. Some other RTKs unique to GBM are tyrosine kinase with immunoglobulin-like and EGF-like domains 1 (TIE1), discoidin domain receptor (DDR1), epoxide hydrolase (EPH1), and Axl. These play an important role in glioma invasion by suppressing the immune system. Recently, researcher identified ring-like contrasts which surround the tumours in MRI images and concluded that these contrasts indicate angiogenic changes during the progression of GBM. Along with VEGF and VEGFR, fibronectin, tenascin-C, and vitronectin along with cathepsin B and matrix metalloproteinases (MMPs) are the key angiogenesis factors in glioblastoma. These prevent ECM degradation via the regulation of collagenase expression. Aberrant activation and overexpression tumour suppressor protein phosphatase and tensin homolog (PTEN), sphingosylphosphorylcholine (SPC), Rho kinases, and GPCRs of RTK/PI3K/Akt cascade (refer Figure 9.1) are prominent in cases of GBM (Sanchez-Vega et al, 2018). Integrins like β1 and β5 are linked with invasion and migration. RAC of the Rho GTPase family is involved in lamellipodial formation which is important in tumour migration. Identifying the mechanisms that contribute to the increase of oncogenic signalling and other pathways during glioma growth is critical. Such researches lead to the discovery of new prospective targets, better therapeutic regimen selection, and personalized medication.

9.3 THERAPEUTIC APPROACHES TO GBM

Every year cancer claims millions of lives globally. Due to this, innovative and novel cancer therapeutic approaches are underway. Diagnostic technologies play an important role in the development of therapeutic strategies for cancer treatment. Surgery, radiation therapy, and chemotherapy are the most common treatment options whereas hormonal therapy and immunotherapy are some of the recent options. GBM is a type of brain cancer that is the cause of many hindrances during the development of a therapeutic approach. Some of them are:

- Intracranial malignancy.
- The danger of irreparable damage to the brain during surgery.
- The blood-brain barrier (BBB) prevents drugs from entering the brain.
- ATP-binding cassettes efflux drugs from the BBB.
- Resistance of tumour cells to chemicals.
- Mutations in isocitrate dehydrogenases inhibit complement-mediated tumour repression.

Thus, the majority of the treatment options are focused on symptomatic relief (palliative treatment) and improving the patient's life expectancy. While designing any treatment option for cancer, certain factors should be taken into consideration. These are tumour location, potential symptoms, potential benefits, and associated limitations.

9.3.1 Surgery

Surgery is one of the oldest yet the go-to options during the initial stages of cancer. Chemotherapy and radiation therapy follow surgery if only surgery is not an option. The surgery is carried out to remove the tumour and some of the surrounding tissue to prevent metastasis. But it is hardly effective when the tumour has undergone aggressive metastasis. There are three stages to surgery. The first is the diagnosis in which a biopsy is used to diagnose the type of cancer. The second is staging where the size, metastasis rate, and other parameters are evaluated. The last step is the surgical removal of the tumour by debulking or palliative surgery. Maximal safe resection is the most effective and safe surgical procedure in which the glioma is excised while ensuring maximum preservation of the surrounding neurological tissues. This can also be achieved by the use of 5-aminolevulinic which defines the tumour region for safe resection. But before this, some preoperative procedures like MRI, diffusion tensor imaging, and navigated transcranial magnetic stimulation need to be conducted to effectively map the surgical procedure.

9.3.2 Radiation Therapy

Radiation therapy is typically administered following surgery. Comparative studies have shown that combining surgery and radiation therapy is more beneficial than either procedure alone. If surgery is not an option due to the location of the tumour, radiation can be considered the sole therapeutic option.

- TMZ is used in conjunction with radiotherapy to improve the tumour's sensitivity to radiation.
- The current standard of therapy includes over six to weeks with three-dimensional conformal beam or intensity-modulated radiotherapy (RT) (2 Gy per day for a total dose of 60 Gy) with concomitant TMZ chemotherapy at a dose of 75 mg per square meter of body-surface area per day, 7 days per week followed by a maintenance dose.
- Proton therapy which uses charged particles (protons), is used to treat deep-brain gliomas.

9.3.3 Chemotherapy

Due to high mortality and limited therapeutic options for GBM, innovative and novel cancer therapeutic approaches are underway. While developing the treatment regime, particularly for GBM, many factors need to be considered. These are as follows:

- BBB permeability.
- ATP-binding cassettes (e.g., P-gp) which efflux drugs from the BBB.
- Resistance of tumour cells to chemicals.
- Mutations in isocitrate dehydrogenases inhibit complement-mediated tumour repression.

Chemotherapy, one of the most widely explored treatment options in cancer, has unfortunately not shown much of a positive effect on GBM. Figure 9.3 illustrates the mechanism of action of TMZ (a prodrug), an alkylating and antineoplastic medication that is taken orally. It has a bioavailability score of 0.55. It is also used during the treatment of recurrent or progressive GBM or anaplastic astrocytoma. About 50% of the patients treated with TMZ do not respond positively due to overexpression of its metabolite MGMT and lowered rate of DNA repair due to overexpression of Mdm2, a p53 antagonist (Jiapaer et al, 2018).

The activity of the gene MGMT is a major reason for the less efficacy of TMZ treatment. MGMT encodes a protein that repairs the alkylated DNA (repair at the alkylated O^6 position of guanine) in turn promoting tumour growth.

About 50% of the patients treated with TMZ do not respond positively due to overexpression of its metabolite MGMT and lowered rate of DNA repair due to overexpression of Mdm2, a p53 antagonist (Jiapaer et al, 2018). One of the most important hindrances in TMZ therapy is the activity of the gene MGMT. MGMT encodes a protein that repairs the alkylated DNA (repair at the alkylated O^6 position of guanine) in turn promoting tumour growth.

Bevacizumab, a humanized monoclonal antibody, is used to suppress angiogenesis (targets VEGF) but has not shown much efficacy in GBM. Similarly, cilengitide, the $\alpha v\beta 3$, and $\alpha v\beta 5$ integrin inhibitor is a pentapeptide that has passed multiple clinical studies in GBM but has not been further explored as an anticancer medicine because

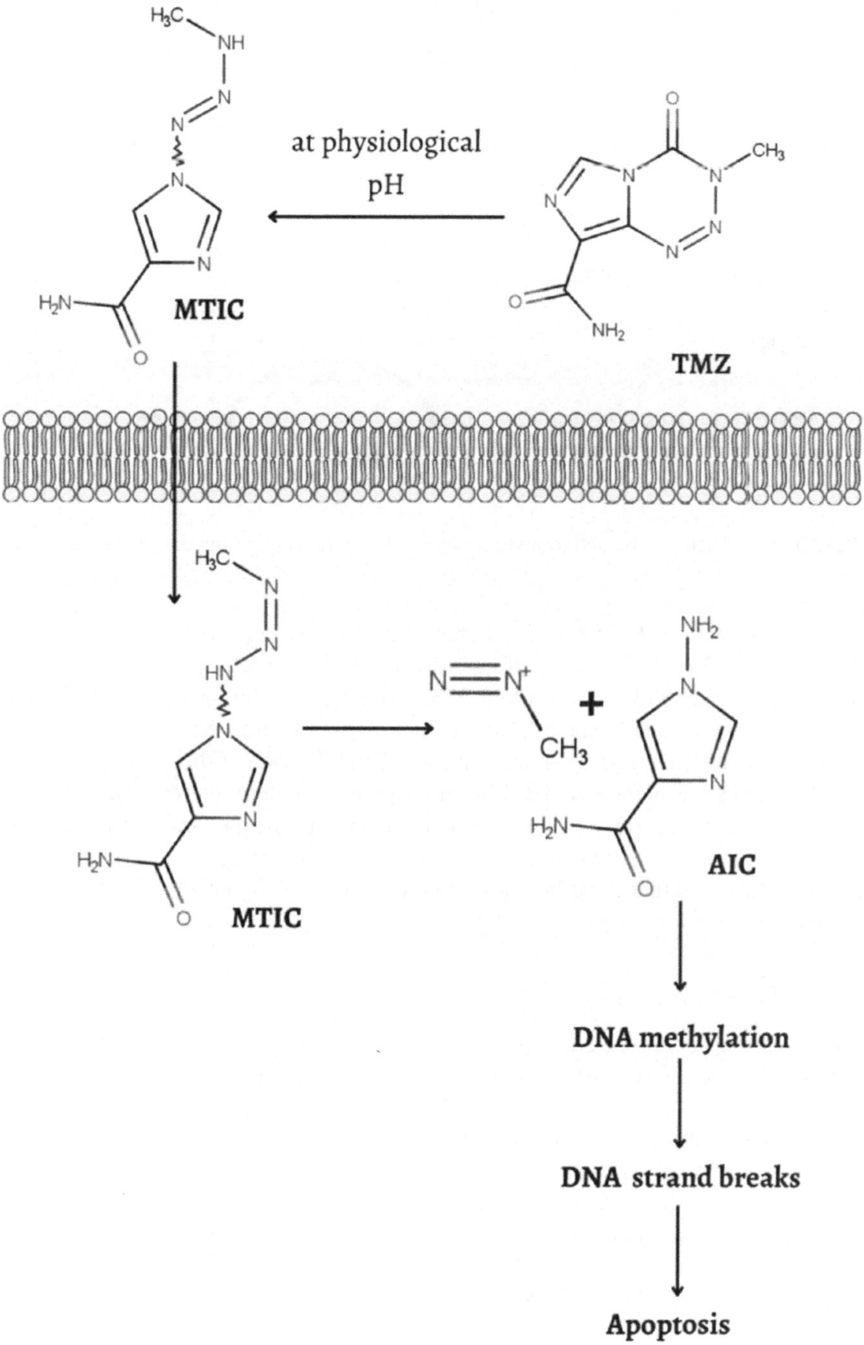

FIGURE 9.3 Mechanism action of TMZ against GBM.

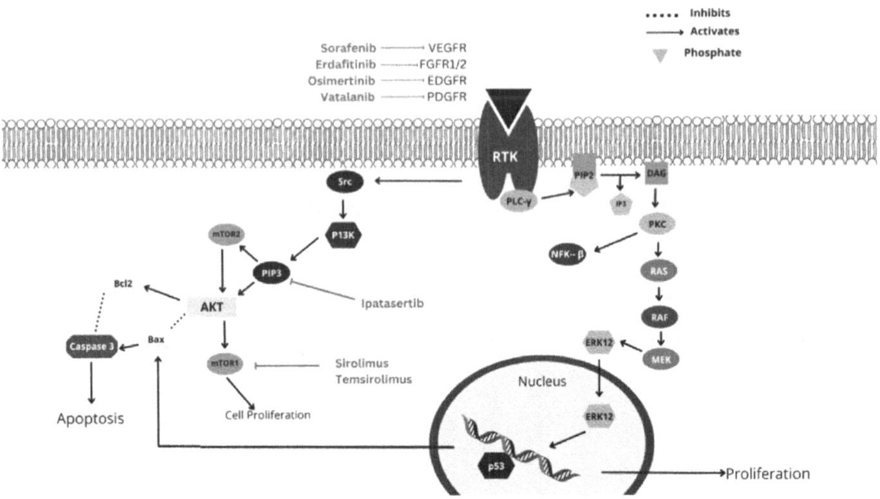

FIGURE 9.4 Different bioactive small molecule compounds targeting peripheral molecular targets.

its combination with TMZ failed to enhance treatment outcomes in a phase III trial (Nabors et al, 2015). Figure 9.4 illustrates molecules targeting peripheral tumour targets like tyrosine kinase receptors, PI3K, and mTOR. But these drugs have relatively high toxicity and adverse effects as the targets are required extensively in normal cellular function. Furthermore, with the rising lifespan of the world population and the higher prevalence of GBM in senior patients, elderly patients have become one of the reasons for poor prognostic factors in GBM. The physical health of senior patients, as well as whether medication damage to the ageing brain is irreversible, are considerations that must be considered during the drug development and optimization process.

9.3.4 Shortcomings of Chemotherapy

In chemotherapy, the type and the concentration of the drug decide the side effects of that drug. These can be acute as well as chronic. Some of these side effects include:

1. Immunosuppression: Depression of the immune system accompanied by myelosuppression and neutropenia.
2. Hair loss: One of the most common side effects. Continuous and high doses of drugs like doxorubicin and paclitaxel can lead to alopecia totalis (complete loss of facial and skull hair).
3. Teratogenicity: Physical abnormalities in the foetus during pregnancy (generally first trimester) are called teratogenicity. For example, arsenic and thalidomide (promyelocytic leukaemia).
4. Organ toxicity: Most common are cardiotoxicity, hepatotoxicity, and nephrotoxicity due to anti-cancer drugs like doxorubicin and epirubicin. Myelotoxicity is a side effect of 5-fluorouracil.

5. Post-chemotherapy cognitive impairment: Characterized by fatigue and an overall decrease in concentration and other neurocognitive issues.

Some of the major limitations of chemotherapy in cancer are as follows:

1. Shows limited application in stage 4 cancers.
2. Permeation through the BBB: Only small and lipophilic molecules like TMZ can cross the BBB and the cerebrospinal fluid. The presence of transporter pumps like p-glycoprotein expels drugs from the BBB into the surrounding capillaries. P-glycoprotein, also known as ATP-binding cassette sub-family B member 1 (ABCB1), expels xenobiotics from the intestinal cells to the lumen, from hepatocytes in the bile ducts, and reabsorbs the drugs from the convoluted tubules in the peritubular capillaries.
3. DNA repair enzymes: These enzymes help in the repair of DNA damage. But, alkylating agents like TMZ also induce DNA damage to induce apoptosis. But the DNA repair enzymes also tend to repair this damage thus inhibiting apoptosis.
4. Gene over-amplification: The anti-cancer drug fails to prevent multiple expressions of the amplified genes. This reduces the drug's efficacy over time.

These limitations have led to the development of various other techniques which can help in increasing the efficacy of chemotherapy. Nanomedicine deals with the development of biocompatible systems which can help in targeted drug delivery. Use of extracellular vesicles like exosomes which act as biologically compatible and low immunogenic nanocarriers for small and low molecular weight drugs. The use of phytopharmaceuticals either coupled or independent has also shown promising effects, for example, curcumin, due to its chemoprotective and cytotoxic effects, has been used as an adjuvant in many drug combinations. But these all techniques are under study and research as much is yet to be elucidated to fully adopt these systems as anti-cancer strategies

9.4 BIOACTIVE PHYTOPHARMACEUTICAL AS ADJUNCTS TO CHEMOTHERAPY

A phytopharmaceutical drug is a purified and standardized fraction of an extract of a medicinal plant or part of it that has a minimum of four bioactive or phytochemical compounds that have been tested for their quality and quantity. This fraction can be used internally or externally in humans or animals to diagnose, treat, mitigate, or prevent any disease or disorder, but parenteral administration is not included. So, it becomes imperative to discuss the bioactives or phytochemicals used in these phytopharmaceuticals. While dealing with cancer, it is important to note that it involves many signalling cascades and protein interactions at the cellular as well as the genetic level. Therefore, it becomes very difficult to develop targeted therapy. But here is where the bioactives have an upper hand. One of the most important properties of natural products is a multi-targeted effect and less toxicity. Several bioactive chemicals derived from medicinal plants have also been found to inhibit angiogenesis and delay metastasis. According to reports, bioactive substances affect carcinogenesis-related intracellular signalling too. Moreover, the effective doses of most phytopharmaceuticals do not exert any toxicities or side effects making them ideal anticancer agents.

9.4.1 Resveratrol

Resveratrol (3,4′,5-trihydroxy-trans-stilbene) is a polyphenolic phytochemical. "French Paradox" was where the role of resveratrol (refer Figure 9.5) present in red wine was first considered to be cardioprotective. Multiple clinical trials have evaluated and confirmed its anti-oxidant, analgesic, cardioprotective, and chemo-preventive properties (Kiskova et al, 2020; Liu et al, 2015; Riba et al, 2017). Its diversified functions are attributed to its glycosylation and half-life which results in its high permeability through the GIT. Resveratrol readily binds to the human serum albumin and various lipoproteins, a step necessary for its transport and cellular absorption. After absorption, it accumulates in tissues that are its target sites namely colorectal tissues, breast, and cardiac tissues (Berman et al, 2017). In the brain tissues, it acts as an antineoplastic agent by suppressing oxidative stress due to reactive oxygen species (ROS) and inflammation.

In 2002, Wang et al proved that resveratrol crosses the BBB via the tight junctions and mainly acts on the NF-κB in the neuronal cells. By inhibiting NF-κB, it significantly reduces the formation of MGMT protein and reduces TMZ degradation (Frémont, 2000). Tumour cells use autophagy to avoid apoptosis-mediated cell death by getting rid of toxic molecules in the neoplasm. Resveratrol has been shown to increase TMZ efficacy by reducing autophagy and inducing ROS and ERK-mediated apoptosis. One of the most important events in apoptosis is mitochondria dysfunction. Figure 9.6 shows various modes of action via which resveratrol induces

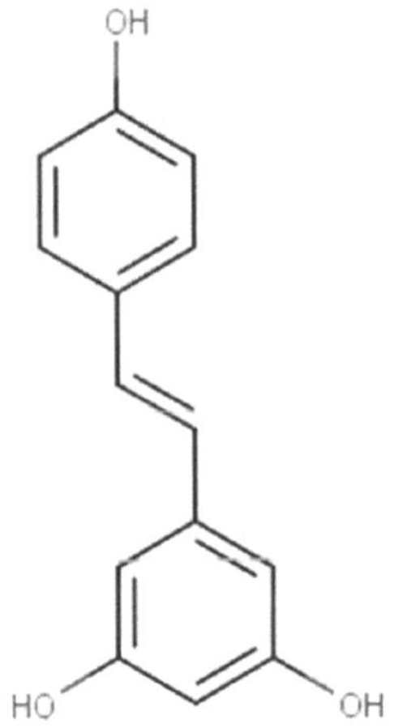

FIGURE 9.5 Structure of resveratrol.

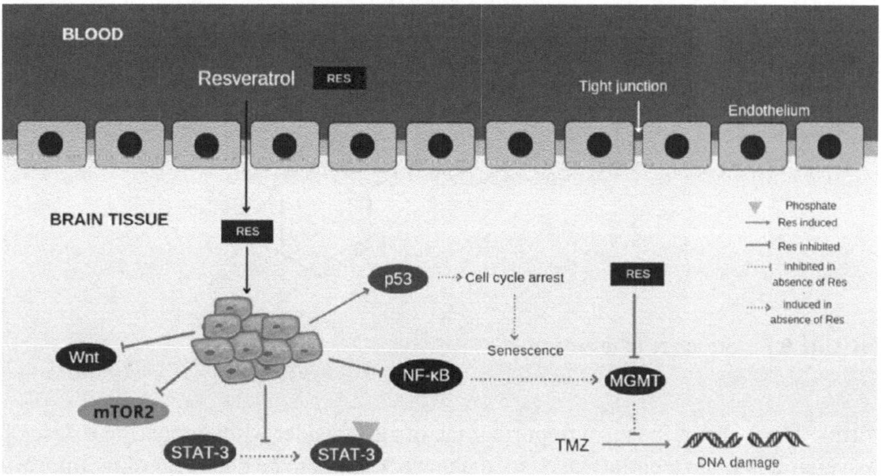

FIGURE 9.6 Mechanism of resveratrol in controlling oncogenic signalling.

apoptosis in cancer cells. Resveratrol treatment results in a loss in membrane potential of cancer cell mitochondria that subsequently leads to the rapid release of cytochrome C from the mitochondrial matrix to the cytoplasm (Ginkel et al, 2007). This induces apoptosis mediated by caspases 3 and 9 (Maria et al, 2009). Inhibition of mTOR signalling, downregulation of Bcl2, and activation of p53-mediated apoptosis are some mechanisms by which resveratrol-induced mitochondrial dysfunction enhances TMZ activity (Goffart et al, 2013). uPAR (urokinase plasminogen activator receptor), a cell membrane lipid anchored protein, when associated with uPA, induces MMP-9 formation. MMP-9 is often associated with tissue remodelling, tumour cell adhesion, migration, and invasion which accounts for its overexpression in GBM. Resveratrol induces ECM degradation by NF-κB activation and uPA downregulation thus validating its anti-invasive properties (Ryu et al, 2015). When NF-κB gets inhibited, a cascade inhibition of Bcl-2, VEGF, and MMP-9 occurs. This downregulation can be traced back to resveratrol-mediated inhibition of TNF-α which otherwise through NF- κB activation activates uPA and subsequent malignant glioma-promoting factors (Filippi-Chiela et al, 2013). Resveratrol also demonstrates p53-independent G2-M arrest thus inducing senescence in the cancer cell. It also acts on Wnt signalling and the JAK-STAT pathway by suppressing the phosphorylation of STAT-1 and STAT-3. In the later stages of cancer, that is, malignancy, it can also function as a cytotoxic agent due to inhibition of angiogenesis and by halting tumour invasion and metastasis.

Rapid metabolism and poor bioavailability when administered orally are the major hindrances in the development of resveratrol as a pharmaceutical formulation (Walle et al, 2004). At lower concentrations, it shows anti-oxidant behaviour, but at prolonged exposure to a high concentration of resveratrol, proteolytic cleavage and DNA damage are observed (Schilder et al, 2009). It also exhibits pro-oxidant effects like increased lipid peroxidation and inhibition of P450 cytochromes (Gadacha et al, 2009). It can also interact with other drugs and either attenuate them or form toxic

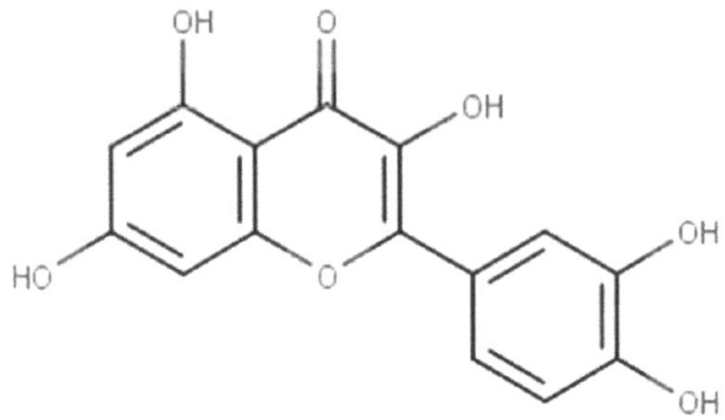

FIGURE 9.7 Structure of quercetin.

forms. Thus, one of the most important factors to consider while preparing a dose of resveratrol is its concentration. Also, much work needs to be done regarding improving the pharmacokinetics of resveratrol.

9.4.2 Quercetin

Quercetin (3,3′,4′,5,7-pentahydroxy-flavone) is a polyphenolic flavonoid and is present in many foods like apples, cherries, onions, (refer Figure 9.7) citrus fruits, and many herbs. It exhibits lipid peroxidation, platelet aggregation, capillary permeability as well as anti-neoplastic and anti-bacterial activities. Quercetin glucoside, the naturally occurring form of quercetin, gets absorbed in ranges only from 3% to 17%. So, limited absorption, high metabolism, and excretion are the reasons for quercetin's low bioavailability (Scholz and Williamson, 2007).

Hsp27, a heat shock protein is overexpressed in cancer and reduces the efficacy of the TMZ regime. Hsp27 is also associated with increased chemoresistance and suppression of apoptosis-mediated death of neoplastic cells (Acunzo et al, 2012). Phosphorylation of Hsp by many protein kinases and MAPKs activates quercetin has been reported to prevent the phosphorylation of Hsp27 which inhibited the SPARC-induced glioma cell invasion (Sang et al, 2014). Quercetin elevates the intracellular levels of ROS and Ca^{2+} leading to the depolarization of mitochondrial membrane potential and subsequent release of cytochrome C. This induces apoptosis mediated by caspases 3 and 9 which support the TMZ regime (Kiekow et al, 2016). Release of the caspase proteins can also be attributed to another activity of quercetin, that is, a change in mitochondrial membrane protection which also leads to an increase in p53 as well as c-Jun kinases. It has also been reported that pre-exposure to quercetin increased the anti-proliferative activity of TMZ. Thus, the quercetin-TMZ treatment regime has been shown to increase the overall efficacy of the treatment of GBM.

Quercetin also arrests the cell cycle, inhibits cell proliferation, and induces apoptosis (refer Figure 9.8). Apart from its function as an adjunct in TMZ therapy, it has various other anti-cancer mechanisms. Yamashita and Kawanishi (2000), reported the DNA topoisomerase inhibitory activity of quercetin. It is also reported to block the NF-κB, Ras-MAPK-ERK, and PI3K-AKT signalling cascade thus inhibiting cancer cell proliferation and angiogenesis (Santos et al, 2015). Quercetin is involved

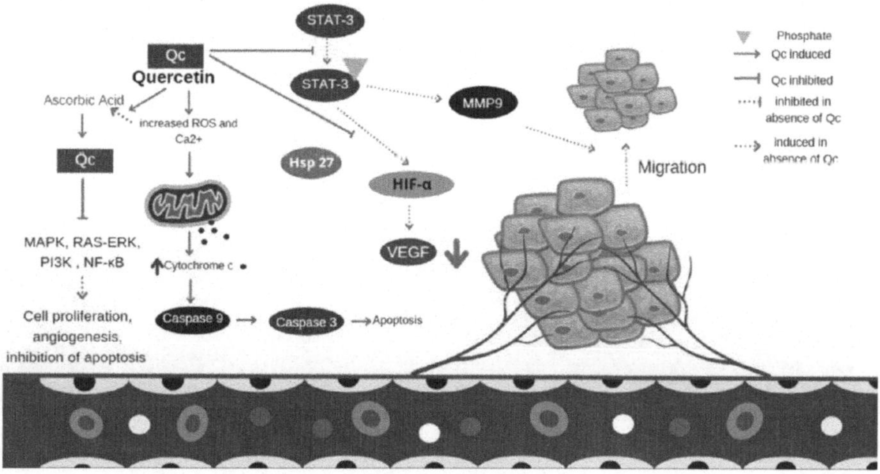

FIGURE 9.8 Anti-cancer activity of quercetin in GBM.

in the deregulation of angiogenic factors like VEGF through inhibition of the VEGFR2-regulated AKT/mTOR cascade. It also causes inhibition of MMP-2 and MMP-9 which helps in lowering matrix degradation and tumour cell metastasis. It also arrests the cell cycle by inhibiting CDK4 and cyclin D along with p53 activation.

But, low bioavailability, rapid degradation, and systemic elimination are some associated limitations. To overcome the problem of quercetin's low bioavailability, quercetin-derived nanoparticles are administered intravenously thus ensuring maximal concentration of quercetin reaching the BBB (Liu et al, 2022). Some other advantages of the nanoparticle mode of delivery are intrinsic tumour vessel targeting, high drug loading, and disruption of pro-angiogenic and angiogenic factors as well as the existing vasculature (Tee et al, 2019; Vafadar et al, 2020). This suggests a vessel-targeting approach as a result of improving the delivery of this bioactive polyphenol.

9.4.3 Curcumin

Curcumin (diferuloylmethane), is a natural sesquiterpene present abundantly in the rhizome *Curcuma longa*, that is, (refer Figure 9.9) turmeric. Curcumin has anti-inflammatory, antioxidant, anticancer, antiseptic, and astringent activities due to which turmeric has been in use for centuries as a therapeutic agent. Rather than using turmeric as a powder, curcumin is isolated and used as an anti-cancer drug due to its ROS scavenging, anti-proliferative, and anti-invasive activity reported in breast cancer, lung metastases, and brain tumours (Bachmeier et al, 2010). It shows such diversified functions due to the presence of three reactive functional groups, namely two phenolic moieties and a 1,3-diketone moiety (Priyadarsini, 2013).

DNA alkylation and ROS generation are the two main mechanisms by which TMZ induces apoptosis in cancer cells. But the cancer cells remodel the electron transport system and enhance the activity of cytochrome c and complex I and II which leads

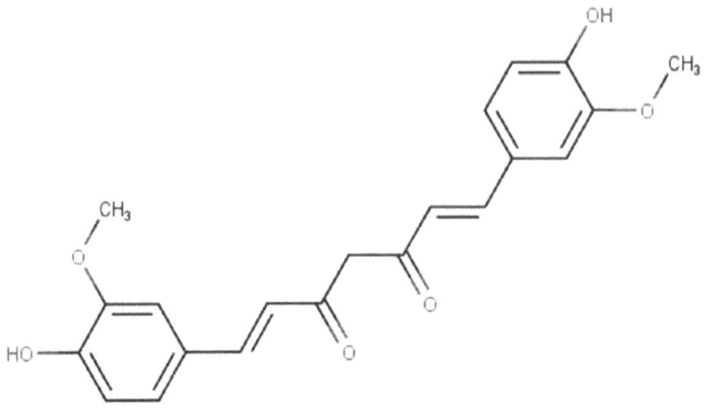

FIGURE 9.9 Structure of curcumin.

to a decrease in oxidative stress which reduces the efficacy of TMZ (Oliva et al, 2011). But in many in-vivo experiments, when resveratrol was used synergistically with TMZ, it showed increased production of ROS and increased TMZ efficacy (Yin et al, 2014). Another hindrance in TMZ treatment is the overexpression of Akt (protein kinase B) which leads to Cdc25C/Cdc2 inactivation and decreased G2 arrest (Hirose et al, n.d.). Figure 9.10 explains the role of curcumin in inhibiting oncogenic signalling molecules. Curcumin inhibits PI3K, thus inhibiting Akt formation which makes cancer cells more prone to TMZ-mediated DNA and subsequent apoptosis (Maiti et al, 2019). Curcumin targets multiple pathways like MAPK, JAK-STAT, Wnt, and PI3K/Akt explains its anti-proliferative, anti-metastatic, and pro-apoptotic activity. Curcumin inhibits the overexpression of PTEN which inhibits the PI3K/Akt

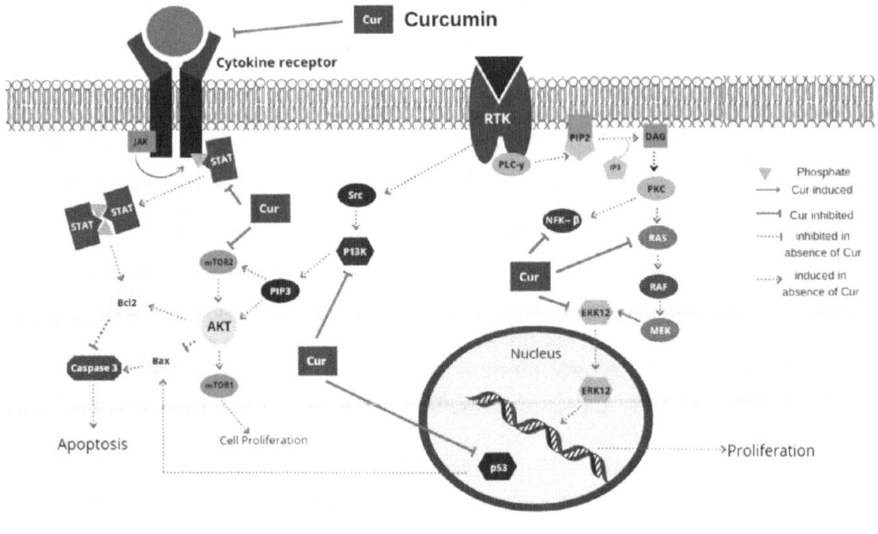

FIGURE 9.10 Role of curcumin in inhibiting oncogenic signalling in GBM.

pathway leading inhibition of Akt and mTOR (Maiti et al, 2019). This leads to the downregulation of Bcl2 causing apoptotic cell death. Curcumin is also reported to reduce cancer cell proliferation and survival via inhibition of NF-κB as it is associated with overexpression of EGFR and c-Jun N-terminal kinases (Dhandapani et al, 2007). Thus, inhibition of NF-κB potentiates the anti-tumour effects in drugs like paclitaxel (Fratantonio et al, 2019). By inhibiting the JAK/STAT pathway, curcumin inhibits tumour cell migration and invasion (Mao et al, 2012). It is also involved in the activation of caspase 3 and 9 mediated apoptotic cascades (Hande et al, 2013).

Water solubility and physicochemical stability limit the bioavailability of curcumin. It has also been determined that curcumin can cross the BBB. Various delivery systems like nanoparticles and liposomes are being experimented upon for more efficient delivery of curcumin. Another problem associated with TMZ/curcumin regime is that it induces autophagy. The use of resveratrol helps reduce autophagy and increases the efficacy of TMZ (Ryskalin et al, 2020). At higher concentrations, curcumin exhibits cytotoxicity leading to the death of the cell. So, monitoring the levels of curcumin is also of utmost importance.

9.4.4 Boswellic Acids

Boswell acids 3-O-acetyl-11-keto-β-boswellic acid (AKBA), is extracted from the resin of *Boswellia serrata* and some other species of the *Boswellia* genus. Many boswellic acid derivates are currently in use. Boswellic acid is a very potent anti-inflammatory molecule that binds to COX-2 and (refer Figure 9.11) inhibits prostaglandin synthesis. It also possesses analgesic, antipyretic, and platelet-inhibitory activities. In cancer, it is used as a cytotoxic and antitumour drug. It is also used for the treatment of tumour-associated oedema (Li et al, 2020). As discussed earlier, autophagy is one of the mechanisms by which cancer cells evade apoptosis.

Figure 9.12 discusses the various mechanisms by which AKBA inhibits autophagy by increasing mTOR and decreasing p53 and ERK expression and regulating the ERK/mTOR and P53/mTOR cascades leading to apoptosis. AKBA also reduces the abnormal accumulation of glucose in cancerous cells (Warburg effect) thus inhibiting the growth of tumour cells. It is also involved in the inhibition of p21 which leads to subsequent inhibition of Cdk1 which stops cell cycle progression. Cdk1 inhibition is also involved in sensitizing the cell to DNA damage thus initiating the signal for apoptosis (Schneider and Weller, 2016). Other mechanisms by which AKBA induces apoptosis are through BAD and Akt inhibition which doesn't allow Bcl2-BAX complex formation thereby inducing caspase 3 and 8 mediated apoptosis (Ravanan et al, 2011). Another mechanism is by blockage of p65, a nuclear protein that normally inhibits apoptosis (Janssen et al, 2000). Inhibition of topoisomerases I and II is another activity by which uncontrolled DNA replication is ceased by boswellic acid which also induces the signal for apoptosis (Ali et al, 2015). AKBA also induces p21 expression which recognizes DNA damage caused by TMZ and induces apoptosis independent of Bcl2 -BAX. NF-κB and Akt play an important role in downregulating TMZ-induced chemoresistance. Boswellic acid downregulates NF-κB and Akt (from Figure 9.12) thereby increasing the sensitivity of the cells to TMZ (Roy et al, 2017).

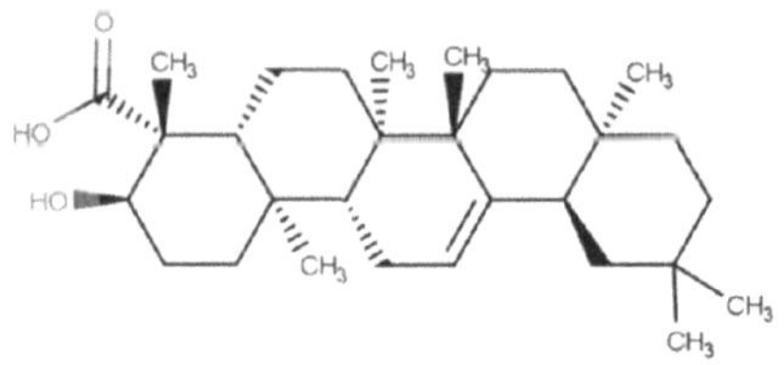

α-boswellic acid

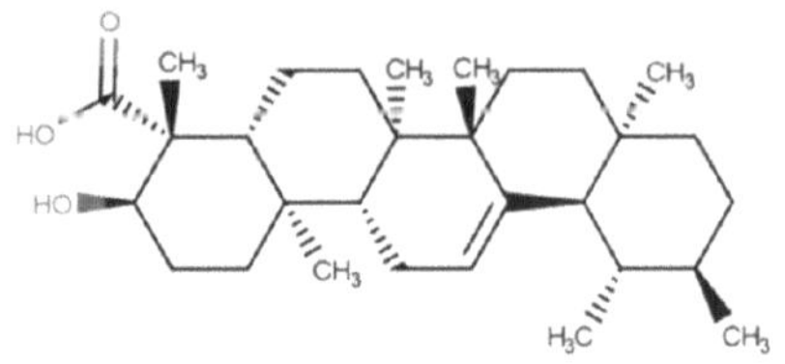

β-boswellic acid

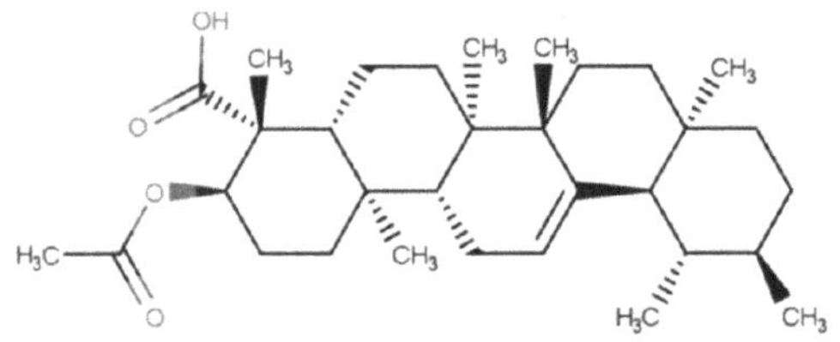

3-Acetyl-beta-boswellic acid

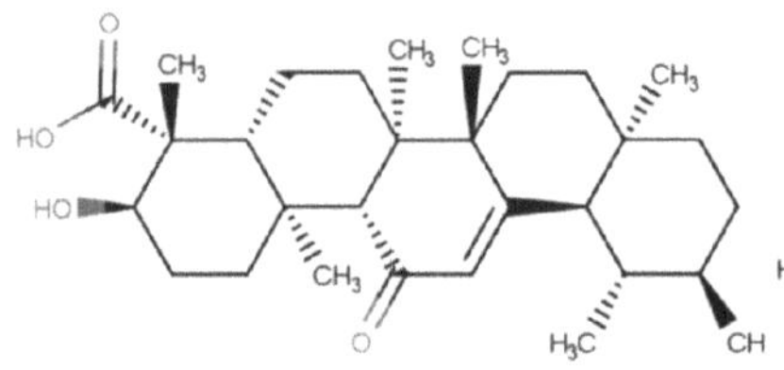

11-Keto-beta-boswellic acid

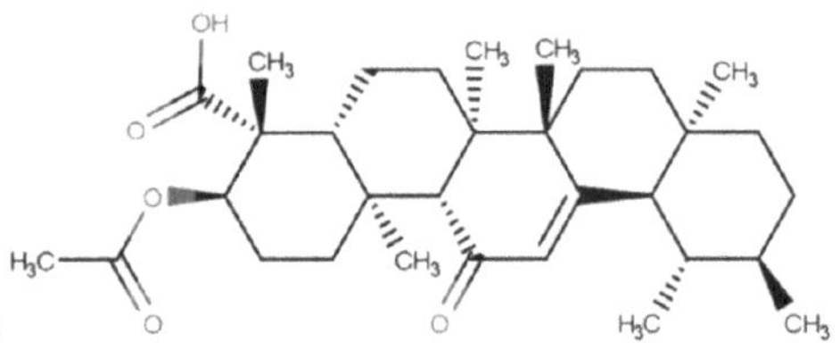

3-Acetyl-11-keto-beta-boswellic acid

FIGURE 9.11 Structures of boswellic acid derivatives.

One of the primary areas of concern is the bioavailability of AKBA which needs to be improved. According to the findings of Ritschel (1980), after oral administration, a stable plasma concentration of AKBA is achieved in thirty hours. This can be attributed to the high lipophilicity of boswellic acids (Krüger et al, 2008). Liposomes, emulsions, solid lipid nanoparticles, nanostructured lipid carriers, micelles, and lactic-co-glycolic acid nanoparticles are all examples of nanoparticle delivery methods that can be used as targeted delivery options (Aqil et al, 2013; Roy et al, 2019).

9.4.5 Epigallocatechin Gallate

Catechins are phenolic compounds belonging to the flavan-3-ols family of flavonoids. Epigallocatechin gallate (EGCG), an ester of epigallocatechin and gallic acid, is the major catechin present in green tea, pomegranates, vinegar, berries, and peaches. It has shown anti-oxidant, anti-inflammatory, anti-diabetic, and antiproliferative effects. Paired with its anti-oxidant activities, EGCG's metal chelation activity helps in preventing neurodegenerative diseases. It has also exhibited anti-tumorigenic activity and is used in combination with TMZ as a chemotherapeutic agent.

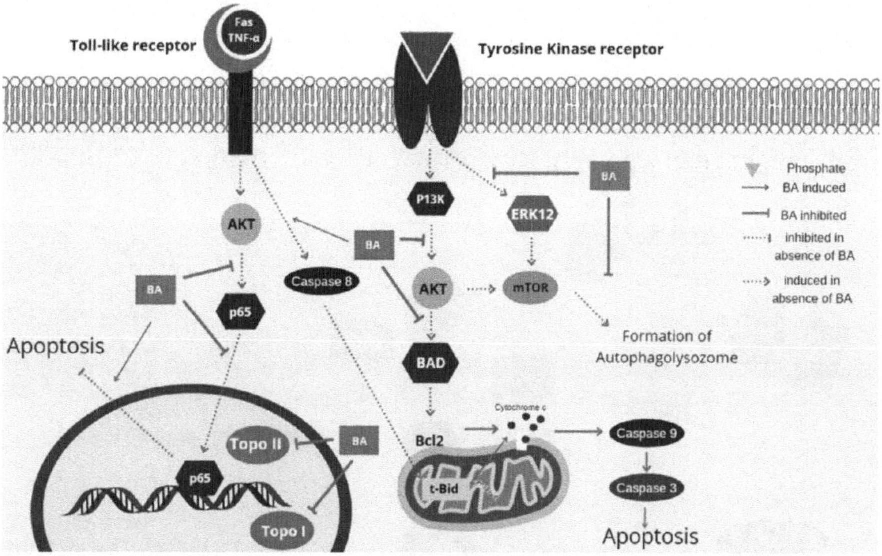

FIGURE 9.12 Mechanism of boswellic acids in inducing apoptosis in tumour cells.

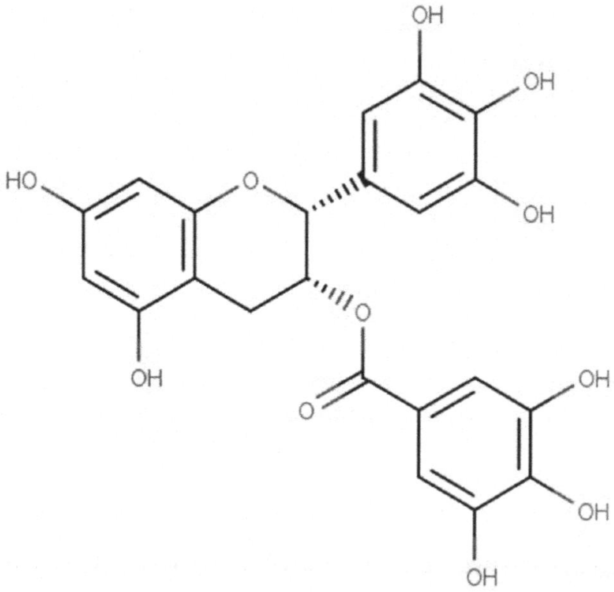

FIGURE 9.13 Structure of epigallocatechin gallate.

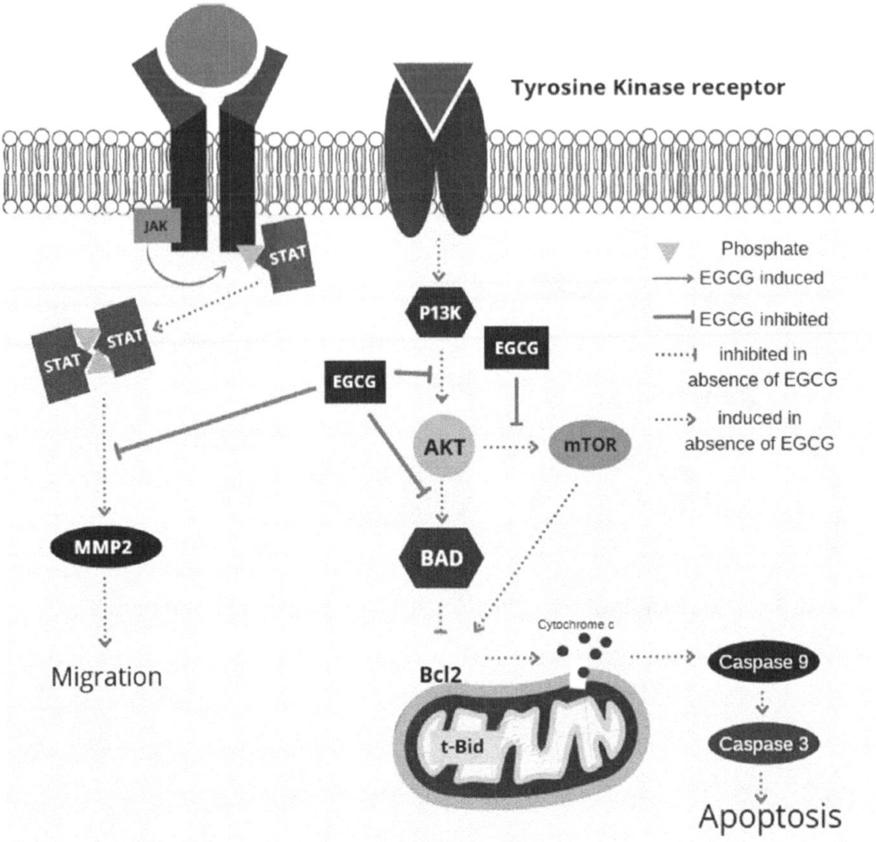

FIGURE 9.14 Effects of epigallocatechin gallate (EGCG) on GBM metastasis.

In cancer such as gliomas, BBB permeation is an important factor that determines a drug's efficacy. EGCG is known to increase BBB permeation of TMZ by inhibiting P-gp expression (Zhang et al, 2015). EGCG inhibits PI3K and/or mTOR kinases and downregulates p-Akt, Bcl-2 promoting the expression of the pro-apoptotic protein Bax which induces apoptosis (Aller et al, 2011). Figure 9.14 depicts the mechanism by which EGCG induces apoptosis by inhibiting NADPH-reducing enzymes. This results in a reduction of the ROS scavenging capacity of the cell leading to cell death via phosphorylation of p38 and JNK (refer Figure 9.14) (Zhang et al, 2015).

EGCG also inhibits the signalling of PDGFR as well as that of various metalloproteinases, cytokines, and chemokines which inhibits cancer cell invasion of tissues. In cancerous cells, high telomerase levels maintain telomere length. It was showed that the addition of EGCG significantly reduced the expression of telomerase mRNA in glioma cells. EGCG inhibits the phosphorylation of PDGFR in glioma cells by blocking PDFG from binding to its receptor which reduces cell proliferation (Weber et al, 2004). MMP-2 is a major protease involved in the invasive behaviour of cells. Inhibition of MMP-2 activation by EGCG leads to reduced invasiveness of glioma cells (Mook et al, 2004). The stability and bioavailability of EGCG are important points of discussion. This is because it is prone to degradation in the body fluids via epimerization and auto-oxidation (Krupkova et al, 2016) Conditions like low temperature and pH of 2–5.5 maintain the stability of EGCG. However, due to a change in pH when it is administered orally, the oral bioavailability of EGCG is low. The bioavailability of EGCG is enhanced when administered along with piperine, that is, co-treatment. Studies based on other strategies like the use of pro-drugs, encapsulation, and chitosan-based nanocarriers have also demonstrated improvement in EGCG's pharmacokinetics.

9.5 EMERGING NANOTECHNOLOGY DELIVERY APPROACH FOR BIOACTIVE PHYTOPHARMACEUTICAL: A POTENTIAL THERAPEUTIC STRATEGY TO GBM

Till date, delivery of bioactive phytopharmaceuticals to brain is a challenging task for scientists. There are numerous barriers such as BBB, poor pharmacokinetic profile, highly hydrophilic compound, less selective that hinder the effective management of GBM. In case of GBM, due to the presence of BBB in brain, the phytopharmaceuticals are unable to reach to tumour site and are restricted to specific targets to bind to the GBM stem cells. Due to this issue, the conventional approaches fail to attend the therapeutic efficacy against various diseases. Despite having a wide spectrum of anticancer actions, bioactives like quercetin have poor water solubility, low bioavailability, prone to oxidative degradation, and severe biotransformation, which makes it difficult to use in vivo. Nano formulations have exhibited significant advances in the delivery of such non-polar molecules. These include high encapsulation efficiency, longer circulation time, tumour-specific biodistribution, controlled release, and increased therapeutic efficacy. As a result, various modified nanoparticles have been developed which are considered as major breakthroughs in quercetin delivery for anticancer therapy. Liposomes, polymeric micelles, PLGA nanoparticles, metal-organic frameworks, inorganic nanoparticles, biomacromolecule-based nanoparticles, and other nanoparticles have been produced since then to deliver quercetin and boost its antitumour action (Zang et al, 2021). Curcumin's efficiency is hampered by its limited absorption, despite its potential anticancer actions. Several study groups have detected very low plasma and extra-intestinal levels of curcumin. Hence, to maximize its solubility and absorption, it is integrated into formulations such as solid dispersion, nanoparticles, micelles, conjugates, and liposomes. The efficacy of nanoformulations for GBM was demonstrated by one study group which used curcumin-loaded poly(lactic-co-glycolic acid)-1,2-distearoyl-glycerol-3-phospho-ethanolamine-N-[methoxy (polyethylene glycol)-2000] ammonium salt (PLGA-DSPE-PEG) hybrid nanoparticles in rats. It reported a significantly lower size of the tumour after five days of injection when compared with the nontreated control group (Orunoğlu et al, 2017). Another study found that curcumin-laden targeted liposomes crossed the BBB twice as much as non-targeted liposomes loaded with curcumin (Gabay et al, 2021). Some of the obstacles to effective GBM treatment are not dependent on the type of tested therapy. Indeed, they are due to: (i) A lack of preclinical models that are close enough to human GBM, (ii) the difficulty in conducting clinical trials on a large enough number of patients to achieve statistical significance of the clinical data, (iii) the design of clinical trials that plan to treat GBM patients at a too advanced stage of the disease, and (iv) the detection of GBM disease at a too late stage.

9.6 REGULATORY ASPECT OF PHYTOPHARMACEUTICALS

People from numerous countries have a long history of utilizing plants and their components in traditional medicine and healing rituals. Among them are India, Africa, Native American nations, Mediterranean nations, China, and Greece. But only a few medicinal herbs have been clinically tested for their potential as a medical treatment, despite the widespread popularity of herbal medicines (Tagde et al, 2021). In the majority of nations, herbal medications are poorly regulated and are frequently neither registered nor regulated by the health authorities. The Food and Drug Administration (FDA) estimates that botanical and other dietary supplements cause over 50,000 adverse events in the United States. On October 24, 2013, the Indian government published a proposed amendment to the Drugs and Cosmetics Act and Rules. Rule 2 of the Drugs and Cosmetics Rules, 1945 states that a "Phytopharmaceutical Drug" is any fraction that has been purified and standardized and has a minimum of four bioactive or phytochemical ingredients. The newly added Appendix I B to Schedule Y outlines the information that must be given with a request to conduct a clinical study, import, or manufacture a phytopharmaceutical drug in the nation. The regulatory requirements for a new medicine's safety and pharmacological information, human research, and confirmatory clinical trials are included in the NDA for the phytopharmaceutical drug. These are:

1. The plant utilized for extraction and fractionation must be properly authenticated, sourced and identified.
2. Extraction procedure followed by fractionation and purification.
3. Information on the manufacturing method for phytopharmaceutical drugs.
4. Data on stability.

Moreover, the applicant is required to send an adequate quantity of phytopharmaceutical, and phytopharmaceutical formulations/products along with adequate quantities of all the identified bioactive/phytochemical compounds to the laboratories when demanded by the CDSCO for testing.

9.7 CONCLUSION AND FUTURE PROSPECTIVES

In recent years, GBM is the common life-threatening tumour associated with the CNS. Primarily, the elderly population are mostly affected due to this disease as it is extremely fatal and still poorly treated. There are many intrinsic problems and side effects exhibited with these existing therapy, and surgical maximal safe resection is the primary most effective and safe option. Besides, issues like poor solubility, low stability, limited bioavailability, chemo-resistance, and toxicity have led to hindrances in developing a chemotherapeutic regime for GBM patients. Bioactive molecules obtained from natural sources like seeds, vegetables, fruits, etc., have numerous therapeutic potentials such as antioxidant, anti-inflammatory, and anticancer properties that may help cancer survivors feel better throughout chemotherapy or other existing approaches to treatment. However, their isolation and purification were the most crucial challenges for the clinical translation against various diseases. Therefore, much of the energy and resources must be used in efficient isolation which is costly due to limited reserves of these bioactive. Thus, the researcher acquired their attention towards the plant tissue culture for maximizing the production of these bioactive. Another important aspect is elucidating the direct target and mechanism of action of any natural product. Chemo proteomics can be used for determining the activity-based protein profiling of lead molecules by making them react with the pre-determined biological targets. In silico techniques like protein docking, target prediction, and ligand screening are important tools of bioinformatics that help in a systemic analysis though the results are not 100% accurate. For the increasing quality of life of GBM patients, nanocarrier-based formulations can be explored for increasing the biodistribution and bioavailability which can help clear malignant glioma cells with minimal side effects. Applications of nanoparticles in tumours are still being studied for reproducibility, stability, and scale production. Optimizing nanomedicine's physical properties (size, zeta potential, surface chemistry, and shape) and compositions (ligand modification and combination with other therapeutic agents) to improve tumor-specific accumulation and antitumour activities while minimizing side effects remains a challenge.

CONFLICT OF INTEREST

The authors declare that there are no conflicts of interest.

REFERENCES

Acunzo, Julie, Maria Katsogiannou, and Palma Rocchi. 2012. "Small Heat Shock Proteins HSP27 (HspB1), Ab-Crystallin (HspB5) and HSP22 (HspB8) as Regulators of Cell Death." *International Journal of Biochemistry and Cell Biology* 44 (10): 1622–31. https://doi.org/10.1016/j.biocel.2012.04.002.

Agnihotri, Sameer, Kelly E. Burrell, Amparo Wolf, Sharzhad Jalali, Cynthia Hawkins, James T. Rutka, and Gelareh Zadeh. 2013. "Glioblastoma, a Brief Review of History, Molecular Genetics, Animal Models and Novel Therapeutic Strategies." *Archivum Immunologiae et Therapiae Experimentalis*. https://doi.org/10.1007/s00005-012-0203-0.

Ali, Shimaa A, Sawsan A Zaitone, and Yasser M Moustafa. 2015. "Boswellic Acids Synergize the Antitumour Activity and Protect Against the Cardiotoxicity of Doxorubicin in Mice Bearing Ehrlich's Carcinoma." *Canadian Journal of Physiology and Pharmacology*. Downloaded from www.nrcresearchpress.com by New York University.

Aller, Glenn S. van, Jeff D. Carson, Wei Tang, Hao Peng, Lin Zhao, Robert A. Copeland, Peter J. Tummino, and Lusong Luo. 2011. "Epigallocatechin Gallate (EGCG), a Major Component of Green Tea, Is a Dual Phosphoinositide-3-Kinase/MTOR Inhibitor." *Biochemical and Biophysical Research Communications* 406 (2): 194–99. https://doi.org/10.1016/j.bbrc.2011.02.010.

Aqil, Farrukh, Radha Munagala, Jeyaprakash Jeyabalan, and Manicka V. Vadhanam. 2013. "Bioavailability of Phytochemicals and Its Enhancement by Drug Delivery Systems." *Cancer Letters*. Elsevier Ireland Ltd. https://doi.org/10.1016/j.canlet.2013.02.032.

Berman, Adi Y., Rachel A. Motechin, Maia Y. Wiesenfeld, and Marina K. Holz. 2017. "The Therapeutic Potential of Resveratrol: A Review of Clinical Trials." *NPJ Precision Oncology* 1 (1). https://doi.org/10.1038/s41698-017-0038-6.

Cloughesy, Timothy F., Webster K. Cavenee, and Paul S. Mischel. 2014. "Glioblastoma: From Molecular Pathology to Targeted Treatment." *Annual Review of Pathology: Mechanisms of Disease* 9: 1–25. https://doi.org/10.1146/annurev-pathol-011110-130324.

Dhandapani, Krishnan M., Virendra B. Mahesh, and Darrell W. Brann. 2007. "Curcumin Suppresses Growth and Chemoresistance of Human Glioblastoma Cells via AP-1 and NFκB Transcription Factors." *Journal of Neurochemistry* 102 (2): 522–38. https://doi.org/10.1111/j.1471-4159.2007.04633.x.

Filippi-Chiela, Eduardo C, Marcos Paulo Thomé, Mardja Manssur Bueno E Silva, Alessandra Luíza Pelegrini, Pitia Flores Ledur, Bernardo Garicochea, Lauren L Zamin, and Guido Lenz. 2013. "Resveratrol Abrogates the Temozolomide-Induced G2 Arrest Leading to Mitotic Catastrophe and Reinforces the Temozolomide-Induced Senescence in Glioma Cells." www.cbtrus.org.

Fratantonio, Deborah, Molonia Maria Sofia, Romina Bashllari, Claudia Muscarà, Guido Ferlazzo, Gregorio Costa, Antonella Saija, Francesco Cimino, and Antonio Speciale. 2019. "Curcumin Potentiates the Antitumour Activity of Paclitaxel in Rat Glioma C6 Cells." *Phytomedicine* 55 (March): 23–30. https://doi.org/10.1016/j.phymed.2018.08.009.

Frkmont, Lucie. 2000. "Minireview Biological Effects of Resveratrol." *Life Sciences*. 663–673.

Gabay, Martin, Abraham Weizman, Nidal Zeineh, Meygal Kahana, Fadi Obeid, Nahum Allon, and Moshe Gavish. 2021. "Liposomal Carrier Conjugated to APP-Derived Peptide for Brain Cancer Treatment." *Cellular and Molecular Neurobiology* 41 (5): 1019–29. https://doi.org/10.1007/s10571-020-00969-1.

Gadacha, Wafa, Mossadok Ben-Attia, Dominique Bonnefont-Rousselot, Ezzedine Aouani, Néziha Ghanem-Boughanmi, and Yvan Touitou. 2009. "Resveratrol Opposite Effects on Rat Tissue Lipoperoxidation: Pro-Oxidant During Day-Time and Antioxidant at Night." *Redox Report* 14 (4): 154–58. https://doi.org/10.1179/135100009X466131.

Ginkel, Paul R. van, Dhruv Sareen, Lalita Subramanian, Quintisha Walker, Soesiawati R. Darjatmoko, Mary J. Lindstrom, Amol Kulkarni, Daniel M. Albert, and Arthur S. Polans. 2007. "Resveratrol Inhibits Tumour Growth of Human Neuroblastoma and Mediates Apoptosis by Directly Targeting Mitochondria." *Clinical Cancer Research* 13 (17): 5162–69. https://doi.org/10.1158/1078-0432.CCR-07-0347.

Goffart, Nicolas, Jérôme Kroonen, and Bernard Rogister. 2013. "Glioblastoma-Initiating Cells: Relationship With Neural Stem Cells and the Micro-Environment." *Cancers*. https://doi.org/10.3390/cancers5031049.

Hande, M. Prakash, Aik Kia Khaw, M. Pradeepa Hande, and Guruprasad Kalthur. 2013. "Curcumin Inhibits Telomerase and Induces Telomere Shortening and Apoptosis in Brain Tumour Cells." *Journal of Cellular Biochemistry* 114 (6): 1257–70. https://doi.org/10.1002/jcb.24466.

Han, Myung Hoon, Kyueng Whan Min, Yung Kyun Noh, Jae Min Kim, Jin Hwan Cheong, Je il Ryu, Yu Deok Won, Seong Ho Koh, and Young Mi Park. 2022. "Identification of Genes from Ten Oncogenic Pathways Associated With Mortality and Disease Progression in Glioblastoma." *Frontiers in Oncology* 12 (August). https://doi.org/10.3389/fonc.2022.965638.

Hirose, Yuchi, Makoto Katayama, Olga K Mirzoeva, Mitchel S Berger, and Russell O Pieper. n.d. "Akt Activation Suppresses Chk2-Mediated, Methylating Agent-Induced G 2 Arrest and Protects from Temozolomide-Induced Mitotic Catastrophe and Cellular Senescence." www.aacrjournals.org.

Jiapaer, Shabierjiang, Takuya Furuta, Shingo Tanaka, Tomohiro Kitabayashi, and Mitsutoshi Nakada. 2018. "Potential Strategies Overcoming the Temozolomide Resistance for Glioblastoma." *Neurologia Medico-Chirurgica*. Japan Neurosurgical Society. https://doi.org/10.2176/nmc.ra.2018-0141.

Kiekow, Cíntia J., Fabrício Figueiró, Fabrícia Dietrich, Luciana Dalla Vechia, Elisa N.S. Pires, Elisa H.F. Jandrey, Simone C.B. Gnoatto, Christianne G. Salbego, Ana Maria O. Battastini, and Grace Gosmann. 2016. "Quercetin Derivative Induces Cell Death in Glioma Cells by Modulating NF-KB Nuclear Translocation and Caspase-3 Activation." *European Journal of Pharmaceutical Sciences* 84 (March): 116–22. https://doi.org/10.1016/j.ejps.2016.01.019.

Kiskova, Terezia, Peter Kubatka, Dietrich Büsselberg, and Monika Kassayova. 2020. "The Plant-Derived Compound Resveratrol in Brain Cancer: A Review." *Biomolecules*. MDPI AG. https://doi.org/10.3390/biom10010161.

Krüger, Phillip, Rambod Daneshfar, Gunter P. Eckert, Jochen Klein, Dietrich A. Volmer, Ute Bahr, Walter E. Müller, Michael Karas, Manfred Schubert-Zsilavecz, and Mona Abdel-Tawab. 2008. "Metabolism of Boswellic Acids in Vitro and in Vivo." *Drug Metabolism and Disposition* 36 (6): 1135–42. https://doi.org/10.1124/dmd.107.018424.

Krupkova, Olga, Stephen J. Ferguson, and Karin Wuertz-Kozak. 2016. "Stability of (-)-Epigallocatechin Gallate and Its Activity in Liquid Formulations and Delivery Systems." *Journal of Nutritional Biochemistry*. Elsevier Inc. https://doi.org/10.1016/j.jnutbio.2016.01.002.

Li, Wan, Liwen Ren, Xiangjin Zheng, Jinyi Liu, Jinhua Wang, Tengfei Ji, and Guanhua Du. 2020. "3-O-Acetyl-11-Keto-β-Boswellic Acid Ameliorated Aberrant Metabolic Landscape and Inhibited Autophagy in Glioblastoma." *Acta Pharmaceutica Sinica B* 10 (2): 301–12. https://doi.org/10.1016/j.apsb.2019.12.012.

Liu, Fuyao, Bin Peng, Miao Li, Junning Ma, Gang Deng, Shenqi Zhang, and Wendy C. Sheu, et al. 2022. "Targeted Disruption of Tumour Vasculature via Polyphenol Nanoparticles to Improve Brain Cancer Treatment." *Cell Reports Physical Science* 3 (1). https://doi.org/10.1016/j.xcrp.2021.100691.

Liu, Fu Chao, Yung Fong Tsai, Hsin I. Tsai, and Huang Ping Yu. 2015. "Anti-Inflammatory and Organ-Protective Effects of Resveratrol in Trauma-Hemorrhagic Injury." *Mediators of Inflammation*. Hindawi Publishing Corporation. https://doi.org/10.1155/2015/643763.

Louis, David N., Arie Perry, Peter Burger, David W. Ellison, Guido Reifenberger, Andreas von Deimling, and Kenneth Aldape, et al. 2014. "International Society Of Neuropathology–Haarlem Consensus Guidelines for Nervous System Tumour Classification and Grading." *Brain Pathology (Zurich, Switzerland)* 24: 429–35. https://doi.org/10.1111/bpa.12171.

Louis, David N., Arie Perry, Guido Reifenberger, Andreas von Deimling, Dominique Figarella-Branger, Webster K. Cavenee, Hiroko Ohgaki, Otmar D. Wiestler, Paul Kleihues, and David W. Ellison. 2016. "The 2016 World Health Organization Classification of Tumours of the Central Nervous System: A Summary." *Acta Neuropathologica*. Springer Verlag. https://doi.org/10.1007/s00401-016-1545-1.

Maiti, Panchanan, Jason Scott, Dipanwita Sengupta, Abeer Al-Gharaibeh, and Gary L. Dunbar. 2019. "Curcumin and Solid Lipid Curcumin Particles Induce Autophagy, but Inhibit Mitophagy and the PI3K-Akt/MTOR Pathway in Cultured Glioblastoma Cells." *International Journal of Molecular Sciences* 20 (2). https://doi.org/10.3390/ijms20020399.

Mao, Hua, Drake G. Lebrun, Jingxuan Yang, Vivian F. Zhu, and Min Li. 2012. "Deregulated Signaling Pathways in Glioblastoma Multiforme: Molecular Mechanisms and Therapeutic Targets." *Cancer Investigation* 30 (1): 48–56. https://doi.org/10.3109/07357907.2011.630050.

Maria, Rosaria, Rita Gangemi, Fabrizio Griffero, Daniela Marubbi, Marzia Perera, Maria Cristina Capra, and Paolo Malatesta, et al. 2009. "SOX2 Silencing in Glioblastoma

Tumour-Initiating Cells Causes Stop of Proliferation and Loss of Tumourigenicity." *Stem Cells* 27: 40–48. https://doi.org/10.1634/stemcells.2008-0493.
Marumoto T, Saya H. Molecular biology of glioma. Adv Exp Med Biol. 2012;746:2-11. doi: 10.1007/978-1-4614-3146-6_1. PMID: 22639155.
Mook, Olaf R.F., Wilma M. Frederiks, and Cornelis J.F. van Noorden. 2004. "The Role of Gelatinases in Colorectal Cancer Progression and Metastasis." *Biochimica Et Biophysica Acta - Reviews on Cancer.* https://doi.org/10.1016/j.bbcan.2004.09.006.
Nabors, L. Burt, Karen L. Fink, Tom Mikkelsen, Danica Grujicic, Rafal Tarnawski, Do Hyun Nam, and Maria Mazurkiewicz, et al. 2015. "Two Cilengitide Regimens in Combination With Standard Treatment for Patients With Newly Diagnosed Glioblastoma and Unmethylated MGMT Gene Promoter: Results of the Open-Label, Controlled, Randomized Phase II CORE Study." *Neuro-Oncology* 17 (5): 708–17. https://doi.org/10.1093/neuonc/nou356.
Nakada, Mitsutoshi, Daisuke Kita, Takuya Watanabe, Yutaka Hayashi, Lei Teng, Ilya V. Pyko, and Jun Ichiro Hamada. 2011. "Aberrant Signaling Pathways in Glioma." *Cancers.* https://doi.org/10.3390/cancers3033242.
Nakada, M., S. Nakada, T. Demuth, N. L. Tran, D. B. Hoelzinger, and M. E. Berens. 2007. "Molecular Targets of Glioma Invasion." *Cellular and Molecular Life Sciences.* https://doi.org/10.1007/s00018-007-6342-5.
Oliva, Claudia R., Douglas R. Moellering, G. Yancey Gillespie, and Corinne E. Griguer. 2011. "Acquisition of Chemoresistance in Gliomas Is Associated With Increased Mitochondrial Coupling and Decreased ROS Production." *PLoS ONE* 6 (9). https://doi.org/10.1371/journal.pone.0024665.
Orunoğlu, Merdan, Abbas Kaffashi, Sibel Bozdağ Pehlivan, Selma Şahin, Figen Söylemezoğlu, Kader Karlı-Oğuz, and Melike Mut. 2017. "Effects of Curcumin-Loaded PLGA Nanoparticles on the RG2 Rat Glioma Model." *Materials Science and Engineering C* 78 (September): 32–38. https://doi.org/10.1016/j.msec.2017.03.292.
Priyadarsini, K Indira. 2013. "Send Orders of Reprints at Reprints@benthamscience.Net Current Pharmaceutical Design."
Ravanan, Palaniyandi, Sanjay K. Singh, G. S.R. Subba Rao, and Paturu Kondaiah. 2011. "Growth Inhibitory, Apoptotic and Anti-Inflammatory Activities Displayed by a Novel Modified Triterpenoid, Cyano Enone of Methyl Boswellates." *Journal of Biosciences* 36 (2): 297–307. https://doi.org/10.1007/s12038-011-9056-7.
Riba, Adam, Laszlo Deres, Balazs Sumegi, Kalman Toth, Eszter Szabados, and Robert Halmosi. 2017. "Cardioprotective Effect of Resveratrol in a Postinfarction Heart Failure Model." *Oxidative Medicine and Cellular Longevity* 2017. https://doi.org/10.1155/2017/6819281.
Roy, N.K., A.B. Kunnumakkara, D. Bordoloi, J. Monisha, G. Padmavathi, J. Kotoky, and R Golla. 2017. "Specific Targeting of Akt Kinase Isoforms: Taking the Precise Path for Prevention and Treatment of Cancer." *Current Drug Targets* 18(4): 421–35.
Roy, Nand Kishor, Dey Parama, Kishore Banik, Devivasha Bordoloi, Amrita Khwairakpam Devi, Krishan Kumar Thakur, and Ganesan Padmavathi, et al. 2019. "An Update on Pharmacological Potential of Boswellic Acids Against Chronic Diseases." *International Journal of Molecular Sciences.* MDPI AG. https://doi.org/10.3390/ijms20174101.
Ryskalin, Larisa, Francesca Biagioni, Carla L. Busceti, Gloria Lazzeri, Alessandro Frati, and Francesco Fornai. 2020. "The Multi-faceted Effect of Curcumin in Glioblastoma from Rescuing Cell Clearance to Autophagy-independent Effects." *Molecules.* MDPI AG. https://doi.org/10.3390/molecules25204839.
Ryu, Jinhyun, Nal Ae Yoon, Hyemin Seong, Joo Yeon Jeong, Seokmin Kang, Nammi Park, and Jungil Choi, et al. 2015. "Resveratrol Induces Glioma Cell Apoptosis Through Activation of Tristetraprolin." *Molecules and Cells* 38 (11): 991–97. https://doi.org/10.14348/molcells.2015.0197.

Salvati, Maurizio, Alessandro Frati, Natale Russo, Emanuela Caroli, Filippo Maria Polli, Giuseppe Minniti, and Roberto Delfini. 2003. "Radiation-Induced Gliomas: Report of 10 Cases and Review of the Literature." *Surgical Neurology* 60 (1): 60–67. https://doi.org/10.1016/S0090-3019(03)00137-X.

Sanchez-Vega, Francisco, Marco Mina, Joshua Armenia, Walid K. Chatila, Augustin Luna, Konnor C. La, and Sofia Dimitriadoy, et al. 2018. "Oncogenic Signaling Pathways in The Cancer Genome Atlas." *Cell* 173 (2): 321–337.e10. https://doi.org/10.1016/j.cell.2018.03.035.

Sang, Dong Ping, Ru Jun Li, and Qing Lan. 2014. "Quercetin Sensitizes Human Glioblastoma Cells to Temozolomide in Vitro via Inhibition of Hsp27." *Acta Pharmacologica Sinica* 35 (6): 832–38. https://doi.org/10.1038/aps.2014.22.

Santos, Balbino L., Mona N. Oliveira, Paulo L.C. Coelho, Bruno P.S. Pitanga, Alessandra B. da Silva, Taís Adelita, and Victor Diógenes A. Silva, et al. 2015. "Flavonoids Suppress Human Glioblastoma Cell Growth by Inhibiting Cell Metabolism, Migration, and by Regulating Extracellular Matrix Proteins and Metalloproteinases Expression." *Chemico-Biological Interactions* 242 (December): 123–38. https://doi.org/10.1016/j.cbi.2015.07.014.

Schilder, Yvonne D.C., Elke H. Heiss, Daniel Schachner, Jürgen Ziegler, Gottfried Reznicek, Dan Sorescu, and Verena M. Dirsch. 2009. "NADPH Oxidases 1 and 4 Mediate Cellular Senescence Induced by Resveratrol in Human Endothelial Cells." *Free Radical Biology and Medicine* 46 (12): 1598–1606. https://doi.org/10.1016/j.freeradbiomed.2009.03.013.

Schneider, Hannah, and Michael Weller. 2016. "Boswellic Acid Activity Against Glioblastoma Stem-Like Cells." *Oncology Letters* 11 (6): 4187–92. https://doi.org/10.3892/ol.2016.4516.

Scholz, Susanne, and Gary Williamson. 2007. "Interactions Affecting the Bioavailability of Dietary Polyphenols in Vivo." *International Journal for Vitamin and Nutrition Research* 77: 224–35. https://doi.org/10.1024/0300-9831.77.3.224.

Tagde, Priti, Pooja Tagde, Sandeep Tagde, Tanima Bhattacharya, Vishal Garg, Rokeya Akter, and Md Habibur Rahman, et al. 2021. "Natural Bioactive Molecules: An Alternative Approach to the Treatment and Control of Glioblastoma Multiforme." *Biomedicine and Pharmacotherapy*. Elsevier Masson s.r.l. https://doi.org/10.1016/j.biopha.2021.111928.

Tee, Jie Kai, Li Xian Yip, Eveline Sheau Tan, Supawan Santitewagun, Arun Prasath, Pu Chun Ke, Han Kiat Ho, and David Tai Leong. 2019. "Nanoparticles' Interactions With Vasculature in Diseases." *Chemical Society Reviews*. Royal Society of Chemistry. https://doi.org/10.1039/c9cs00309f.

Vafadar, Asma, Zahra Shabaninejad, Ahmad Movahedpour, Farzaneh Fallahi, Mona Taghavipour, Younes Ghasemi, and Maryam Akbari, et al. 2020. "Quercetin and Cancer: New Insights into Its Therapeutic Effects on Ovarian Cancer Cells." *Cell and Bioscience*. BioMed Central Ltd. https://doi.org/10.1186/s13578-020-00397-0.

Walle, Thomas, Faye Hsieh, Mark H. DeLegge, John E. Oatis, and U. Kristina Walle. 2004. "High Absorption but Very Low Bioavailability of Oral Resveratrol in Humans." *Drug Metabolism and Disposition* 32 (12): 1377–82. https://doi.org/10.1124/dmd.104.000885.

Weber, Artur Aron, Thomas Neuhaus, Romanita Adriana Skach, Jürgen Hescheler, Hee Yul Ahn, Karsten Schrör, Yon Ko, and Agapios Sachinidis. 2004. "Mechanisms of the Inhibitory Effects of Epigallocatechin-3 Gallate on Platelet-Derived Growth Factor-BB-Induced Cell Signaling and Mitogenesis." *The FASEB Journal : Official Publication of the Federation of American Societies for Experimental Biology* 18 (1): 128–30. https://doi.org/10.1096/fj.03-0007fje.

Wen, Patrick Y, and Santosh Kesari. 2008. "Medical Progress Malignant Gliomas in Adults." *New England Journal of Medicine* 359. www.nejm.org.

Yin, Haitao, Yun Zhou, Cuixia Wen, Chong Zhou, Wei Zhang, Xiang Hu, Lifeng Wang, Chuanwen You, and Junfei Shao. 2014. "Curcumin Sensitizes Glioblastoma to Temozolomide by Simultaneously Generating ROS and Disrupting AKT/MTOR Signaling." *Oncology Reports* 32 (4): 1610–16. https://doi.org/10.3892/or.2014.3342.

Zang, Xinlong, Mingyang Cheng, Xiaoxu Zhang, and Xuehong Chen. 2021. "Quercetin Nanoformulations: A Promising Strategy for Tumour Therapy." *Food and Function*. Royal Society of Chemistry. https://doi.org/10.1039/d1fo00851j.

Zhang, Yong, Shao Xiang Wang, Ji Wei Ma, Hai Ying Li, Jie Cheng Ye, Si Ming Xie, Bin Du, and Xue Yun Zhong. 2015. "EGCG Inhibits Properties of Glioma Stem-Like Cells and Synergizes With Temozolomide Through Downregulation of P-Glycoprotein Inhibition." *Journal of Neuro-Oncology* 121 (1): 41–52. https://doi.org/10.1007/s11060-014-1604-1.

10 Eugenol as Neuro-Phytomedicine

Recent Trends Pertaining to the Treatment of Neurological Disorders

Bhagawati Saxena

10.1 INTRODUCTION

Neurological disorders are medically described as conditions that impact the brain, spinal cord, and body's nerves. The most frequent neurological conditions include Alzheimer's disease (AD), Parkinson's disease (PD), brain and spinal cord injury, epilepsy, migraines, multiple sclerosis, stroke, etc. Recently, there has been an upsurge in the frequency of neurological disorders. Treatment of various neurological illnesses presents a myriad of challenges, one of which is the delivery of drugs to the brain (Bhavsar et al, 2022). According to reports, phytochemicals derived from medicinal plants have therapeutic and pharmacological effects that were neuroprotective, making them possible treatments for neurological illnesses. The blood-brain barrier (BBB), which encircles and protects the brain, hinders a significant number of drugs from entering the brain. Eugenol is a widely used researched molecule. Eugenol appears to pass through the BBB and act in situ when consumed orally due to its hydrophobic characteristic. Eugenol demonstrates a variety of biological effects while having a rather simplistic chemical composition. It is renowned for its anti-microbial (Marchese et al, 2017), antioxidant, and anti-inflammatory properties (Barboza et al, 2018). According to earlier experimental findings, eugenol has been proven to be protective against the neurotoxicity caused by chlorpyrifos (Singh and Panwar, 2014), aluminium (Mesole et al, 2020; Said and Rabo, 2017), acrylamide (Prasad and Muralidhara 2013), 6-hydroxydopamine (6-OHDA) (Kabuto et al, 2007; Kabuto and Yamanushi, 2011) and scopolamine (Garabadu and Sharma, 2019). Eugenol can be utilised to treat several neurological or neurodegenerative problems due to its chemical properties such as hydrophobicity as well as a wide range of pharmacological features, such as antioxidative, anti-inflammatory activities, etc. Eugenol continues to be of great interest to researchers because of its multidirectional impact, which makes it a practical component in therapies with therapeutic potential against many brain disorders. The present level of knowledge on the neuroprotective properties and uses of eugenol and its derivatives in neurological disorders such as

DOI: 10.1201/9781003389781-10

depression, AD, Parkinson's diseases, stress and stress-related disorders, stroke, and traumatic brain injury (TBI) are emphasised in this chapter in the following sections.

10.2 EUGENOL AND ITS PROPERTIES

Eugenol ($C_{10}H_{12}O_2$) is chemically 4-allyl-2-methoxy phenol. It is an aromatic phenolic compound. It is frequently used as a scent in fragrances and soaps. Eugenol has an oily viscosity, a clear to light yellow colour, and a pungent scent. Inorganic solvents are the best solubilisers of eugenol, while it is sparingly soluble in water. Eugenol is unstable chemically and vulnerable to oxidation and other chemical reactions. It is quickly absorbed by several organs when taken orally and metabolised by the liver. Common sources of it are plants from the Lauraceae, Lamiaceae, Myristicaceae, and Myrtaceae families' natural essential oils. It is reported to occur in a variety of concentrations depending on the species (Table 10.1). However, eugenol is the primary ingredient in the clove oil extract, that is, 70%–90% (*Eugenia caryophyllata, Syzygium aromaticum*). Cloves, in addition to being a common spice used to flavour foods and drinks, can also be applied locally to alleviate toothaches. It is also taken orally, however less frequently to treat digestive and respiratory issues. Cinnamon, nutmeg, and other aromatic spices also contain eugenol, albeit in smaller amounts (Charan Raja, 2015; Khalil et al, 2017). Since eugenol is a key component of the three spices like clove, cinnamon, nutmeg, etc., it is commonly regarded as safe and employed as a natural food flavouring and preservative by the U.S. Food and Drug Administration (FDA) (Burt, 2004; National Toxicology Program, 1983). Eugenol's antimicrobial (Hu et al, 2018; Marchese et al, 2017) properties have led to its usage as a pesticide. Eugenol has broad-spectrum antibacterial activity against

TABLE 10.1
Eugenol Occurrence and Concentration in Various Plants (Charan Raja, 2015; Khalil et al, 2017).

Plant	Common Name	Family	Part	Concentration (mg/g)
Syzygium aromaticum	Clove, clovetree	Myrtaceae	Flower, flower buds, leaf, stem	1809
Syzygium aromaticum	Clove pepper	Myrtaceae	Fruit	36
Pepper betel	Betel pepper	Piperaceae.	Leaf	17.85
Daucus carota	Carrot	Apiaceae or Umbelliferae	Seed	7.0
Ocimum tenuiflorum	Tulsi	Lamiaceae	Leaf	4.2–4.97
Cinnamomum verum	Ceylon cinnamon, cinnamon	Lauraceae	Bark	3.52
Curcuma longa	Turmeric	Zingiberaceae	Leaf, essential oil	2.1
Laurus nobilis	Bay, bay laurel	Lauraceae	Leaf	1.34
Zingiber officinale	Chinese ginger	Zingiberaceae	Rhizome	0.4
Myristica fragrans	Nutmeg	Myristicaceae	Seed	0.32
Origanum vulgare	Small-flowered oregano	Lamiaceae	Shoot	0.055–0.125

microorganisms (Burt, 2004). Eugenol is traditionally been used in dentistry as a dental analgesic. It is used as an analgesic and an antimicrobial in a variety of root canal filling products (Diaz and Sembrano, 1985; Pramod et al, 2016).

10.3 NEUROPROTECTIVE EFFECTS AND MECHANISMS OF EUGENOL IN VARIOUS NEUROLOGICAL DISORDERS

10.3.1 Depression

The exceedingly high frequency of depression in the community qualifies it as one of the most significant neurological illnesses. Moreover, depression frequently coexists with neurodegenerative illnesses such as PD and AD (Chung et al, 2003; Lyketsos and Lee, 2004). Depression is a mood condition that affects many people and frequently results in major psychosocial issues including suicide (Wong and Licinio, 2001). The "monoamine theory" has been the most significant of the pathogenic explanations for depression. Monoamine levels (serotonin, dopamine (DA), and noradrenaline) in post-synaptic areas are markedly diminished in individuals with depression for a plethora of reasons. Symptoms associated with depression should improve if these monoamines are restored. The current line of medications for the treatment of depression includes inhibitors of monoamine oxidase (MAO), serotonin noradrenaline receptor inhibitors (SNRIs), and selective serotonin receptor inhibitors (SSRIs). However, chronic and recurrent depression is still prevalent despite the introduction of these medicines (Gelenberg et al, 2003). This raises the need of developing a new therapy for depression. The anti-depressive effects of eugenol on rats were assessed in the earlier study. Results show that oral administration of eugenol to animals reduced the depression comparable to imipramine administration (Irie et al, 2004).

In patients suffering from depression, the hippocampus area is relatively smaller (Bremner et al, 2000; Brosch et al, 2022). This decrease in the hippocampal region is recovered as the patients receive antidepressant therapy and their clinical condition improves (Malberg et al, 2000). One of the neuronal growth factors in the brain that is crucial for the process of neurogenesis is called brain-derived neurotrophic factor (BDNF). Previous animal research shows that BDNF supplementation can alleviate depression associated with BDNF deficiency (Shirayama et al, 2002). Additionally, recent research showed that the neurogenic benefits of antidepressants were eliminated when this neurogenesis was disrupted by X-ray irradiation in the brain (Santarelli et al, 2003). Thus, it is determined that regenerated neurons are responsible for an antidepressant's therapeutic benefit (Malberg et al, 2000). The elevation of BDNF expression in the hippocampus promotes the antidepressant-induced regeneration of neurons (Björkholm and Monteggia, 2016; Shirayama et al, 2002). It was discovered that eugenol elevated BDNF expression in the hippocampus as with imipramine (Irie et al, 2004). Eugenol also promotes hippocampus neurogenesis, which is seen with other antidepressants (Akbar et al, 2021; Irie, 2006).

In contrast to treatment with imipramine, eugenol enhances the expression of the metallothionein-III (MT-III) gene in the hippocampus (Irie et al, 2004). MT-III is a member of the MT family that is particular to the brain and shields cells and organs from numerous harmful events. It is thought that MT-III functions in the brain to promote neurogenesis and neuroprotection (Irie and Keung, 2001; Tanji et al, 2003). These results imply that eugenol's ability to induce neurogenesis may be mediated through MT-III expression. Certain study shows that eugenol and its analogues inhibit MAO, preferentially MAO-A (Tao et al, 2005); however, another study reported that eugenol neither inhibits MAO-A nor inhibits MAO-B (Kong et al, 2004). This difference in observation may be owing to various experimental circumstances.

10.3.2 Alzheimer's Disease

AD is a neurodegenerative disease that occurs due to the atrophy of cholinergic neurons mainly in the cerebral cortex and hippocampus region. The most prevalent type of dementia is AD. The prevalence of AD is rising globally as the proportion of elderly people increases worldwide (Citron, 2002). The majority of AD cases occur sporadically, that is, without genetic origins, however, some familial cases of AD have been well-described, that is, genetic anomalies in the genes (PS1, PS2, APP, etc.) (Selkoe and Schenk, 2003). Pathogenesis of AD is multifactorial. The two hallmarks of AD's pathogenesis include interneuronal accumulation of amyloid β (Aβ) plaques and intra-neuronal accumulation of neurofibrillary tangles (NFTs). Studies have demonstrated that Aβ plaques result in the death of neuronal cells through several mechanisms (Choi et al, 2019; Gschwind and Huber, 1995; Orellana et al, 2011). Aβ plaque and NFT exert their neurotoxicity through a range of mechanisms. First, the reactive oxygen species (ROS) are generated as a consequence of Aβ plaque deposition. Generated ROS results in direct damage to cell membranes and DNA to cause atrophy of neurons (Behl et al, 1994; Cheignon et al, 2018; Wang et al, 2010). Second, NFT and Aβ plaque have been linked to promoting excitotoxicity by allowing an excessive influx of calcium ions via NMDA receptor into the neuronal cells (Pallo et al, 2016). Insoluble deposits of Aβ plaques between neurons result in the activation of microglia, an innate immune cell within the brain. Activated microglia have two distinct tasks in the pathogenesis of AD. Through phagocytic action, it initially provides neuroprotective effects by eliminating Aβ deposits, but with time, it releases cytotoxic chemicals and pro-inflammatory cytokines, causing gradual neurodegeneration (Mandrekar-Colucci and Landreth, 2010). Although microglia express all toll-like receptor (TLR) subtypes, TLR2 and TLR4 are particularly abundant (Fiebich et al, 2018). TLR4 is crucial in the identification of the Aβ plaques by microglia, which is believed to be a part of inflammatory reactions in AD pathogenesis (Papageorgiou et al, 2016). In AD, TLR4 is crucial for the initiation of neuroinflammation which in turn results in neurodegeneration (Chavali et al, 2020; Saxena and Chavali, 2019; Walter et al, 2007).

Memantine, an NMDA inhibitor of the glutamate receptor, three different cholinesterase blockers (donepezil, galantamine, and rivastigmine), and most recently Aducanumab, a monoclonal antibody, were the only medications that the US-FDA approved for treating AD patients (Chauhan et al, 2022; Long and Holtzman, 2019; Yang and Sun, 2021). Thus, excitotoxicity via glutamate acting on NMDA, as well as cholinesterase, is found promising target for the new molecule.

Eugenol is well known for its antioxidant as well as anti-inflammatory activity. Eugenol can directly neutralise hydroxyl free radicals and lessen oxidative stress (Taira et al, 1992). Additionally, it was revealed that eugenol's free radicals scavenging activity (antioxidative effect) is linked to its capacity as a metal transition (Ito et al, 2005). It is suggested that eugenol prevents lipid peroxidation caused by lipoxygenase (Naidu, 1995). Additionally, earlier reports show that eugenol has inhibitory activity on calcium ion channels (Chung et al, 2008; Seo et al, 2013). Eugenol prevents excitotoxicity induced by NMDA in cultured neuronal cells by inhibiting Ca^{2+} absorption (Wie et al, 1997). It has also been proven that NMDA receptors mediate the neurotoxic effects of Aβ deposits (Parks et al, 2001). Hence, it stands to reason that eugenol might counteract the harmful action of Aβ, which is mediated through the NMDA receptor. TLR4 is a key player in the innate immune responses in neurodegenerative diseases like AD, PD, etc. We docked eugenol against the human myeloid differentiation factor-2 (hMD2) co-receptor of TLR4 to better comprehend the mechanism underlying the anti-inflammatory effect. Eugenol has a −6.897 kcal/mol binding affinity towards the hMD2 active site (Vora et al, 2022). Eugenol dramatically decreased the gene expression of TLR4 indicators, according to a prior study (Choudhury et al, 2014). Lipopolysaccharide (LPS) is the natural agonist of TLR4, and eugenol inhibits the production of pro-inflammatory mediators generated by LPS in human macrophages (Lee et al, 2007). Thus, eugenol's antagonistic action at TLR4 might be partially responsible for its anti-inflammatory effects. Further research is necessary to corroborate this conclusion. Thus, eugenol is supposed to be effective against AD by simultaneously acting on multiple pathways.

Eugenol inhibits the aggregation of amyloid protein (Dubey et al, 2017). Eugenol administration prevented the loss in memory and learning performance against insulin and Aβ-induced AD rat models (Taheri et al, 2019). In vitro study shows that eugenol has antioxidant activities as well as inhibits MAO, butyrylcholinesterase (BChE), acetylcholinesterase (AChE), and activities dose-dependently (Adefegha et al, 2021). Eugenol was also found to alleviate the impairment in the learning-memory ability and augmented oxidative stress in the hippocampus of rats with AD (Liu et al, 2013). The report shows that eugenol enhanced learning and memory in the mouse by augmenting neurogenesis as well as dendritic complexity of neurons in the dentate gyrus (DG) and cornu ammonis 1 (CA1) basal regions of the mouse brain (Akbar et al, 2021). Additionally, eugenol reduced the memory loss brought on by scopolamine. Eugenol also lessened the effects of scopolamine on the rat's hippocampus cholinergic and mitochondrial dysfunction and glutamate neurotoxicity (Garabadu and Sharma, 2019). Aluminium administration results in dementia, hastening Aβ plaque formation and accumulation, neuroinflammation, and oxidative stress in the rat's brain (Chavali et al, 2020; Saxena and Chavali, 2019). It was discovered that eugenol protected against aluminium-induced neurotoxicity (Said and Rabo, 2017; Mesole et al, 2020). Thus, these pieces of literature show that eugenol is found effective for treating AD via acting through multiple pathways which are shown in Figure 10.1.

Eugenol analogue, that is, acetoxy eugenol acetate has shown neuroprotective activity against cognitive impairment in mice caused by Aβ (Jayasingh Chellammal et al, 2019). Eugenol-rich grass leaf sweet flag rhizome shields PC-12 cells against

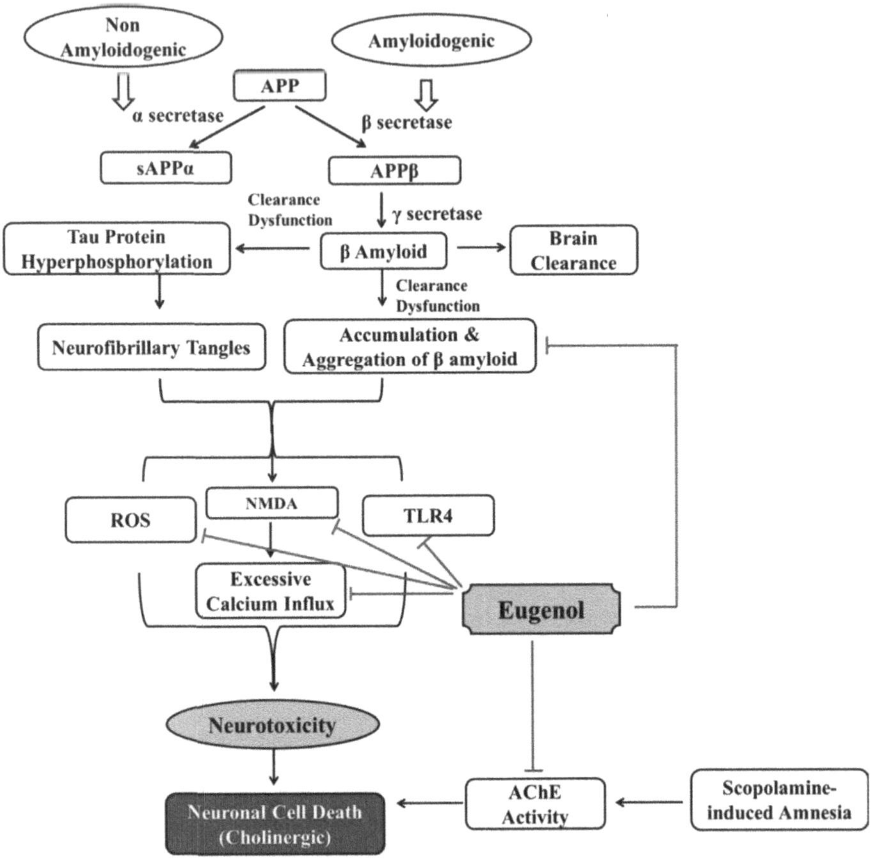

FIGURE 10.1 Neuroprotective mechanisms of eugenol in Alzheimer's disease. Proteolysis of amyloid precursor protein (APP), a transmembrane glycoprotein occurs by two pathways, that is, non-amyloidogenic pathway and amyloidogenic pathway. In the non-amyloidogenic pathway, APP is first cleaved by alpha-secretase and then by gamma-secretase and thus leads to the generation of soluble fragments sAPPα. On the other hand, in the amyloidogenic pathway, neurotoxic Aβ formed through cleavage of APP by beta-secretase followed by gamma-secretase and formed sAPPβ and then Aβ oligomers. Deposition of Aβ plaque leads to hyperphosphorylation of tau proteins and ultimately leads to the formation of NFTs. Formation and deposition of Aβ plaque and NFTs results in ROS generation, excessive influx of calcium ion by activation of NMDA receptors, and activation of neuroinflammatory receptors like toll-like receptor 4 (TLR4), which further play role in neurotoxicity and cause neuronal death of cholinergic nerve fibres. In scopolamine-induced amnesia, there is an increase in AchE activity which decreases the level of acetylcholine. Neurodegeneration or decreased activity of cholinergic system leads to memory loss associated with Alzheimer's disease. Eugenol inhibits the formation of ROS, NMDA receptor as well as TLR4. It also blocks the influx of calcium ions and thus inhibits excitotoxicity. It also inhibits AchE activity and thus increases the concentration of Ach within the synapse. Thus, eugenol exerts its neuroprotection against Alzheimer's disease by acting on multiple pathways.

the Aβ peptide's harmful effects (Liang et al, 2015). Moreover, eugenol containing *Syzygium aromaticum* was found to be protective against $A\beta_{25\text{-}35}$-induced neurotoxicity in neuronal cells (Shekhar et al, 2018). Certain analogues of eugenol, that is, asaronic acid, o-eugenol, 4-hydroxy-3-methoxybenzylamine, ethyl 4-hydroxy-3-methyoxycinnamate, ethyl homovanillate, and 2-methoxy5-nitrophenol greatly shielded PC-12 cells from the harmful effect of Aβ 1–40 (2.5 mM) (Irie, 2006).

10.3.3 Parkinson's Disease

After AD, PD is the second most typical neurodegenerative condition. PD is caused by the depletion of DA in the nigrostriatal pathway due to the gradual degeneration of neurons (dopaminergic) in the substantia nigra pars compacta (SNpc) (Bergman and Deuschl, 2002). Characteristic features of PD include impairment of the motor and nonmotor (limbic and cognitive) functions. DA deficiency in the SNpc results in motor symptoms that include resting tremors, stiff joints, akinesia/bradykinesia, etc (Jankovic, 2008)). The majority of the time, nevertheless, there are no apparent genetic backgrounds; however, rarely, PD develops as a hereditary condition brought on by a gene defect (Bergman and Deuschl, 2002). Treatment for PD has advanced significantly. Many prospective treatments for PD are developing as a consequence of advancements in experimental medicines (Ellis and Fell, 2017). The most efficient drug for treating the signs and symptoms of PD is currently levodopa, but it is also linked to serious side effects such as the "wearing off" effect including dyskinesias and other motor problems (Bargiotas and Konitsiotis, 2013). Alternative treatments for PD include DA agonists, nondopaminergic therapy, and catechol-o-methyl-transferase inhibitors. These treatments can be taken in conjunction with levodopa or alone (Jankovic and Aguilar, 2008).

6-OHDA and 1-methly-4-phenyl-1,2,3,6-tetrahydropyridine (MPTP) are widely used classic toxins that induce PD in animals (Schober, 2004). Eugenol was proven to be effective in the 6-OHDA-induced rat model of PD (Moreira Vasconcelos et al, 2020). Further research suggests that eugenol inhibits 6-OHDA-induced DA depletion in the striatum both directly by inhibiting lipid peroxidation and indirectly through the activation of glutathione (GSH) and L-ascorbate-producing systems. Additionally, additional GSH by conjugating with the p-quinone created during 6-OHDA auto-oxidation could prevent cell death (Kabuto et al, 2007). According to a report, eugenol pre-treatment inhibits 6-OHDA-induced DA reduction by halting lipid peroxidation. On the contrary, eugenol post-treatment enhanced the DA decline caused by 6-OHDA. According to the study, eugenol pre-treatment reduces lipid peroxidation, which in turn prevents 6-OHDA's ability to cause DA depletion in the striatum. Eugenol post-treatment, however, further enhanced the DA decline. Thus, depending on when it was administered concerning when the model PD started, eugenol showed negative consequences (Kabuto and Yamanushi, 2011).

The synthetic neurotoxin MPTP has shown to cause PD-like symptoms in mice, including dyskinesia (Mustapha and Taib, 2021; Schober, 2004). MPTP, that is, less toxic is converted to MPP+ (highly toxic) in the mitochondria of the brain by MAO, particularly MAO-B. MPP+ inhibits the function of mitochondrial complex I, which causes the production of ROS (Mustapha and Taib, 2021). These facts suggest that

MAO-B inhibitors are currently used in therapeutic settings. In the MPTP-induced PD mouse model, our research demonstrates that pre-treatment with eugenol has positive effects on behavioural and pathological development. The protective effect of eugenol against the MPTP-induced PD model is because of its antioxidant effect. However, post-treatment with eugenol further deteriorates the MPTP-induced alteration in behaviour and pathological progression (Vora et al, 2022).

10.3.4 Stress and Stress-Related Disorders

The level of stress that people sense nowadays appears to have significantly increased. Stress leads to numerous other conditions or disorders, including anxiety, depression, cognitive impairment, diabetes, immunosuppression, irritable bowel syndrome, ulcerative colitis, gastric ulcer, hypertension, and male impotence. Stress is described as the body's non-specific reaction to demands placed on it, which are typically unpleasant (Selye, 1936). The body responds to stress through the process of adaption (also known as allostasis) by maintaining physiological processes within a predetermined range. However, excessive stress or ineffective allostasis control results in an allostatic load, which can cause diseases and must thus be managed (McEwen, 2000). The sympathoadrenal system (SAS), brain monoaminergic systems (BMS), and hypothalamic-pituitary-adrenal (HPA) all have a role in the physiological and psychological reactions to stress. In response to stress, the ulcer index, plasma corticosterone, and norepinephrine (NE) levels were elevated. Increased plasma corticosterone and NE are a marker of the stimulated HPA axis and the SAS, respectively. The earlier report shows that eugenol exerts an anti-stress effect against restraint-induced stress (Sen et al, 1992). Another study shows that the seven-day pre-treatment with eugenol showed a selective impact on the HPA axis and thus, reduced the ulcer scoring and increased corticosterone level in plasma induced by stress but did not influence the stress-induced altered levels of NE in plasma (Garabadu et al, 2011). Additionally, eugenol restored the stress-related alterations in 5-HT levels in the various region of the brain while NE levels were reversed in the various region of the brain except for the hippocampus. Results further demonstrated that eugenol has strong anti-stress activity in the four-hour restraint stress paradigm and that the impact is a consequence of HPA and BMS regulation. Eugenol-rich bark of *Cassia zelynicum* (cinnamon) (Saxena and Saxena, 2012) and buds of eugenia caryophyllus (clove) (Singh et al, 2009) were also found to be effective as an anti-stress agent.

People today experience inescapable chronic stress due to the difficult and extremely demanding life rhythm. Numerous major health issues, such as anxiety, depression, gastrointestinal dysfunctions consequently irritable bowel syndrome, dysregulations of the neuroendocrine system, etc., are linked to chronic stress. Studies also show that eugenol is also effective against stress-induced disorders. Study shows that eugenol reduces anxiety brought on by stress in rodents by inhibiting the expression of glyoxalase 1 (Siyal et al, 2021) and downregulation of neurokinin-1 receptors (Siyal et al, 2020). The manifestation of emotional behaviours, such as pathologic anxiety disorders, is critically influenced by the central amygdala (CeA). CeA-specific infusion or intraperitoneal injection of methyl eugenol, a derivative of eugenol decreased anxiety-like behaviours in mice. This is most likely a result of the methylleugenol's ability to activate A-type GABA receptors (GABAARs). Stimulation of GABAARs increases gamma-aminobutyric acid (GABA)ergic inhibitory influence in the CeA, which lowers anxiety (Liu et al, 2019). Eugenol also showed a protective effect against gastrointestinal dysfunction caused by constraint stress. The results of this investigation suggest that eugenol may be used to treat irritable bowel syndrome (Garabadu et al, 2015). Chronic stress-related hormonal variations or irregularities and associated psychological and physical effects, including reproductive dysfunction, are more common in women. Stress activates the HPA axis, elevating blood cortisol levels, which attenuates the release of anti-Mullerian and oestradiol hormones from the ovary while augmenting anterior pituitary synthesis of luteinising and follicle-stimulating hormones. Stress also results in insulin resistance, oxidative stress and consequently molecular damage in the ovaries. These alterations cumulatively result in fertility issues (Bethea et al, 2008). Another study shows the preventive effects of eugenol against stress-induced infertility in female rats (Helmy et al, 2022). It's interesting to note that the injection of eugenol dramatically reduced all of these negative consequences by modifying animal behaviour, ovarian damage, corticosterone upsurge hormonal abnormalities and thus, ultimately reproductive disturbances.

10.3.5 Eugenol for the Treatment of Stroke

Stroke is one of the most dangerous illnesses endangering people's health and life. A stroke occurs when the blood flow to a certain area of the brain is interrupted. This may result in permanent brain damage, chronic disability, or even death. Ischemic stroke and haemorrhagic stroke are the two forms of stroke. Ischemic stroke happens when blood arteries get blocked, cutting off the brain's blood supply whereas eruption of the blood vessels, results in haemorrhagic stroke, resulting in an influx of blood into the brain cavity. Up to 60%–80% of all strokes are ischemic, making them more prevalent (Chugh, 2019). Stroke is the most frequent long-term disability, the second-leading cause of death, and dementia in those over sixty (Chamorro et al, 2016). The financial burden of stroke is rising globally due to an ageing population (Feigin et al, 2016; Forouzanfar et al, 2015). Recombinant tissue plasminogen activator (rt-PA) is the only thrombolytic FDA-approved to date for the treatment of stroke (Lekoubou et al, 2017). rt-PA may not be used as widely as it might due to its short therapy windows and adverse effects. Research has indicated that eugenol may protect against brain ischemia damage (Ahmad et al, 2018; Won et al, 1998). Autophagy is a phenomenon which is induced as a result of ischemic stroke, its significance for stroke is yet debatable. Numerous studies indicate that autophagy triggered in response to ischemia acts as a double-edged sword that can have both advantageous and detrimental effects on brain neurons (Shi et al, 2021; Wang et al, 2018). However, certain studies show that autophagy prevents neuronal damage caused by cerebral ischemia followed by reperfusion. The molecules having the capacity to induce autophagy are found to prevent cerebral ischemia (Wang et al, 2018; Yao et al, 2019). The study by Sun and the team shows that by enhancing autophagy, eugenol reduces the effects of cerebral ischemia-reperfusion injury (Sun et al, 2020). Moreover, it has been shown that the analogous of eugenol, that is, methyleugenol can reduce brain ischemia damage by preventing oxidative stress, neuro-inflammation, and consequently apoptotic neuronal death (Choi et al, 2010).

10.3.6 Traumatic Brain Injury

TBI can be described as the disruption of the brain's physiological and regular activities as a result of an external mechanical impact (Khellaf et al, 2019). TBI is one of the leading causes of disability and death, globally (Najem et al, 2018). Pathogenesis of TBI consists of two phases: primary injury and subsequent secondary injury. A direct mechanical impact to the head causes primary injuries such as cortical contusions, axonal shearing hematomas at epidural or subdural regions and microvascular injuries (Najem et al, 2018). An initial blow results in a delayed, long-lasting secondary injury that lasts for hours to months to years. Mechanisms involved in secondary injuries include excitotoxicity, mitochondrial dysfunction, neuroinflammation, oxidative stress, axon deterioration, and consequently neuronal death (Ng and Lee, 2019). TBI also causes cerebral oedema as a result of the obstruction of the outflow. This in turn results in rise in a intracranial pressure (ICP), which ultimately leads to herniation, a primary cause of death. To prevent the swollen brain from being compressed, a section of the cranium is removed in the decompressive craniectomy (DC) (Bor-Seng-Shu et al, 2013). However, literature reported that DC lowers mortality, but it also leads to brain damage (Cole et al, 2011; Lad et al, 2019) and turns fatalities into survivors with significant disabilities (Honeybul, 2017). Our results show that pre-treatment with eugenol in rats ameliorated TBI-induced functional disabilities (motor coordination, locomotor activity, and short-term memory). Additionally, eugenol pre-treatment reduced trauma-induced enhanced neuronal cell death, lipid peroxidation, oedema, and impaired BBB integrity (Barot and Saxena, 2021).

10.4 CONCLUSION

Numerous spices and therapeutic plants contain eugenol. It is frequently used as a food flavouring and preservation as well as a dental analgesic. Eugenol, a very simple chemical, has been established to have a wide array of neuroprotective properties. Recent research indicates that eugenol may potentially have neurogenerative properties. Eugenol has a variety of effects that cannot be attributed to just one or two of its functions. The potential therapeutic effects of eugenol are likely the result of at least many multiple processes. Eugenol can now be used to treat various neurological illnesses and diseases, including AD, depression, PD, and other conditions. Future developments in more potent analogues might be promised.

REFERENCES

Adefegha, Stephen A., Bathlomew M. Okeke, and Ganiyu Oboh. 2021. "Antioxidant Properties of Eugenol, Butylated Hydroxylanisole, and Butylated Hydroxyl Toluene With Key Biomolecules Relevant to Alzheimer's Diseases-In Vitro." *Journal of Food Biochemistry* 45 (3). https://doi.org/10.1111/JFBC.13276.

Ahmad, Niyaz, Rizwan Ahmad, Md Aftab Alam, and Farhan Jalees Ahmad. 2018. "Quantification and Brain Targeting of Eugenol-Loaded Surface Modified Nanoparticles Through Intranasal Route in the Treatment of Cerebral Ischemia." *Drug Research* 68 (10): 584–95. https://doi.org/10.1055/A-0596-7288.

Akbar, Latiful, Berry Juliandi, Arief Boediono, Irmanida Batubara, and Mawar Subangkit. 2021. "Effects of Eugenol on Memory Performance, Neurogenesis, and Dendritic Complexity of Neurons in Mice Analyzed by Behavioral Tests and Golgi Staining of Brain Tissue." *Journal of Stem Cells & Regenerative Medicine* 17 (1): 35–41. https://doi.org/10.46582/JSRM.1701005.

Barboza, Joice Nascimento, Carlos da Silva Maia Bezerra Filho, Renan Oliveira Silva, Jand Venes R. Medeiros, and Damião Pergentino de Sousa. 2018. "An Overview on the Anti-Inflammatory Potential and Antioxidant Profile of Eugenol." *Oxidative Medicine and Cellular Longevity* 2018. https://doi.org/10.1155/2018/3957262.

Bargiotas, Panagiotis, and Spyridon Konitsiotis. 2013. "Levodopa-Induced Dyskinesias in Parkinson's Disease: Emerging Treatments." *Neuropsychiatric Disease and Treatment* 9 (October): 1605–17. https://doi.org/10.2147/NDT.S36693.

Barot, Jeetprakash, and Bhagawati Saxena. 2021. "Therapeutic Effects of Eugenol in a Rat Model of Traumatic Brain Injury: A Behavioral, Biochemical, and Histological Study." *Journal of Traditional and Complementary Medicine* 11 (4): 318–27. https://doi.org/10.1016/J.JTCME.2021.01.003.

Behl, C., J. B. Davis, R. Lesley, and D. Schubert. 1994. "Hydrogen Peroxide Mediates Amyloid Beta Protein Toxicity." *Cell* 77 (6): 817–27. https://doi.org/10.1016/0092-8674(94)90131-7.

Bergman, Hagai, and Günther Deuschl. 2002. "Pathophysiology of Parkinson's Disease: From Clinical Neurology to Basic Neuroscience and Back." *Movement Disorders: Official Journal of the Movement Disorder Society* 17 Suppl 3 (SUPPL. 3). https://doi.org/10.1002/MDS.10140.

Bethea, Cynthia L., Maria Luisa Centeno, and Judy L. Cameron. 2008. "Neurobiology of Stress-Induced Reproductive Dysfunction in Female Macaques." *Molecular Neurobiology* 38 (3): 199–230. https://doi.org/10.1007/S12035-008-8042-Z.

Bhavsar, Nikhil, Bhagawati Saxena, and Jigna Shah. 2022. "Blood-Brain Barrier and Central Nervous System Drug Delivery: Challenges and Opportunities." *Nanocarriers for Drug-Targeting Brain Tumors* 31–48. https://doi.org/10.1016/B978-0-323-90773-6.00005-1.

Björkholm, Carl, and Lisa M. Monteggia. 2016. "BDNF - a Key Transducer of Antidepressant Effects." *Neuropharmacology* 102 (March): 72–79. https://doi.org/10.1016/J.NEUROPHARM.2015.10.034.

Bor-Seng-Shu, Edson, Eberval G. Figueiredo, Erich Talamoni Fonoff, Yasunori Fujimoto, Ronney B. Panerai, and Manoel Jacobsen Teixeira. 2013. "Decompressive Craniectomy and Head Injury: Brain Morphometry, ICP, Cerebral Hemodynamics, Cerebral Microvascular Reactivity, and Neurochemistry." *Neurosurgical Review* 36 (3): 361–70. https://doi.org/10.1007/S10143-013-0453-2.

Bremner, J. Douglas, Meena Narayan, Eric R. Anderson, Lawrence H. Staib, Helen L. Miller, and Dennis S. Charney. 2000. "Hippocampal Volume Reduction in Major Depression." *The American Journal of Psychiatry* 157 (1): 115–17. https://doi.org/10.1176/AJP.157.1.115.

Brosch, Katharina, Frederike Stein, Simon Schmitt, Julia Katharina Pfarr, Kai G. Ringwald, Florian Thomas-Odenthal, and Tina Meller, et al. 2022. "Reduced Hippocampal Gray Matter Volume Is a Common Feature of Patients With Major Depression, Bipolar Disorder, and Schizophrenia Spectrum Disorders." *Molecular Psychiatry* 27 (10). https://doi.org/10.1038/S41380-022-01687-4.

Burt, Sara. 2004. "Essential Oils: Their Antibacterial Properties and Potential Applications in Foods - A Review." *International Journal of Food Microbiology* 94 (3): 223–53. https://doi.org/10.1016/j.ijfoodmicro.2004.03.022.

Chamorro, Ángel, Ulrich Dirnagl, Xabier Urra, and Anna M. Planas. 2016. "Neuroprotection in Acute Stroke: Targeting Excitotoxicity, Oxidative and Nitrosative Stress, and Inflammation." *The Lancet. Neurology* 15 (8): 869–81. https://doi.org/10.1016/S1474-4422(16)00114-9.

Charan Raja, Mamilla R. 2015. "Versatile and Synergistic Potential of Eugenol: A Review." *Pharmaceutica Analytica Acta* 06 (05). https://doi.org/10.4172/2153-2435.1000367.

Chauhan, Heena, Pawan Gupta, Bhagawati Saxena, Heena Chauhan, Pawan Gupta, and Bhagawati Saxena. 2022. "MicroRNAs as Future Treatment Tools and Diagnostic Biomarkers in Alzheimer's Disease." *Alzheimer's Disease*. https://doi.org/10.5772/INTECHOPEN.103173.

Chavali, Vijaya Durga, Milee Agarwal, Vivek Kumar Vyas, and Bhagawati Saxena. 2020. "Neuroprotective Effects of Ethyl Pyruvate Against Aluminum Chloride-Induced Alzheimer's Disease in Rats via Inhibiting Toll-Like Receptor 4." *Journal of Molecular Neuroscience: MN* 70 (6): 836–50. https://doi.org/10.1007/S12031-020-01489-9.

Cheignon, C., M. Tomas, D. Bonnefont-Rousselot, P. Faller, C. Hureau, and F. Collin. 2018. "Oxidative Stress and the Amyloid Beta Peptide in Alzheimer's Disease." *Redox Biology* 14: 450–64. https://doi.org/10.1016/J.REDOX.2017.10.014.

Choi, Yoo Keum, Geum Sil Cho, Sunyoung Hwang, Byung Woo Kim, Ji H. Lim, Jae Chul Lee, Hyoung Chun Kim, Won Ki Kim, and Yeong Sik Kim. 2010. "Methyleugenol Reduces Cerebral Ischemic Injury by Suppression of Oxidative Injury and Inflammation." *Free Radical Research* 44 (8): 925–35. https://doi.org/10.3109/10715762.2010.490837.

Choi, Heesun, Chaeyoung Kim, Hyundong Song, Moon Yong Cha, Hyun Jin Cho, Sung Min Son, Haeng Jun Kim, and Inhee Mook-Jung. 2019. "Amyloid β-Induced Elevation of O-GlcNAcylated c-Fos Promotes Neuronal Cell Death." *Aging Cell* 18 (1). https://doi.org/10.1111/ACEL.12872.

Choudhury, Sudhansu S., Leena Bashyam, Nalini Manthapuram, Prasanth Bitla, Padmasree Kollipara, and Sarada D. Tetali. 2014. "Ocimum Sanctum Leaf Extracts Attenuate Human Monocytic (THP-1) Cell Activation." *Journal of Ethnopharmacology* 154 (1): 148–55. https://doi.org/10.1016/J.JEP.2014.03.049.

Chugh, Chandril. 2019. "Acute Ischemic Stroke: Management Approach." *Indian Journal of Critical Care Medicine : Peer-Reviewed, Official Publication of Indian Society of Critical Care Medicine* 23 (Suppl 2): S140–46. https://doi.org/10.5005/JP-JOURNALS-10071-23192.

Chung, T. H., K. H.O. Deane, S. Ghazi-Noori, H. Rickards, and C. E. Clarke. 2003. "Systematic Review of Antidepressant Therapies in Parkinson's Disease." *Parkinsonism and Related Disorders* 10 (2): 59–65. https://doi.org/10.1016/S1353-8020(03)00108-1.

Chung, G, J. N Rhee, S. J Jung, J. S Kim, and S. B Oh. 2008. "Modulation of CaV2.3 Calcium Channel Currents by Eugenol." *Journal of Dental Research* 87 (2): 137–41. https://doi.org/10.1177/154405910808700201.

Citron, Martin. 2002. "Alzheimer's Disease: Treatments in Discovery and Development." *Nature Neuroscience* 5 Suppl (11s): 1055–57. https://doi.org/10.1038/NN940.

Cole, Jeffrey T., Angela Yarnell, William S. Kean, Eric Gold, Bobbi Lewis, Ming Ren, and David C. McMullen, et al. 2011. "Craniotomy: True Sham for Traumatic Brain Injury, or a Sham of a Sham?" *Journal of Neurotrauma* 28 (3): 359–69. https://doi.org/10.1089/NEU.2010.1427.

Diaz, MR, and JM Sembrano. 1985. "A Comparative Study of the Efficacy of Garlic and Eugenol as Palliative Agents Against Dental Pain of Pulpal Origin - PubMed." *The Journal of the Philippine Dental Association* 35 (1): 3–10. https://pubmed.ncbi.nlm.nih.gov/3869252/.

Dubey, Kriti, Bibin G. Anand, Dolat Singh Shekhawat, and Karunakar Kar. 2017. "Eugenol Prevents Amyloid Formation of Proteins and Inhibits Amyloid-Induced Hemolysis." *Scientific Reports* 7: 40744. https://doi.org/10.1038/SREP40744.

Ellis, J. Michael, and Matthew J. Fell. 2017. "Current Approaches to the Treatment of Parkinson's Disease." *Bioorganic & Medicinal Chemistry Letters* 27 (18): 4247–55. https://doi.org/10.1016/J.BMCL.2017.07.075.

Feigin, Valery L., Gregory A. Roth, Mohsen Naghavi, Priya Parmar, Rita Krishnamurthi, Sumeet Chugh, and George A. Mensah, et al. 2016. "Global Burden of Stroke and Risk Factors in 188 Countries, During 1990-2013: A Systematic Analysis for the Global Burden of Disease Study 2013." *The Lancet. Neurology* 15 (9): 913–24. https://doi.org/10.1016/S1474-4422(16)30073-4.

Fiebich, Bernd L., Carla Ribeiro Alvares Batista, Soraya Wilke Saliba, Nizar M. Yousif, and Antonio Carlos Pinheiro de Oliveira. 2018. "Role of Microglia TLRs in Neurodegeneration." *Frontiers in Cellular Neuroscience* 12. https://doi.org/10.3389/FNCEL.2018.00329.

Forouzanfar, M. H., L. Alexander, V. F. Bachman, S. Biryukov, M. Brauer, D. Casey, and M. M. Coates, et al. 2015. "Global, Regional, and National Comparative Risk Assessment of 79 Behavioural, Environmental and Occupational, and Metabolic Risks or Clusters of Risks in 188 Countries, 1990-2013: A Systematic Analysis for the Global Burden of Disease Study 2013." *Lancet (London, England)* 386 (10010): 2287–2323. https://doi.org/10.1016/S0140-6736(15)00128-2.

Garabadu, Debapriya, Ankit Shah, Ausaf Ahmad, Vijaya B. Joshi, Bhagawati Saxena, Gautam Palit, and Sairam Krishnamurthy. 2011. "Eugenol as an Anti-Stress Agent: Modulation of Hypothalamic-Pituitary-Adrenal Axis and Brain Monoaminergic Systems in a Rat Model of Stress." *Stress (Amsterdam, Netherlands)* 14 (2): 145–55. https://doi.org/10.3109/10253890.2010.521602.

Garabadu, Debapriya, Ankit Shah, Sanjay Singh, and Sairam Krishnamurthy. 2015. "Protective Effect of Eugenol Against Restraint Stress-Induced Gastrointestinal Dysfunction: Potential Use in Irritable Bowel Syndrome." *Pharmaceutical Biology* 53 (7): 968–74. https://doi.org/10.3109/13880209.2014.950674.

Garabadu, Debapriya, and Mahima Sharma. 2019. "Eugenol Attenuates Scopolamine-Induced Hippocampal Cholinergic, Glutamatergic, and Mitochondrial Toxicity in Experimental Rats." *Neurotoxicity Research* 35 (4): 848–59. https://doi.org/10.1007/S12640-019-0008-6.

Gelenberg, Alan J., Madhukar H. Trivedi, A. John Rush, Michael E. Thase, Robert Howland, Daniel N. Klein, and Susan G. Kornstein, et al. 2003. "Randomized, Placebo-Controlled Trial of Nefazodone Maintenance Treatment in Preventing Recurrence in Chronic Depression." *Biological Psychiatry* 54 (8): 806–17. https://doi.org/10.1016/S0006-3223(02)01971-6.

Gschwind, Martin, and Gerda Huber. 1995. "Apoptotic Cell Death Induced by Beta-Amyloid 1-42 Peptide Is Cell Type Dependent." *Journal of Neurochemistry* 65 (1): 292–300. https://doi.org/10.1046/J.1471-4159.1995.65010292.X.

Helmy, Hebatullah, Nermin Abdel Hamid Sadik, Laila Badawy, and Noha H. Sayed. 2022. "Mechanistic Insights into the Protective Role of Eugenol Against Stress-Induced Reproductive Dysfunction in Female Rat Model." *Chemico-Biological Interactions* 367: 110181. https://doi.org/10.1016/J.CBI.2022.110181.

Honeybul, Stephen. 2017. "Decompressive Craniectomy for Severe Traumatic Brain Injury Reduces Mortality but Increases Survival With Severe Disability." *Evidence-Based Medicine* 22 (2): 61. https://doi.org/10.1136/EBMED-2016-110616.

Hu, Qiao, Meifang Zhou, and Shuyong wei. 2018. "Progress on the Antimicrobial Activity Research of Clove Oil and Eugenol in the Food Antisepsis Field." *Journal of Food Science* 83 (6): 1476–83. https://doi.org/10.1111/1750-3841.14180.

Irie, Yoshifumi. 2006. "Effects of Eugenol on the Central Nervous System: Its Possible Application to Treatment of Alzheimers Disease, Depression, and Parkinsons Disease." *Current Bioactive Compounds* 2 (1): 57–66. https://doi.org/10.2174/157340721060201057.

Irie, Yoshifumi, Nanae Itokazu, Naoko Anjiki, Atsushi Ishige, Kenji Watanabe, and Wing Ming Keung. 2004. "Eugenol Exhibits Antidepressant-Like Activity in Mice and Induces Expression of Metallothionein-III in the Hippocampus." *Brain Research* 1011 (2): 243–46. https://doi.org/10.1016/j.brainres.2004.03.040.

Irie, Yoshifumi, and Wing Ming Keung. 2001. "Metallothionein-III Antagonizes the Neurotoxic and Neurotrophic Effects of Amyloid Beta Peptides." *Biochemical and Biophysical Research Communications* 282 (2): 416–20. https://doi.org/10.1006/BBRC.2001.4594.

Ito, Masae, Keiko Murakami, and Masataka Yoshino. 2005. "Antioxidant Action of Eugenol Compounds: Role of Metal Ion in the Inhibition of Lipid Peroxidation." *Food and Chemical Toxicology : An International Journal Published for the British Industrial Biological Research Association* 43 (3): 461–66. https://doi.org/10.1016/J.FCT.2004.11.019.

Jankovic, J. 2008. "Parkinson's Disease: Clinical Features and Diagnosis." *Journal of Neurology, Neurosurgery, and Psychiatry* 79 (4): 368–76. https://doi.org/10.1136/JNNP.2007.131045.

Jankovic, Joseph, and L. Giselle Aguilar. 2008. "Current Approaches to the Treatment of Parkinson's Disease." *Neuropsychiatric Disease and Treatment* 4 (4): 743–57. https://doi.org/10.2147/NDT.S2006.

Jayasingh Chellammal, Hanish Singh, Alagarsamy Veerachamy, Dhani Ramachandran, Sridhar Babu Gummadi, Mohamed Mansor Manan, and Narsimha Reddy Yellu. 2019. "Neuroprotective Effects of 1`δ-1`-Acetoxyeugenol Acetate on Aβ(25-35) Induced Cognitive Dysfunction in Mice." *Biomedicine & Pharmacotherapy = Biomedecine & Pharmacotherapie* 109: 1454–61. https://doi.org/10.1016/J.BIOPHA.2018.10.189.

Kabuto, Hideaki, Mika Tada, and Masahiro Kohno. 2007. "Eugenol [2-Methoxy-4-(2-Propenyl)Phenol] Prevents 6-Hydroxydopamine-Induced Dopamine Depression and Lipid Peroxidation Inductivity in Mouse Striatum." *Biological & Pharmaceutical Bulletin* 30 (3): 423–27. https://doi.org/10.1248/BPB.30.423.

Kabuto, Hideaki, and Tomoko T. Yamanushi. 2011. "Effects of Zingerone [4-(4-Hydroxy-3-Methoxyphenyl)-2-Butanone] and Eugenol [2-Methoxy-4-(2-Propenyl)Phenol] on the Pathological Progress in the 6-Hydroxydopamine-Induced Parkinson's Disease Mouse Model." *Neurochemical Research* 36 (12): 2244–49. https://doi.org/10.1007/S11064-011-0548-5.

Khalil, Anees Ahmed, Ubaid Ur Rahman, Moazzam Rafiq Khan, Amna Sahar, Tariq Mehmood, and Muneeb Khan. 2017. "Essential Oil Eugenol: Sources, Extraction Techniques and Nutraceutical Perspectives." *RSC Advances* 7 (52): 32669–81. https://doi.org/10.1039/C7RA04803C.

Khellaf, Abdelhakim, Danyal Zaman Khan, and Adel Helmy. 2019. "Recent Advances in Traumatic Brain Injury." *Journal of Neurology* 266 (11): 2878–89. https://doi.org/10.1007/s00415-019-09541-4.

Kong, L. D., Christopher H.K. Cheng, and R. X. Tan. 2004. "Inhibition of MAO a and B by Some Plant-Derived Alkaloids, Phenols and Anthraquinones." *Journal of Ethnopharmacology* 91 (2–3): 351–55. https://doi.org/10.1016/j.jep.2004.01.013.

Lad, Krishna A., Anurag Maheshwari, and Bhagawati Saxena. 2019. "Repositioning of an Anti-Depressant Drug, Agomelatine as Therapy for Brain Injury Induced by Craniotomy." *Drug Discoveries & Therapeutics* 13 (4): 189–97. https://doi.org/10.5582/DDT.2019.01056.

Lee, Ya Yun, Shan Ling Hung, Sheng Fang Pai, Yuan Ho Lee, and Shue Fen Yang. 2007. "Eugenol Suppressed the Expression of Lipopolysaccharide-Induced Proinflammatory Mediators in Human Macrophages." *Journal of Endodontics* 33 (6): 698–702. https://doi.org/10.1016/J.JOEN.2007.02.010.

Lekoubou, Alain, Jean Joë L. Awoumou, and André Pascal Kengne. 2017. "Incidence of Seizure in Stroke Patients Treated With Recombinant Tissue Plasminogen Activator: A Systematic Review and Meta-Analysis." *International Journal of Stroke: Official*

Journal of the International Stroke Society 12 (9): 923–31. https://doi.org/10.1177/1747493017729239.

Liang, Zi Hao, Xiao Hui Cheng, Zhi Gang Ruan, Han Wang, Shan Shan Li, Jing Liu, Guo Ying Li, and Su Min Tian. 2015. "Protective Effects of Components of the Chinese Herb Grassleaf Sweetflag Rhizome on PC12 Cells Incubated With Amyloid-Beta42." *Neural Regeneration Research* 10 (8): 1292–97. https://doi.org/10.4103/1673-5374.162762.

Liu, Yan Mei, Hui Ran Fan, Shining Deng, Tailin Zhu, Yuhua Yan, Wei Hong Ge, Wei Guang Li, and Fei Li. 2019. "Methyleugenol Potentiates Central Amygdala GABAergic Inhibition and Reduces Anxiety." *The Journal of Pharmacology and Experimental Therapeutics* 368 (1): 1–10. https://doi.org/10.1124/JPET.118.250779.

Liu, Zhibin, Wenmin Niu, Xiaohang Yang, and Yuan Wang. 2013. "Effects of Combined Acupuncture and Eugenol on Learning-Memory Ability and Antioxidation System of Hippocampus in Alzheimer Disease Rats via Olfactory System Stimulation." *Journal of Traditional Chinese Medicine = Chung i Tsa Chih Ying Wen Pan* 33 (3): 399–402. https://doi.org/10.1016/S0254-6272(13)60186-7.

Long, Justin M., and David M. Holtzman. 2019. "Alzheimer Disease: An Update on Pathobiology and Treatment Strategies." *Cell* 179 (2): 312–39. https://doi.org/10.1016/J.CELL.2019.09.001.

Lyketsos, Constantine G., and Hochang B. Lee. 2004. "Diagnosis and Treatment of Depression in Alzheimer's Disease. A Practical Update for the Clinician." *Dementia and Geriatric Cognitive Disorders* 17 (1–2): 55–64. https://doi.org/10.1159/000074277.

Malberg, Jessica E., Amelia J. Eisch, Eric J. Nestler, and Ronald S. Duman. 2000. "Chronic Antidepressant Treatment Increases Neurogenesis in Adult Rat Hippocampus." *The Journal of Neuroscience : The Official Journal of the Society for Neuroscience* 20 (24): 9104–10. https://doi.org/10.1523/JNEUROSCI.20-24-09104.2000.

Mandrekar-Colucci, Shweta, and Gary E. Landreth. 2010. "Microglia and Inflammation in Alzheimer's Disease." *CNS & Neurological Disorders Drug Targets* 9 (2): 156–67. https://doi.org/10.2174/187152710791012071.

Marchese, Anna, Ramona Barbieri, Erika Coppo, Ilkay Erdogan Orhan, Maria Daglia, Seyed Fazel Nabavi, Morteza Izadi, Mohammad Abdollahi, Seyed Mohammad Nabavi, and Marjan Ajami. 2017. "Antimicrobial Activity of Eugenol and Essential Oils Containing Eugenol: A Mechanistic Viewpoint." *Critical Reviews in Microbiology* 43 (6): 668–89. https://doi.org/10.1080/1040841X.2017.1295225.

McEwen, Bruce S. 2000. "The Neurobiology of Stress: From Serendipity to Clinical Relevance." *Brain Research* 886 (1–2): 172–89. https://doi.org/10.1016/S0006-8993(00)02950-4.

Mesole, Samuel Bolaji, Okpanachi Omachonu Alfred, Uthman Ademola Yusuf, Lwiindi Lukubi, and Dailesi Ndhlovu. 2020. "Apoptotic Inducement of Neuronal Cells by Aluminium Chloride and the Neuroprotective Effect of Eugenol in Wistar Rats." *Oxidative Medicine and Cellular Longevity* 2020: 8425643. https://doi.org/10.1155/2020/8425643.

Moreira Vasconcelos, Carlos Franciney, Nívea Maria da Cunha Ferreira, Nayanne Hardy Lima Pontes, Thomas Dominik de Sousa dos Reis, Ricardo Basto Souza, Francisco Eduardo Aragão Catunda Junior, Lissiana Magna Vasconcelos Aguiar, and Rodrigo Maranguape Silva da Cunha. 2020. "Eugenol and Its Association With Levodopa in 6-Hydroxydopamine-Induced Hemiparkinsonian Rats: Behavioural and Neurochemical Alterations." *Basic & Clinical Pharmacology & Toxicology* 127 (4): 287–302. https://doi.org/10.1111/BCPT.13425.

Mustapha, Musa, and Che Norma Mat Taib. 2021. "MPTP-Induced Mouse Model of Parkinson's Disease: A Promising Direction of Therapeutic Strategies." *Bosnian Journal of Basic Medical Sciences* 21 (4): 422–33. https://doi.org/10.17305/BJBMS.2020.5181.

Naidu, K. A. 1995. "Eugenol–an Inhibitor of Lipoxygenase-Dependent Lipid Peroxidation." *Prostaglandins, Leukotrienes, and Essential Fatty Acids* 53 (5): 381–83. https://doi.org/10.1016/0952-3278(95)90060-8.

Najem, Dema, Kerry Rennie, Maria Ribecco-Lutkiewicz, Dao Ly, Julie Haukenfrers, Qing Liu, Munyao Nzau, Douglas D. Fraser, and Mahmud Bani-Yaghoub. 2018. "Traumatic Brain Injury: Classification, Models, and Markers." *Biochemistry and Cell Biology = Biochimie et Biologie Cellulaire* 96 (4): 391–406. https://doi.org/10.1139/BCB-2016-0160.

National Toxicology Program. 1983. "Carcinogenesis Studies of Eugenol (CAS No. 97-53-0) in F344/N Rats and B6C3F1 Mice (Feed Studies) – PubMed." *National Toxicology Program Technical Report Series* 223: 1–159. https://pubmed.ncbi.nlm.nih.gov/12778213/.

Ng, Si Yun, and Alan Yiu Wah Lee. 2019. "Traumatic Brain Injuries: Pathophysiology and Potential Therapeutic Targets." *Frontiers in Cellular Neuroscience* 13: 1–23. https://doi.org/10.3389/fncel.2019.00528.

Orellana, Juan A., Kenji F. Shoji, Verónica Abudara, Pascal Ezan, Edwige Amigou, Pablo J. Sáez, Jean X. Jiang, Christian C. Naus, Juan C. Sáez, and Christian Giaume. 2011. "Amyloid β-Induced Death in Neurons Involves Glial and Neuronal Hemichannels." *The Journal of Neuroscience : The Official Journal of the Society for Neuroscience* 31 (13): 4962–77. https://doi.org/10.1523/JNEUROSCI.6417-10.2011.

Pallo, Susanne P., John Dimaio, Alexis Cook, Bradley Nilsson, and Gail V.W. Johnson. 2016. "Mechanisms of Tau and Aβ-Induced Excitotoxicity." *Brain Research* 1634: 119–31. https://doi.org/10.1016/J.BRAINRES.2015.12.048.

Papageorgiou, Ismini E., Andrea Lewen, Lukas v. Galow, Tiziana Cesetti, Jörg Scheffel, Tommy Regen, Uwe Karsten Hanisch, and Oliver Kann. 2016. "TLR4-Activated Microglia Require IFN-γ to Induce Severe Neuronal Dysfunction and Death in Situ." *Proceedings of the National Academy of Sciences of the United States of America* 113 (1): 212–17. https://doi.org/10.1073/PNAS.1513853113.

Parks, Janice K., Trisha S. Smith, Patricia A. Trimmer, James P. Bennett, and Davis Parker Jr. 2001. "Neurotoxic Abeta Peptides Increase Oxidative Stress in Vivo Through NMDA-Receptor and Nitric-Oxide-Synthase Mechanisms, and Inhibit Complex IV Activity and Induce a Mitochondrial Permeability Transition in Vitro." *Journal of Neurochemistry* 76 (4): 1050–56. https://doi.org/10.1046/J.1471-4159.2001.00112.X.

Pramod, Kannissery, M. R. Aji Alex, Manisha Singh, Shweta Dang, Shahid H. Ansari, and Javed Ali. 2016. "Eugenol Nanocapsule for Enhanced Therapeutic Activity Against Periodontal Infections." *Journal of Drug Targeting* 24 (1): 24–33. https://doi.org/10.3109/1061186X.2015.1052071.

Prasad, Sathya N., and Muralidhara. 2013. "Neuroprotective Efficacy of Eugenol and Isoeugenol in Acrylamide-Induced Neuropathy in Rats: Behavioral and Biochemical Evidence." *Neurochemical Research* 38 (2): 330–45. https://doi.org/10.1007/S11064-012-0924-9.

Said, Mahmoud M., and Marwa M. Abd Rabo. 2017. "Neuroprotective Effects of Eugenol Against Aluminiuminduced Toxicity in the Rat Brain." *Arhiv Za Higijenu Rada i Toksikologiju* 68 (1): 27–37. https://doi.org/10.1515/AIHT-2017-68-2878.

Santarelli, Luca, Michael Saxe, Cornelius Gross, Alexandre Surget, Fortunato Battaglia, Stephanie Dulawa, and Noelia Weisstaub, et al. 2003. "Requirement of Hippocampal Neurogenesis for the Behavioral Effects of Antidepressants." *Science (New York, N.Y.)* 301 (5634): 805–9. https://doi.org/10.1126/SCIENCE.1083328.

Saxena, Bhagawati, and Vijaya Durga Chavali. 2019. "The Role of Toll Like Receptor 4 in Pathogenesis of Alzheimer's Disease Induced by Aluminum Chloride." *International Journal of Emerging Technologies and Innovative Research* 6 (4): 96–99. https://www.jetir.org/view?paper=JETIRBF06019.

Saxena, Bhagawati, and Urmila Saxena. 2012. "View of Anti-Stress Effects of Cinnamon (Cassia Zelynicum) Bark Extract in Cold Restraint Stress Rat Model." *International Journal of Research and Development in Pharmacy and Life Sciences* 1 (1): 28–31. https://ijrdpl.com/index.php/ijrdpl/article/view/155/159.

Schober, Andreas. 2004. "Classic Toxin-Induced Animal Models of Parkinson's Disease: 6-OHDA and MPTP." *Cell and Tissue Research* 318 (1): 215–24. https://doi.org/10.1007/S00441-004-0938-Y.

Selkoe, Dennis J., and Dale Schenk. 2003. "Alzheimer's Disease: Molecular Understanding Predicts Amyloid-Based Therapeutics." *Annual Review of Pharmacology and Toxicology* 43: 545–84. https://doi.org/10.1146/ANNUREV.PHARMTOX.43.100901.140248.

Selye, Hans. 1936. "A Syndrome Produced by Diverse Nocuous Agents." *Nature* 138 (3479): 32–32. https://doi.org/10.1038/138032a0.

Sen, P, P C Maiti, S Puri, A Ray, N A Audulov, and A v Valdman. 1992. "Mechanism of Anti-Stress Activity of Ocimum Sanctum Linn, Eugenol and Tinospora Malabarica in Experimental Animals – PubMed." *Indian Journal of Experimental Biology* 30 (7): 592–96. https://pubmed.ncbi.nlm.nih.gov/1459632/.

Seo, Haengsoo, Hai Ying Li, Edward Perez-Reyes, and Jung Ha Lee. 2013. "Effects of Eugenol on T-Type Ca^{2+} Channel Isoforms." *The Journal of Pharmacology and Experimental Therapeutics* 347 (2): 310–17. https://doi.org/10.1124/JPET.113.207936.

Shekhar, Shashank, Yudhishthir Yadav, Amrendra Pratap Singh, Rashmita Pradhan, Gaurav Rajesh Desai, A. B. Dey, and Sharmistha Dey. 2018. "Neuroprotection by Ethanolic Extract of Syzygium Aromaticum in Alzheimer's Disease Like Pathology via Maintaining Oxidative Balance Through SIRT1 Pathway." *Experimental Gerontology* 110: 277–83. https://doi.org/10.1016/J.EXGER.2018.06.026.

Shi, Qianyu, Quancheng Cheng, and Chunhua Chen. 2021. "The Role of Autophagy in the Pathogenesis of Ischemic Stroke." *Current Neuropharmacology* 19 (5): 629. https://doi.org/10.2174/1570159X18666200729101913.

Shirayama, Yukihiko, Andrew C.H. Chen, Shin Nakagawa, David S. Russell, and Ronald S. Duman. 2002. "Brain-Derived Neurotrophic Factor Produces Antidepressant Effects in Behavioral Models of Depression." *The Journal of Neuroscience : The Official Journal of the Society for Neuroscience* 22 (8): 3251–61. https://doi.org/10.1523/JNEUROSCI.22-08-03251.2002.

Singh, Anand Kumar, Sunil S. Dhamanigi, and Mohammed Asad. 2009. "Anti-Stress Activity of Hydro-Alcoholic Extract of Eugenia Caryophyllus Buds (Clove)." *Indian Journal of Pharmacology* 41 (1): 28–31. https://doi.org/10.4103/0253-7613.48889.

Singh, Varsha, and Rupali Panwar. 2014. "In Vivo Antioxidative and Neuroprotective Effect of 4-Allyl-2-Methoxyphenol Against Chlorpyrifos-Induced Neurotoxicity in Rat Brain." *Molecular and Cellular Biochemistry* 388 (1–2): 61–74. https://doi.org/10.1007/S11010-013-1899-9.

Siyal, FahadJibran, Zahida Memon, Rehan Ahmed Siddiqui, Zara Aslam, Uzair Nisar, Rehan Imad, and Muhammad Raza Shah. 2020. "Eugenol and Liposome-Based Nanocarriers Loaded With Eugenol Protect Against Anxiolytic Disorder via Down Regulation of Neurokinin-1 Receptors in Mice." *Pakistan Journal of Pharmaceutical Sciences* 33 (5): 2275–84. https://pubmed.ncbi.nlm.nih.gov/33832901/.

Siyal, F. J., R. A. Siddiqui, Z. Memon, Z. Aslam, U. Nisar, R. Imad, and M. R. Shah. 2021. "Eugenol and Its Liposome-Based Nano Carrier Reduce Anxiety by Inhibiting Glyoxylase-1 Expression in Mice." *Brazilian Journal of Biology = Revista Brasleira de Biologia* 83. https://doi.org/10.1590/1519-6984.251219.

Sun, Xiaowei, Dongyan Wang, Tingting Zhang, Xuejian Lu, Fangfang Duan, Lili Ju, Xiaotong Zhuang, and Xicheng Jiang. 2020. "Eugenol Attenuates Cerebral Ischemia-Reperfusion Injury by Enhancing Autophagy via AMPK-MTOR-P70S6K Pathway." *Frontiers in Pharmacology* 11. https://doi.org/10.3389/FPHAR.2020.00084.

Taheri, P., P. Yaghmaei, H. Sepasi Tehrani, and A. Ebrahim-Habibi. 2019. "Effects of Eugenol on Alzheimer's Disease-Like Manifestations in Insulin- and Aβ-Induced Rat Models." *Neurophysiology* 51 (2): 114–19. https://doi.org/10.1007/s11062-019-09801-z.

Taira, Junsei, Takeshi Ikemoto, Tohru Yoneya, Akifumi Hagi, Akira Murakami, and Keisuke Makino. 1992. "Essential Oil Phenyl Propanoids. Useful as. OH Scavengers?" *Free Radical Research Communications* 16 (3): 197–204. https://doi.org/10.3109/10715769209049172.

Tanji, Kunikazu, Yoshifumi Irie, Yoko Uchida, Fumiaki Mori, Kei Satoh, Yutaka Mizushima, and Koichi Wakabayashi. 2003. "Expression of Metallothionein-III Induced by Hypoxia Attenuates Hypoxia-Induced Cell Death in Vitro." *Brain Research* 976 (1): 125–29. https://doi.org/10.1016/S0006-8993(03)02633-7.

Tao, Guoxin, Yoshifumi Irie, Dian Jun Li, and Ming Keung Wing. 2005. "Eugenol and Its Structural Analogs Inhibit Monoamine Oxidase A and Exhibit Antidepressant-Like Activity." *Bioorganic & Medicinal Chemistry* 13 (15): 4777–88. https://doi.org/10.1016/J.BMC.2005.04.081.

Vora, Urmi, Vivek Kumar Vyas, Pranay Wal, and Bhagawati Saxena. 2022. "Effects of Eugenol on the Behavioral and Pathological Progression in the MPTP-Induced Parkinson's Disease Mouse Model." *Drug Discoveries & Therapeutics* 16 (4): 154–63. https://doi.org/10.5582/DDT.2022.01026.

Walter, Silke, Maryse Letiembre, Yang Liu, Holger Heine, Botond Penke, Wenlin Hao, and Barbara Bode, et al. 2007. "Role of the Toll-Like Receptor 4 in Neuroinflammation in Alzheimer's Disease." *Cellular Physiology and Biochemistry : International Journal of Experimental Cellular Physiology, Biochemistry, and Pharmacology* 20 (6): 947–56. https://doi.org/10.1159/000110455.

Wang, Hongmei, Jianfang Ma, Yuyan Tan, Zhiquan Wang, Chengyu Sheng, Shengdi Chen, and Jianqing Ding. 2010. "Amyloid-Beta1-42 Induces Reactive Oxygen Species-Mediated Autophagic Cell Death in U87 and SH-SY5Y Cells." *Journal of Alzheimer's Disease : JAD* 21 (2): 597–610. https://doi.org/10.3233/JAD-2010-091207.

Wang, Pei, Bo Zong Shao, Zhiqiang Deng, Shi Chen, Zhenyu Yue, and Chao Yu Miao. 2018. "Autophagy in Ischemic Stroke." *Progress in Neurobiology* 163–164: 98–117. https://doi.org/10.1016/J.PNEUROBIO.2018.01.001.

Wie, Myung Bok, Moo Ho Won, Keun Ho Lee, Joon Hyun Shin, Jae Chul Lee, Hong Won Suh, Dong Keun Song, and Yung Hi Kim. 1997. "Eugenol Protects Neuronal Cells from Excitotoxic and Oxidative Injury in Primary Cortical Cultures." *Neuroscience Letters* 225 (2): 93–96. https://doi.org/10.1016/S0304-3940(97)00195-X.

Wong, Ma Li, and Julio Licinio. 2001. "Research and Treatment Approaches to Depression." *Nature Reviews. Neuroscience* 2 (5): 343–51. https://doi.org/10.1038/35072566.

Won, Moo Ho, Jae Chul Lee, Yung Hi Kim, Dong Keun Song, Hong Won Suh, Yang Seok Oh, Jung Hoon Kim, Tae Kyun Shin, Young Jae Lee, and Myung Bok Wie. 1998. "Postischemic Hypothermia Induced by Eugenol Protects Hippocampal Neurons from Global Ischemia in Gerbils." *Neuroscience Letters* 254 (2): 101–4. https://doi.org/10.1016/S0304-3940(98)00664-8.

Yang, Pan, and Fusheng Sun. 2021. "Aducanumab: The First Targeted Alzheimer's Therapy." *Drug Discoveries & Therapeutics* 15 (3): 166–68. https://doi.org/10.5582/DDT.2021.01061.

Yao, Xiaoxi, Rui Yao, Fengzhen Huang, and Jiping Yi. 2019. "LncRNA SNHG12 as a Potent Autophagy Inducer Exerts Neuroprotective Effects Against Cerebral Ischemia/Reperfusion Injury." *Biochemical and Biophysical Research Communications* 514 (2): 490–96. https://doi.org/10.1016/J.BBRC.2019.04.158.

11 Deciphering the Deep Learning and Machine Learning Tactics in Advancement of Neuroprotection by Phytochemicals

Gadde Shareena and Dileep Kumar

11.1 INTRODUCTION

Recent decades have witnessed tremendous breakthroughs and research advancements in the medical field that are changing the way numerous chronic diseases are being treated. With an increasing amount of therapeutics being developed in the current pharmaceutical marketplace, several instances of drug toxicity, poor tolerability, curability, and high healthcare costs are increasing. Dementia-related neurodegenerative disorders (NDD) and neurotoxicity notably cause chronic disorder states worldwide (Hajat and Stein, 2018; Shin, 2022). Dementia-related NDD ranks as the seventh leading mortality cause, affecting nearly 55.2 million individuals worldwide (Ponjoan et al, 2019).

The primary constituents of natural products, terpenes, flavonoids, and carotenoids (CT), are often extracted from several plants, including *Aloe barbadensis* (Asphodelaceae), *Berberis vulgaris (Berberidaceae), Curcuma longa* (Zingiberaceae), and *Ginkgo biloba* (Ginkgoaceae), and are often distinguished for their structural similarity (Veeresham, 2012). Throughout the decades of phytochemical research, plant-derived extracts have undergone extensive preclinical experimentation to determine their pharmacological impact, bioactive phytochemicals, medicinal values, and therapeutic targets (Dzobo, 2022). The principal plant components, polyphenols (anthocyanins, flavonoids, phenolic acid, stilbenes, lignans), vitamins (C, E), CT (carotenes, xanthophylls) exhibit a broad range of biological effects such as robust antioxidant, anti-apoptotic, anti-platelet, anti-fibrotic, and anti-inflammatory bioactivities (Manach et al, 2004).

Since the emergence of elevated drug adversity, poor tolerability, and clinical drug trial failure of conventional drug therapies have affected patients worldwide, the clinical application of plant-derived compounds, therapeutics, and experimental drugs for multiple disease associations to signaling pathways, and correlation

DOI: 10.1201/9781003389781-11

with other conditions, has been surging (Cragg and Newman, 2013). More recently, several research institutions have been displaying interest in plant extracts due to advancements in natural product screening techniques, including liquid chromatography (LC)/mass spectrometry (MS), LC/ultraviolet spectroscopy, LC/nuclear magnetic resonance spectroscopy, high performance liquid chromatograpy–MS/MS, high-resolution Fourier transform mass spectrometry, and photo-diode arrays, providing ample structural configurations, elucidation for lead compound identification (Hostettmann et al, 2001)

Drug discovery strategies include phenotypic- and target-based computer-aided drug design approaches, with the target-based strategy involving target identification related to disease state. While these investigations drive potent chemical optimization of lead compounds, clinical trials may have limitations (Swinney, 2013; Zheng et al, 2013). According to a recent study, invalidated illness targets result in many unsuccessful medication candidates in Phase II and III clinical trials. Computational technologies such as machine learning (ML) tools have been widely used to improve the hit rate in drug development molecules (Jaén-Oltra et al, 2000; Marrero-Ponce et al, 2005).

Singh and colleagues established a Bayesian classification model based on structural fingerprints and physicochemical property descriptors. They applied it to virtually screen an independent data set of 200k molecules, exhibiting that the model can screen top hits of PubChem Bioassay actives with up to 76% accuracy. Computational approaches such as deep learning and ML have been highly employed in improving drug discovery hit rates for many synthetic and naturally derived drugs (Singh et al, 2012). Ekins and colleagues developed Bayesian models to predict compound activity toward *Mycobacterium tuberculosis*, then computationally screened 82,403 molecules and chose 550 for in vitro testing, yielding 124 actives against *Mycobacterium tuberculosis*. However, there has been little study on categorization predictions for phenotypic screening of neuroprotective drugs so far (Ekins et al, 2014).

The objective of this book chapter is to provide an overview of multiple natural products and their diverse phytochemical constituent profile against neurotoxicity, oxidative stress, and inflammation. In addition, recent tactics of ML and deep learning in natural product screening against neuroprotection have also been emphasized.

11.2 PHYTOCHEMICAL CONSTITUENTS OF NATURAL PLANT PRODUCTS

The majority of the bioactive components found in different parts (leaves, seeds, flowers, roots, and stems) of natural plants are reported to be flavonoids, terpenes, polyphenols, and CT, at varying concentrations. Phytochemical compounds display multiple therapeutic mechanisms, including kinase-mediated aberrant signal transduction and antioxidant and anti-inflammatory stimulation. Flavonoids are classified as secondary metabolites of polyphenolic nature, found mainly in plants, with dietary consumption values. They have a generalized 15-carbon skeletal structure consisting of a heterocyclic pyrane ring connecting (with an embedded O group) two benzene rings, C6-C3-C6. Presently, nearly 6000 types of flavonoids have been identified from several plant species, including proanthocyanins, flavones, biflavonoids, flavonol glycosides, and acylated flavonol glycosides (Ullah et al, 2020). Alteration in the structural skeleton affects and defines the structural basis for its related derivatives. They exert several beneficial effects in the prevention of cancer, cardiovascular disorders, NDD, and diabetes (Safe et al, 2021). CT are a sub-class of terpenoids, tetraterpenoids, producing organic pigments for plants, algae, and certain fungi. They are further divided into two categories, carotenes (containing hydrocarbons with no O-atoms) and xanthophylls (containing O-atom). They are composed of four terpene units with ten carbon atoms each, providing a total of 40 carbon atoms, with their absorbance wavelengths ranging from 400 to 550 nM. Data from several epidemiological studies suggest that carotenoid consumption in humans and animal models appears to be protective against breast, prostate, head, and neck cancer and Parkinson's disease (PD) (Tan and Norhaizan, 2019). Terpenes are unsaturated hydrocarbon-containing natural products produced by plants and contain nearly 30,000 compounds. They are further classified by the number of carbon present, such as monoterpenes, diterpenes, and sesquiterpenes, and are regarded as beneficial phytocompounds for their anti-apoptotic, anti-platelet, anti-microbial, and anti-inflammatory activities (Ninkuu et al, 2021).

11.3 ALZHEIMER'S DISEASE – A CRITICAL PROSPECT OF NEUROTOXICITY

Alzheimer's disease (AD) is a neurological condition that gradually deteriorates cognition and perception prior to progressing to advanced dementia. The underlying pathogenic cause of AD is considered to be neurofibrillary tangles (NFTs) and amyloid-β (Aβ) fibrils. The clinical effectiveness of natural plants for treating neural conditions can be achieved by the evaluation of these two pathological protein aggregates in the brain tissue. Numerous studies suggest that bioactive plants and plant extracts such as curcumin, *Ginkgo biloba*, and resveratrol are known to possess potent anti-AD activities for their varied constitution of bioactive phytochemicals (Shareena and Kumar, 2022). EGB 761-treated P301S TG mice depicted declining levels of p-tau, enhanced learning retention, spatial memory, mitigated p38, glycogen synthase kinase-3-beta (GSK-3β) signaling, rescued cyclic AMP-response element binding protein phosphorylation, and synaptophysin loss in cultured cortical neurons (Qin et al, 2018). Further, dietary EGb-treated-TF/CRND8 AD mice were shown to mitigate AD pathology via downregulating Aβ aggregate formation and modulation of β-secretase enzyme activity while reducing Aβ-induced proinflammatory cytokine release, such as tumor necrosis factor alpha (TNF-α), interleukin (IL)-1β, preventing Aβ-mediated aberrant microglial activation (Colciaghi et al, 2004). In human SH-SY5Y neuroblastoma cell lines, curcumin suppresses Aβ-mediated tau and GSK-3β phosphorylation at Ser396, Thr231, and Ser9 residues and ameliorates HDAC6 overexpression and Aβ toxicity. However, the neuroprotective impact of curcumin on Aβ-mediated GSK-3β dephosphorylation is not directly associated with oxidative stress. It abolishes Aβ-induced downregulation of Ak strain transforming (AKT) and PDPK1 phosphorylation at Thr308, Ser473, and Ser241, respectively, suggesting that the second response, phosphatidylinositol (3,4,5)-trisphosphate (PIP3), consists of curcumin-protective signaling. Likewise, insulin receptor (IR)/phosphatidylinositol-3-kinase (PI3K) pathway, as a regulatory response signal of PIP3, does not involve Aβ-mediated AKT deactivation (Thr308, Ser473 dephosphorylation). In addition, Aβ expression leads to increased levels of phosphatase and tensin homolog (PTEN), a negative PIP3 regulator, which is suppressed by curcumin treatment, implying that curcumin is effectively Aβ-mediated tau phosphorylation engaging PTEN/AKT/GSK-3β pathway (Huang et al, 2014). In another study, curcumin, in combination with resveratrol, downregulates ROS generation, ameliorates oxidative stress, prevents tau hyperphosphorylation at T181 and T205 residue sites, and exhibits neuroprotective actions on SH-SY5Y cells from Aβ oligomer damage (Yu et al, 2022). In P301L/rTg4510 mice, cornel iridoid glycoside (CIG) treatment improves memory capabilities, spatial learning, and cognitive impairment, represses synapse, and neuronal loss, mitigates brain atrophy, enhances synaptic proteins, preserves cytoskeleton, prevents tau hyperphosphorylation, aggregation in rTg4510 mice's cerebral tissues. Mechanistically, CIG enhances the PP2A activity, upregulates PP2Ac methylation at Leu309, reduces PP2Ac phosphorylation at Tyr307, and promotes protein expressions of PTPA, PTP1B, and LCMT-1 in the brain (Various plant products and their extract's potent anti-AD activity is represented in Figure 11.1) (Ma et al, 2019). Novel osmotin

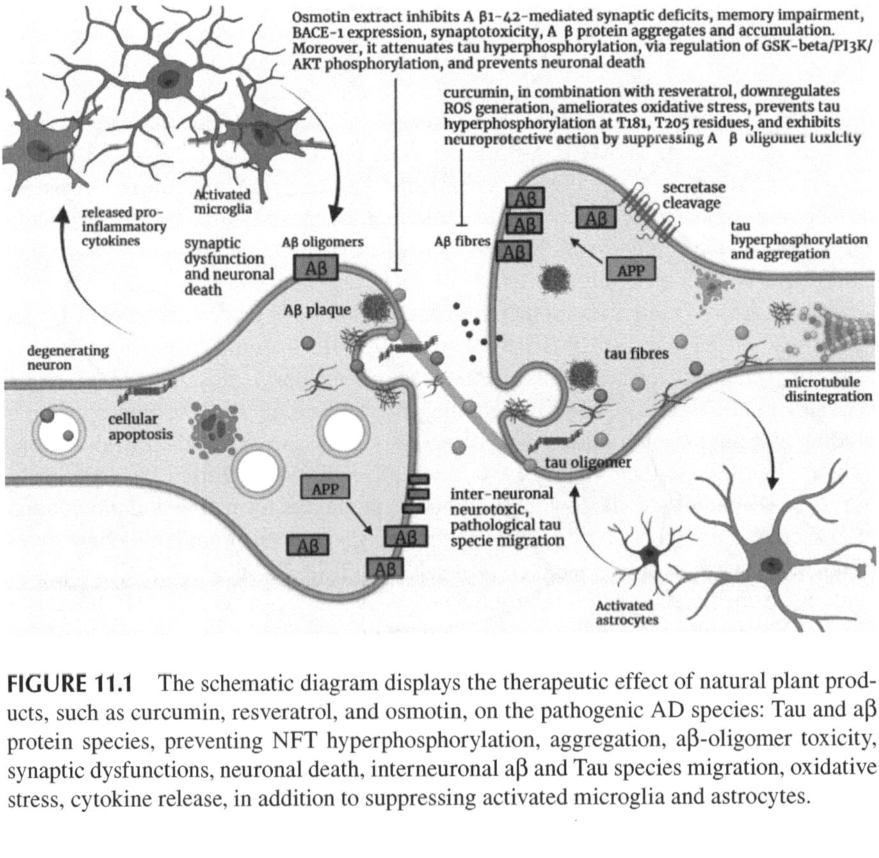

FIGURE 11.1 The schematic diagram displays the therapeutic effect of natural plant products, such as curcumin, resveratrol, and osmotin, on the pathogenic AD species: Tau and aβ protein species, preventing NFT hyperphosphorylation, aggregation, aβ-oligomer toxicity, synaptic dysfunctions, neuronal death, interneuronal aβ and Tau species migration, oxidative stress, cytokine release, in addition to suppressing activated microglia and astrocytes.

extract significantly inhibits Aβ1-42-mediated synaptic deficits, memory impairment, BACE-1 expression, synaptotoxicity, and Aβ protein aggregates and accumulation. Moreover, osmotin-treated Aβ1-42-mice attenuates Aβ1-42-mediated tau protein hyperphosphorylation at Ser413 via regulation of aberrant p-GSK-3β (Ser9), p-PI3K, p-AKT (Ser473) phosphorylation, prevents neuronal apoptosis via p53-induced caspase-associated apoptotic pathway and neurodegeneration along with alleviating cellular neurotoxicity (Ali et al, 2015).

11.4 COGNITION

A common characteristic of many pathological conditions is cognitive decline or impairment. Memory, learning, thinking, and information processing are primarily affected by the physiological changes that result in the body. For their abilities to improve cognition, many plant extracts have been used extensively in conditions like dementia, AD, and PD related to memory loss as the body ages (Eckert, 2010). *Angelica gigas* (AG), in combination with *Bombyx mori* silkworm (SW), notably protects hippocampal neuronal cells from H_20_2-mediated cell death in HT22 mouse cell lines. It recovers Scopolamine-induced cognitive impairment and spatial learning. In specific, it upregulates mRNA and protein levels of phosphorylated p38 and extracellular signal-regulated kinase (ERK)1/2 and reduces Bax/Bcl2 apoptotic index expression (Guo et al, 2021). Gypenoside LXXV, derived from *Gynostemma pentaphyllum*, substantially alleviated cognitive deficits in db/db mice and enhanced lipid metabolism and glucose tolerance. It drastically improved glucose uptake by the brain and markedly elevated p-AKT/total AKT, GLUT4, and PPARγ expressions with a negative impact on p-IRS-1/total IRS-1 levels (Meng et al, 2022). Dihydromyricetin (DHM) decreased protein carbonyl, lipid peroxidation levels, augmented catalase, and superoxide dismutase in the brain tissue in Pb-mediated mice. It mitigated Pb-mediated cell apoptosis, with an evident reduction in cleaved caspase-3, Bax. DHM facilitated AMP-activated protein kinase (AMPK) phosphorylation, prevented activation of TLR4, p38, MyD88, and GSK-3, inhibited inflammatory cytokines (iNOS, IL-6, TNF-α, and COX-2), Aβ levels, and NF-κB nuclear translocation, indicating that DHM could enhance cognitive functional profile via prevention of oxidative stress, inflammation, and apoptosis (Meng et al, 2022). Salidroside, obtained from *Rhodiola rosea*, prevents cognitive impairment due to hypoperfusion, and cerebral hypoxia, which was attributed to alteration of PI3K, NF-κB, HIFα, mitogen-activated protein kinase (MAPK), and matrix metalloproteinases (MMPs) signaling pathways. It enhances lipopolysaccharides (LPS)-mediated learning impairment, memory deficit, and neuroinflammation via modulating the SIRT-1dependent nuclear factor erythroid 2–related factor 2 (NRF-2)/HO-1/NF-κB pathway. Salidroside improves cognitive profile in orthopedic surgery-mediated cognitive impairment mice, decreasing the expressions of M1 marker genes (CCL5, CD16, CXCL1, CXCL10, IFN-γ, and TNF-α), elevating M2 marker genes (Argl, TGF-β, YM1, IL-4, and IL-10) and activating AMPK kinase/PPARγ pathway (Pan et al, 2022).

11.5 BIOACTIVITIES OF NATURALLY DERIVED PHYTOCONSTITUENTS

11.5.1 Polyphenols

Numerous polyphenolic compounds are a class of natural products that can exhibit potent anti-tumor, antioxidant, and anti-inflammatory properties through modulation of ROS levels, ER stress, Ca^{+2} levels, and autophagy. B16-F10 cells, on treatment with P2Et, (*Caesalpinia spinosa* extract), mediated ER stress and apoptosis in melanoma cells. P2Et extract mediated ER stress signaling via (PKR)-like endoplasmic reticulum kinase (PERK) kinase phosphorylation, which can modulate Ca+2 levels and promote anti-tumor responses. It induced expression of ICD-associated DAMPs (ATP, Ecto-CRT, and HMGB1) release in a PERK-dependent manner (Prieto et al, 2019). Dietary supplementation of green tea-derived EGCG rescues synaptic development and dendritic defect in CDKL5-KO mice neurons. It restored the spine and brain maturation density, along with attenuating defective spine maturation and PSD95+ puncta cell population (Trovò et al, 2020). EGCG-treated BAEC cells resulted in increased NO production and phosphorylation of AKT, eNOS, and Fyn, depicting that ECGC has potent endothelial-dependent vasodilatory effects mediated by ROS and Fyn-dependent intracellular pathways, which lead to PI3K/AKT, eNOS activation. Mulberry polyphenol extracts (MPE) substantially suppressed phosphorylated ERK, RAS, and B-galactosidase levels and enhanced NO and iNOS synthase levels. NO increased the activation of AMPK and reduced HMG-CoA reductase activity via its phosphorylation. It promoted the association of cyclins to their CDK kinases and hyperphosphorylated RB. MPE prevented K-Ras-mediated A7r5 vascular smooth muscle cell senescence via enhancing AMPK and iNOS-dependent pathways, suggesting its powerful role in mitigating age-induced atherosclerosis (Chen et al, 2022). In p53/wild-type HCT116 colorectal cancer cells, *Artemisia annua* L. polyphenols (pKAL) induced late apoptosis, morphological changes in cancer cells, associated with elevated acidic vesicles, loss of Golgi integrity, reduced DNA content via upregulation and downregulation of γ-H2AX/p53/p21/Bak cleavage/phospho-c-Jun N-terminal kinases (JNK)/p62/MAPK1/LC3B-I axis and AKT/β-catenin/cyclophilin A/GM130, respectively, implying that p-KAL plays an essential survival role in HCT116 cell lines, via p-JNK/p62 signaling and by inhibiting ROS-independent p53-dependent cell death signaling (Jung et al, 2021). *Annurca apple* polyphenols selectively prevent malondialdehyde (MDA)-MB-231 cell proliferation and viability. APE promoted G2/M phase cell cycle arrest, correlated with p-cdc25C and p27 upregulation, along with p21 decrement. It also reduced the expression of several oncoproteins, including β-catenin, NF-κB, and c-myc, and suppressed AKT activation. Moreover, it lowered Dusp-1 levels, enhanced ROS production, and JNK-c-Jun phosphorylation, portraying the underlying pathways of APE-mediated cell death in MDA-MB-231 TNBC cell lines (Martino et al, 2019).

11.5.2 Saponin

Hederoside C (HedC), a pentacyclic triterpene saponin isolated from *P. Koreana*, demonstrated potent anti-cancer properties toward MG63, and U20S osteosarcoma cell lines wherein HedC prevented proliferative activity of both U20S and MG63 cell lines, and induced apoptosis (downregulation of cleaved poly(ADP-ribose) polymerase (PARP), terminal deoxynucleotidyl transferase dUTP nick end labeling–positive cells, cleaved caspase-9 and -3), in time- and dose-dependent manner. HedC-treated cells exhibited elevated p21, p53, Bax, and lowered Bcl-2, whereas HedC-induced apoptosis was followed by reduced STAT3, MAPKs (JNK, ERK1/2, and p38) phosphorylation. In *in vivo* xenograft mouse model, HedC showed an anti-metastatic effect via suppression of neoplastic migration and invasion, ameliorated and enhanced expressions of p-STAT3, PCNA, and p53, cleaved caspase-3, respectively, asserting that HedC has anti-metastatic, anti-tumor potentials (Park et al, 2021). Ginsenoside Ro (GRo) (Panax ginseng's oleanolic saponin) notably attenuates LPS-mediated lung injury and proinflammatory mediators, including IL-6, IL-1β, TNF-α. In addition, it inhibits the binding of LPS488 (fluorescence-labeled LPS) to membranes of RAW264.7 macrophages, suppresses MAPK, NF-κB phosphorylation, p65 nuclear translocation, dose-dependently. Moreover, docking and molecular dynamics (MD) simulation analysis suggest that GRo is effectively docked into LPS binding regions of MD2/TLR4 complex with stable binding conformation, implying that GRo serves as a therapeutic anti-inflammatory agent (Xu et al, 2022). Findings display that cycloastragenol-treated mice brains were evaluated for their effects against Aβ-mediated neurogenic dysfunctions, mitochondrial apoptosis, and oxidative stress. Treatment with cycloastragenol effectively inhibited expressions of MAP kinases such as p-p38, p-JNK, and ERK1/2, reduced inflammatory markers (Iba-1, IL-1β, TNF-α, and GFAP), mitigated activated microglia and astrocytes. It improved Aβ-mediated cognitive impairment, regulated anti-apoptotic effects, upregulating and downregulating Bcl-2, Bax, Bim, and Casp-3, respectively, and promoted neuronal nuclei (NeuN), p-CREB expressions in the frontal cortex and hippocampal regions of mice's brains, confirming its neuronal survival effect (Ikram et al, 2021).

11.5.3 Carotenoids (CT)

CTs are a class of tetraterpenoids, constituting conjugated polyene chains with eight units of C5 isoprenoids. CTs are found in several sources, such as photosynthetic and non-photosynthetic plants, animals, and aquatic sources. With the presence of varied functional groups, number of conjugated double bonds, and cumulative polarities, CT exerts potent ROS scavenging, antioxidant, anti-inflammatory, neuroprotective, anti-drug resistance, and anti-cancer properties (Kabir et al, 2022). Carotenoid, identified from *Spondias mombin*, exhibits potent anti-angiogenic, anti-proliferative activities in 7,12-dimethylbenz[a]-anthracene breast cancer models in female Wistar rats. Carotenoid isolate treatment remarkably downregulates the expressions of MMP-2, HIF-1, VEGF, VEGFR, and EGFR in mammary tumors while upregulating mRNA levels of CHD-1. The binding of astaxanthin produced the DFG-out conformation, 7,7′,8,8′-tetrahydro-β, -β-carotene, and β-carotene-15,15′-epoxide to the ATP binding domain, with interaction energies of −8.2, −10.3, and −10.5 kcal/mol, respectively (Metibemu et al, 2021). Synergistic dosing of doxorubicin, in tandem with dietary carotenoid fucoxanthin (FUC), drastically enhances DOX's cytotoxicity profile and limits the dose of DOX (FR) in DOX-resistant cancer cell lines, including HepG-2/

ADR (HCC), MCF-7/ADR (BC) and SKOV-3/ADR (ovarian cancer) by 6.28, 8.42, and 4.56-fold, respectively, FUC elevates DOX accumulation in resistant cells, in contrast to verapamil, FUC, and DOX together, increase Rho123 build-up, p53 caspases (CASP8, CASP3) activity, decrease the expressions of ABCB1, ABCC1, and ABCG2 along with PXR, GST, and CUP3A4 in resistant cancer cells, respectively (Eid et al, 2020). A low DOX dose with carotenoid substantially promoted anti-proliferation and cytotoxicity in treated MCF-7 and MDA-MV-231 cell lines, compared with high-dose DOX treatment alone. It caused G0/G1 phase cell cycle arrest, promoted selective ROS-induced apoptosis, prevented mitochondrial dysfunctions, and downregulated protein expressions of p21, p27, p53, Bcl-2, Bax, and cyclin D1, thereby improving cancer therapy. In MDA-MB-231 cell lines, β-carotenes were reported to display significant antitumoral actions by suppressing cell proliferation, migration, and invasion ability. β-carotene promotes apoptosis by interfering with S-phase cell arrest and upregulates 3H-deoxy-D-glucose uptake, while it does not affect either 3H-folic acid or 3H-glutamine uptake and oxidative stress. B-carotene's anti-proliferative effect involves JNK intracellular signaling and did not impact cell viability, cell cycle, proliferation, and migration capabilities of MCF12-A non-tumoral cell lines, asserting that β-carotene exhibits cancer cell-selective actions (Antunes et al, 2022).

11.5.4 Flavonoids

Flavonoids are a group of polyphenolic secondary bioactive metabolites that are found in plants and are commonly consumed as dietary supplements due to their immense therapeutic benefits. A citrus flavonoid, diosmetin displays a range of therapeutic impacts for its antioxidant, anti-bacterial, and anti-inflammatory effects. In DNCB-mediated atopic dermatitis (AD) mouse models, diosmetin remarkably lowers epidermis thickness, dermatitis score, and mast cell population compared with untreated groups. It mitigates the macrophage infiltration into AD lesions, with the observed reduction in IL-4, IL-1β, and TNF-α. Further, diosmetin hampers nitric oxide production, downregulates iNOS expression, suppresses MAPKs (JNK, p38, ERK1/2), and JAK/STAT3 signaling phosphorylation and activation, respectively, in raw 264.7 mouse macrophage cell line (Lee et al, 2020). Oral dosing of fisetin (FIS) in HFD-fed mice ameliorates HFD-induced cardiac dysregulations and metabolic disorders by reducing insulin levels, insulin resistance, body weight, and fasting blood glucose. FIS supplementation substantially prevents the pathology of dyslipidemia in both mouse cardiomyocytes and the heart, triggered by metabolic stress. Moreover, FIS treatment represses HFD-mediated inflammatory responses in cardiac tissues via lowered Tnfr-1/Tnfr-2 signaling and downstream target expressions. It promotes a robust reduction in fibrosis-associated genes, attributing to attenuation of fibrosis by inactivating TGF-β1/Smads/ERK1/2 signaling (Hu et al, 2020). Dietary flavonoid galangin prevents metastatic features such as TPA-mediated HepG2 cell migration and invasion via inhibiting TPA-mediated PKC-δ, PKC-α, phospho-IkBα, c-Jun, C-Fos, and NF-κB, thus showcasing its applications in anti-metastatic clinical therapy. Myricetin portrays potent anti-cancer, antioxidant, and anti-inflammatory activities, serving as a potential therapeutic agent. In U-86 MG cell lines, myricetin treatment depicts anti-glioblastoma, anti-proliferative effects, hampering cellular migration and invasion, and PTEN status independently. The cytotoxic actions of myricetin are cell-sensitive, wherein it shows minimal cytotoxicity to normal astrocytes, in contrast to the GBM cell line. Additionally, it suppresses the synthesis of focal adhesions, membrane ruffles, lamellipodia, and vasculogenic mimicry and blocks the ROCK2 phosphorylation, cortactin, paxillin, JNK, and PI3K/AKT signaling. Myricetin binds to various kinases and scaffold proteins such as 3-phosphoinositide-dependent kinase 1 (PDK1), c-Jun, JNK, vinculin, VE-cadherin, and PI3K catalytical isoforms (p110α-γ) (Zhao et al, 2018).

11.6 MACHINE LEARNING AND DEEP LEARNING TACTICS IN NATURAL PRODUCT DRUG DISCOVERY AND DEVELOPMENT

11.6.1 Machine Learning Approaches

ML involves algorithms analyzing curated data to make decisions and build efficient models for different functions. So far, several ML tools and development kits have been utilized, including TensorFlow, Random Forest, Waikato Environment for Knowledge Analysis, Scikit-Learn, and Support vector machine. The benefits of using ML tools in a quantitative structure activity relationship (QSAR) study include an accurate prediction of the influence of a compound's chemical structure on its bioactivity and the modeling of drug toxicity and metabolism (Stitou et al, 2019). Therefore, ML is considered an essential resource for discovering new drugs (Dara et al, 2022).

Yang et al reported developing a neuroprotective ML model for identifying potent neuroprotective compounds in Xiaxuming decoction (XXMD) against H202-mediated and Hypoxia-induced brain cell damage. The TCM prescription Xiaoxuming decoction (XXMD) (made up of 12 herbs) has been a beneficial remedy for the therapy of stroke in clinical settings, as documented in the Tang Dynasty's Beiji Qianjin Yaofang. They constructed stacked naïve Bayesian models based on molecular fingerprint descriptors and four distinct single classifiers (AB, CT, kNN, and RF). The finalized models were used for the virtual screening of neuroprotective agents in XXMD and further selected for cell-based assay. Among the compounds, two compounds substantially inhibited $Na_2S_2O_4$-induced and H_2O_2-mediated neurotoxicity (Yang et al, 2019). Baicalein was discovered to be a potent neuroprotective drug in this work, with effects on hypoxia and oxidative stress phenotypes. Baicalein has gained significant focus for its antioxidant and anti-inflammatory effects as a bioactive phenolic flavonoid molecule. The validation of baicalein against two damage phenotypes suggests that it might be used as a promising neuroprotective agent (Dinda et al, 2017).

Another study by Fang J and colleagues reported 28 compounds exhibiting neuroprotective properties against H_2O_2-mediated and monosodium glutamate-induced neurotoxicity by ML approaches. The results suggested that techniques that integrated single classifiers into combined Bayesian models could be a viable approach to predicting neuroprotective compounds. On monosodium glutamate-induced and H_2O_2-induced PC12 cells, three selected molecules (J14572, J27152, and J27114) demonstrate a favorable dose-response relation and cell viability. Cell survival for model groups affected by 40 mM monosodium glutamate or 300 mM H_2O_2 was significantly lower than in the control group (P 0.01) (Fang et al, 2016).

11.6.2 Deep Learning Approaches

Deep learning is a sub-field of ML that utilizes artificial neural cells to process data in decision-making. It also applies to the drug discovery process besides computational fields such as pattern recognition. It is highly implemented in QSAR studies; virtual screening; *absorption, distribution, metabolism, excretion, and toxicity* properties; and lead optimization. Earlier deep learning implementations propose a wide range of disease detection and prediction screening models, including a Densely connected convolutional neural network (CNN)-based automated COVID screening model, drug-target interaction CNN models, and CNN RF models to analyze drug functions from chemical structure (Chakravarti and Alla, 2019; Kumari and Subbarao, 2021). This novel approach has led to the discovery of several impactful techniques that could be effectively utilized in phytochemical drug screening and their therapeutic and pharmacological targets in several chronic states, such as neuroprotection, neurotoxicity, apoptosis, oxidative stress, and inflammation. Hence, several studies surrounding such have been described, showing a vast diversity of models throughout its development phases (Shanmuganathan, 2016).

The algorithm for all the naturally derived drugs and compounds functions in three steps: (A) Collection of natural compounds and drug information from the public database. (B). Producing molecular interaction and chemical features from collected data by text mining, network analysis, and chemical property analysis. (C) Training the deep learning models on the elements of approved drugs and predicting the medicinal properties based on the trained model. Deep learning can increase prediction performance when the input characteristics are extensive and diversified by extracting high-level interpretation using low-level features. The mentioned approach has four consecutive levels: Input, partially linked hidden layers, fully coupled hidden layers, and output. The models are developed to forecast the likely effects list utilizing the input attributes. Theoretical knowledge, molecular interaction, and chemical property data are generated for each drug or natural component and used as inputs to the model. Hidden layers extended their outputs by detecting nonlinear connections between low- and high-level data to provide a higher-level representation than the last layer (Yoo et al, 2020).

Rodríguez FR and colleagues developed a computational strategy involving deep learning, structural bioinformatics, and signaling pathway manual reconstruction to predict eight novel nicotine analogs' neuroprotective activity based on the PI3KAKT pathway in the context of PD therapeutics. The model predicted the potential neuroprotective efficacy of seven novel nicotine analogs based on the binomial Bcl-2 response regulated by PI3K/AKT activation (Rojas-Rodríguez et al, 2020).

Further studies by Wang H and co-workers identified sclareol as a naturally derived Cav1.3-neuroprotective antagonist in PD. While L-type voltage-gated calcium channel blockers treatment selectively alleviates Cav1.3 for PD, drug development is impeded due to a shortfall in high-throughput screening techniques allowing

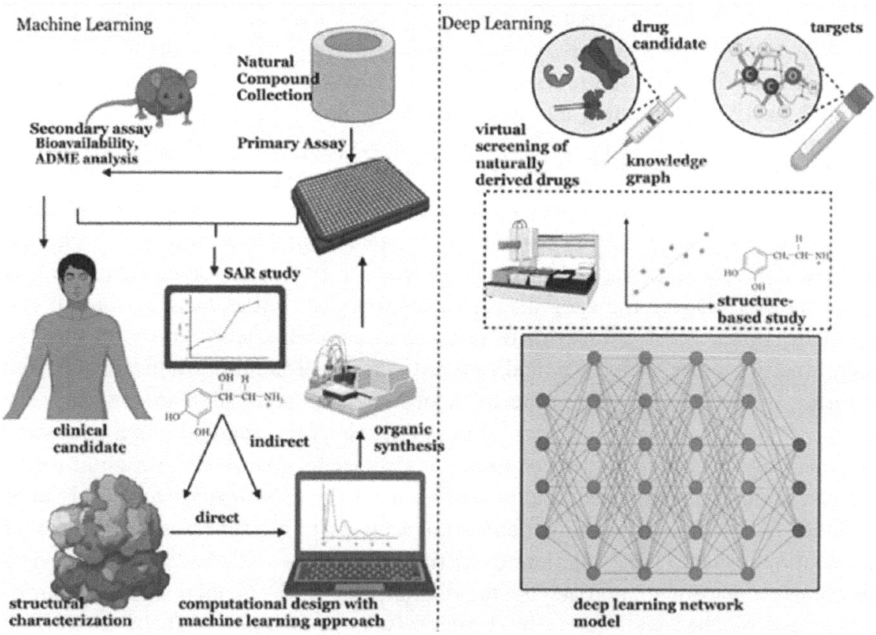

FIGURE 11.2 The above illustration depicts a contrasting feature between ML approaches and deep learning techniques, utilized in drug discovery and development, virtual screening, and SAR study of phytochemicals and natural products.

isoform-specific assessment of Cav-antagonistic activities. Integrated in silico virtual screening and deep learning models enabled the discovery of (6)-Gingerol and sclareol as novel Cavl.3 antagonists. In silico analysis of 198 candidate molecules produced 14 hits as the most promising Cavl.3 inhibitors.

Moreover, structure clustering analysis enabled a selection of five phytoconstituents as potential structures. In addition, parallel artificial intelligence–based validation showcased similar results, with a *receiver operating characteristic curve–area under curve* value of 97.78%. Experimentations on the five compounds with CaB-A assay confirmed that sclareol and (6)-gingerol portrayed robust inhibition on Cavl.3. Both molecules showed a more substantial antagonistic effect on Cavl.3-induced reporter gene expression than the Cavl.3-dependent CaB system. Sclareol-treated PD mouse models observed minimal loss in DA neurons, exhibited in vivo neuroprotective effects toward 6-ODHA-mediated neurodegeneration, prevented oversynchronization and locomotion deficits of striatal neurons (the contrast between ML and deep learning approaches are described in Figure 11.2) (Wang et al, 2022).

11.7 CONCLUSION

The emergence of expanding disease conditions and the shortage of disease-modulating or disease-alleviating therapies pose alarming challenges to research institutions. Although diagnosing and treating the symptomatic events of chronic diseases aids in the prevention of advancing the disease state, the genetic predisposition and the onset of the disorder should be mitigated beforehand. Thus, the advancements in molecular research technology have enhanced the identification of potent bioactive compounds and their underlying biochemical mechanism of action on target conditions. Along with it, several thousands of natural target molecules are being screened for varied in silico analysis, hoping to find feasible inhibitor molecules to discover potent drug candidates that involve phytochemical-based signaling pathways. Their optimization becomes an essential discovery tool in chronic disease management. Natural products serve as a potential source of the potential drug candidate in the drug discovery pipeline, as well as their nutritional and cosmetic value. Most clinical drug trials and research studies conclude that the increased potencies, efficacies, and lowered toxicity of natural products have been asserted and found to produce synergistic positive effects in the treatment and management of a spectrum of diseases, such as AD. However, maintaining a consistent rate of drug discovery and development is crucial to guaranteeing a good medication profile and improved healthcare. As a result, several discovery tools have included ML and deep learning methodologies to assure lower error margins, exact calculations, and a viable solution to existing difficulties. This book chapter review summarized a critical approach to studying natural products, their derivative properties, and biochemical factors in chronic diseases, including AD. In addition, It highlighted the recent trends in ML and deep learning techniques providing a basis to synthesize further lead compounds.

ACKNOWLEDGMENT

All the schematic diagrams were made using online illustration software, Biorender.com.

CONFLICT OF INTEREST

The authors have no conflicts of interest to declare that are relevant to the content of this article.

Approval of ethics: Not applicable.
Participation consent: Yes.
Publication consent: Yes
Availability of data and materials: Not applicable.
Competing interests: None.
Funding: None.
Authors' contributions: All authors contributed to the review study conception and design.
Acknowledgements: Authors acknowledge support from principal Poona college of Pharmacy, Pune.
Disclosure of potential conflicts of interest: Not applicable.
Research involving Human Participants and/or Animals: Not applicable.
Informed consent: Not applicable.

ABBREVIATIONS

AD	Alzheimer's disease
ADP	Adenosine di-phosphate
AMPK	AMP-activated protein kinase
DL	Deep learning
EGFR	Epidermal growth factor receptor
ERK	Extracellular signal-regulated kinase
JNK	c-Jun N-terminal kinases
LDH	*Lactate dehydrogenase*
LPS	Lipopolysaccharides
MAPK	Mitogen-activated protein kinase
MCP1	Monocyte chemoattractant protein-1
MD simulation	Molecular dynamics simulations
MDA	Malondialdehyde
MET	Mesenchymal-epithelial transition factor
ML	Machine learning
MMP	Matrix metalloproteinases
mTOR	Mammalian target of rapamycin
NADPH	Nicotinamide adenine dinucleotide phosphate
NDD	Neurodevelopmental disorders
NeuN	Neuronal nuclei
NF-kB	Nuclear factor kappa light chain enhancer of activated B cells
NO	Nitric oxide
NRF-2	Nuclear factor erythroid 2–related factor 2
OA	Osteoarthritis
PD	Parkinson's disease
PDK1	3-Phosphoinositide-dependent kinase 1
ROS	Reactive oxygen species
PARP	Poly(ADP-ribose) polymerase
PDPK1	3-Phosphoinositide-dependent protein kinase-1
PERK	(PKR)-like endoplasmic reticulum kinase
PI3K	Phosphatidylinositol-3-kinase
PIP3	Phosphatidylinositol (3,4,5)-trisphosphate
PKB	Protein kinase B/AKT

REFERENCES

Ali, T., G. H. Yoon, S. A. Shah, H. Y. Lee, and M. O. Kim. 2015. "Osmotin attenuates amyloid beta-induced memory impairment, tau phosphorylation and neurodegeneration in the mouse hippocampus." *Scientific Reports* 5: 11708.

Antunes, A., F. Carmo, S. Pinto, N. Andrade, and F. Martel. 2022. "The anti-proliferative effect of β-carotene against a triple-negative breast cancer cell line is cancer cell-specific and JNK-dependent." *PharmaNutrition* 22: 100320.

Chakravarti, S. K., and S. R. M. Alla. 2019. "Descriptor free QSAR modeling using deep learning with long short-term memory neural networks." *Frontiers in Artificial Intelligence* 2: 17.

Chen, C. P., K. C. Chan, H. H. Ho, H. P. Huang, S. Hsu, and C. J. Wang. 2022. "Mulberry polyphenol extracts attenuated senescence through inhibition of Ras/ERK via promoting Ras degradation in VSMC." *International Journal of Medical Sciences 19* (1): 89–97.

Colciaghi, F., B. Borroni, and M. Zimmermann, et al. 2004. "Amyloid precursor protein metabolism is regulated toward alpha-secretase pathway by Ginkgo biloba extracts." *Neurobiology of Disease 16* (2): 454–460.

Cragg, G. M., and D. J. Newman. 2013. "Natural products: A continuing source of novel drug leads." *Biochimica Et Biophysica Acta 1830* (6): 3670–3695.

Dara, S., S. Dhamercherla, S. S. Jadav, C. M. Babu, and M. J. Ahsan. 2022. "Machine learning in drug discovery: A review." *Artificial Intelligence Review 55* (3): 1947–1999.

Dinda, B., S. Dinda, S. DasSharma, R. Banik, A. Chakraborty, and M Dinda. 2017. "Therapeutic potentials of baicalin and its aglycone, baicalein against inflammatory disorders." *European Journal of Medicinal Chemistry 131*: 68–80.

Dzobo, K. 2022. "The role of natural products as sources of therapeutic agents for innovative drug discovery." *Comprehensive Pharmacology 2*: 408–422.

Eckert, G. P. 2010. "Traditional used plants against cognitive decline and Alzheimer disease." *Frontiers in Pharmacology 1*: 138.

Eid, S. Y., M. A. Althubiti, M. E. Abdallah, M. Wink, and M. Z. El-Readi. 2020. "The carotenoid fucoxanthin can sensitize multidrug resistant cancer cells to doxorubicin via induction of apoptosis, inhibition of multidrug resistance proteins and metabolic enzymes." *Phytomedicine: International Journal of Phytotherapy and Phytopharmacology 77*: 153280.

Ekins, S., A. C. Casey, D. Roberts, T. Parish, and B. A. Bunin. 2014. "Bayesian models for screening and TB Mobile for target inference with Mycobacterium tuberculosis." *Tuberculosis (Edinburgh, Scotland) 94* (2): 162–169.

Fang, J., X. Pang, and R. Yan, et al. 2016. "Discovery of neuroprotective compounds by machine learning approaches." *RSC Advance 6*: 9857–9871.

Guo, Y., H. Jin, and M. Kim, et al. 2021. "Synergistic neuroprotective effects of mature silkworm and *Angelica gigas* against scopolamine-induced mild cognitive impairment in mice and H_2O_2-induced cell death in HT22 mouse hippocampal neuronal cells." *Journal of Medicinal Food 24* (5): 505–516.

Hajat, C., and E. Stein. 2018. "The global burden of multiple chronic conditions: A narrative review." *Preventive Medicine Reports 12*: 284–293.

Hostettmann, K., J. L. Wolfender, and C. Terreaux. 2001. "Modern screening techniques for plant extracts." *Pharmaceutical Biology 39* (Suppl 1): 18–32.

Huang, H., D. Tang, K. Xu, and Z. F. Jiang. 2014. "Curcumin attenuates amyloid-β-induced tau hyperphosphorylation in human neuroblastoma SH-SY5Y cells involving PTEN/Akt/GSK-3β signaling pathway." *Journal of Receptor and Signal Transduction Research 34* (1): 26–37.

Hu, L. F., J. Feng, and X. Dai, et al. 2020. "Oral flavonoid fisetin treatment protects against prolonged high-fat-diet-induced cardiac dysfunction by regulation of multicombined signaling." *The Journal of Nutritional Biochemistry 77*: 108253.

Ikram, M., M. H. Jo, and K. Choe, et al. 2021. "Cycloastragenol, a triterpenoid saponin, regulates oxidative stress, neurotrophic dysfunctions, neuroinflammation and apoptotic cell death in neurodegenerative conditions." *Cells 10* (10): 2719.

Jaén-Oltra, J., M. T. Salabert-Salvador, F. J. García-March, F. Pérez-Giménez, and F. Tomás-Vert. 2000. "Artificial neural network applied to prediction of fluorquinolone antibacterial activity by topological methods." *Journal of Medicinal Chemistry 43* (6): 1143–1148.

Jung, E. J., A. Paramanantham, and J. Kim, et al. 2021. "*Artemisia annua* L. polyphenol-induced cell death is ROS-independently enhanced by inhibition of JNK in HCT116 colorectal cancer cells." *International Journal of Molecular Sciences 22* (3): 1366.

Kabir, M.T., M. H. Rahman, and M. Shah, et al. 2022. "Therapeutic promise of carotenoids as antioxidants and anti-inflammatory agents in neurodegenerative disorders." *Biomedicine & Pharmacotherapy* 146: 112610,

Kumari, M., and N. Subbarao. 2021. "Deep learning model for virtual screening of novel 3C-like protease enzyme inhibitors against SARS coronavirus diseases." *Computers in Biology and Medicine 132*: 104317.

Lee, D. H., J. K. Park, J. Choi, H. Jang, and J. W. Seol. 2020. "Anti-inflammatory effects of natural flavonoid diosmetin in IL-4 and LPS-induced macrophage activation and atopic dermatitis model." *International Immunopharmacology 89* (Pt A): 107046.

Ma, D., Y. Luo, and R. Huang, et al. 2019. "Cornel iridoid glycoside suppresses tau hyperphosphorylation and aggregation in a mouse model of tauopathy through increasing activity of PP2A." *Current Alzheimer Research 16* (14): 1316–1331.

Manach, C., A. Scalbert, C. Morand, C. Rémésy, and L. Jiménez. 2004. "Polyphenols: Food sources and bioavailability." *The American Journal of Clinical Nutrition 79* (5): 727–747.

Marrero-Ponce, Y., M. Iyarreta-Veitía, and A. Montero-Torres, et al. 2005. "Ligand-based virtual screening and in silico design of new antimalarial compounds using nonstochastic and stochastic total and atom-type quadratic maps." *Journal of Chemical Information and Modeling 45*(4): 1082–1100.

Martino, E., D. C. Vuoso, and S. D'Angelo, et al. 2019. "Annurca apple polyphenol extract selectively kills MDA-MB-231 cells through ROS generation, sustained JNK activation and cell growth and survival inhibition." *Scientific Reports 9*(1): 13045.

Meng, X., Y. Zhang, and Z. Li, et al. 2022. "A novel natural PPARγ agonist, gypenoside LXXV, ameliorates cognitive deficits by enhancing brain glucose uptake via the activation of Akt/GLUT4 signaling in db/db mice." *Phytotherapy Research 36* (4): 1770–1784.

Metibemu, D, O. A. Akinloye, I. O. Omotuyi, J. O. Okoye, M. A. Popoola, and A. J. Akamo. 2021. "Carotenoid-enriched fractions from *Spondias mombin* demonstrate HER2 ATP kinase domain inhibition: Computational and *in vivo* animal model of breast carcinoma studies." *Frontiers in Oncology 11*: 687190.

Ninkuu, V., L. Zhang, J. Yan, Z. Fu, T. Yang, and H. Zeng. 2021. "Biochemistry of terpenes and recent advances in plant protection." *International Journal of Molecular Sciences 22* (11): 5710.

Pan, C. L., G. L. Dai, and H. W. Zhang, et al. 2022. "Salidroside ameliorates orthopedic surgery-induced cognitive dysfunction by activating adenosine 5'-monophosphate-activated protein kinase signaling in mice." *European Journal of Pharmacology 929*: 175148.

Park, K. R., H. H. Leem, J. Lee, I. K. Kwon, J. T. Hong, and H. M. Yun. 2021. "Anti-cancer effects of Hederoside C, a pentacyclic triterpene saponin, through the intrinsic apoptosis and STAT3 signaling pathways in osteosarcoma." *American Journal of Cancer Research 11* (9): 4541–4550.

Ponjoan, A., J. Garre-Olmo, and J. Blanch et al. 2019. "Epidemiology of dementia: Prevalence and incidence estimates using validated electronic health records from primary care." *Clinical Epidemiology* 11: 217–228.

Prieto, K., Y. Cao, and E. E. Mohamed et al. 2019. "Polyphenol-rich extract induces apoptosis with immunogenic markers in melanoma cells through the ER stress-associated kinase PERK." *Cell Death Discovery* 5: 134

Qin, Y., Y. Zhang, and I. Tomic, et al. 2018. "Ginkgo biloba extract EGb 761 and its specific components elicit protective protein clearance through the autophagy-lysosomal pathway in tau-transgenic mice and cultured neurons." *Journal of Alzheimer's Disease: JAD 65* (1): 243–263.

Rojas-Rodríguez, F., C. Morantes, and A. Pinzón, et al. 2020. "Machine learning neuroprotective strategy reveals a unique set of Parkinson therapeutic nicotine analogs." *The Open Bioinformatics Journal 13*: 1–14.

Safe, S., A. Jayaraman, R.S. Chapkin, M. Howard, K. Mohankumar, and R. Shrestha. 2021. "Flavonoids: Structure-function and mechanisms of action and opportunities for drug development." *Toxicology Research 37* (2): 147–162.

Shanmuganathan, S. 2016. "Artificial Neural Network Modelling: an Introduction." In *Artificial Neural Network Modelling*, edited by Subana Shanmuganathan and Sandhya Samarasinghe, 1–14. Berlin: Springer.

Shareena, G., and D. Kumar. 2022. "Traversing through half a century research timeline on Ginkgo biloba, in transforming a botanical rarity into an active functional food ingredient." *Biomedicine & Pharmacotherapy = Biomedecine & Pharmacotherapie 153*: 113299.

Shin, J. H. 2022. "Dementia epidemiology fact sheet 2022." *Annals of Rehabilitation Medicine 46* (2): 53–59.

Singh, N., S. Chaudhury, R. Liu, M. D. AbdulHameed, G. Tawa, and A. Wallqvist. 2012. "QSAR classification model for antibacterial compounds and its use in virtual screening." *Journal of Chemical Information and Modeling 52* (10): 2559–2569.

Stitou, M., H. Toufik, M. Bouachrine, H. Bih, and F. Lamchouri. 2019. "Machine learning algorithms used in Quantitative structure-activity relationships studies as new approaches in drug discovery," *2019 International Conference on Intelligent Systems and Advanced Computing Sciences (ISACS)*. pp. 1–8

Swinney, D. C. 2013. "Phenotypic vs. target-based drug discovery for first-in-class medicines." *Clinical Pharmacology and Therapeutics 93* (4): 299–301.

Tan, B. L., and M. E. Norhaizan. 2019. "Carotenoids: How effective are they to prevent age-related diseases?" *Molecules* 24 (9): 1801.

Trovò, L., C. Fuchs, and R De Rosa, et al. 2020. "The green tea polyphenol epigallocatechin-3-gallate (EGCG) restores CDKL5-dependent synaptic defects in vitro and in vivo." *Neurobiology of Disease 138*: 104791.

Ullah, A., S. Munir, and S. L. Badshah, et al. 2020. "Important flavonoids and their role as a therapeutic agent." *Molecules 25* (22): 5243.

Veeresham, C. 2012. "Natural products derived from plants as a source of drugs." *Journal of Advanced Pharmaceutical Technology & Research 3* (4): 200–201.

Wang, H., M. Xie, G. Rizzi, X. Li, K. Tan, and M. Fussenegger. 2022. "Identification of sclareol as a natural neuroprotective $Ca_v1.3$-antagonist using synthetic Parkinson-mimetic gene circuits and computer-aided drug discovery." *Advance Science* 9: 2102855.

Xu, H. L., G. H. Chen, and Y. T. Wu, et al. 2022. "Ginsenoside Ro, an oleanolic saponin of Panax ginseng, exerts anti-inflammatory effect by direct inhibiting toll like receptor 4 signaling pathway." *Journal of Ginseng Research 46* (1): 156–166.

Yang, S., Y. Shen, and W. Lu, et al. 2019. "Evaluation and identification of the neuroprotective compounds of Xiaoxuming decoction by machine learning: A novel mode to explore the combination rules in traditional Chinese medicine prescription." *BioMed Research International 2019*: 6847685.

Yoo, S., H. C. Yang, and S. Lee, et al. 2020. "A deep learning-based approach for identifying the medicinal uses of plant-derived natural compounds." *Frontiers in Pharmacology 11*: 584875.

Yu, H., T. Yamashita, and X. Hu, et al. 2022. "Protective and anti-oxidative effects of curcumin and resveratrol on aβ-oligomer-induced damage in the SH-SY5Y cell line." *Journal of the Neurological Sciences 441*: 120356.

Zhao, H. F., G Wang, and C. P. Wu, et al. 2018. "A multi-targeted natural flavonoid myricetin suppresses lamellipodia and focal adhesions formation and impedes glioblastoma cell invasiveness and abnormal motility." *CNS & Neurological Disorders Drug Targets 17* (7): 557–567.

Zheng, W., N. Thorne, and J. C. McKew, et al. 2013. "Phenotypic screens as a renewed approach for drug discovery." *Drug Discovery Today 18* (21–22): 1067–1073.

12 Future Trends and Directives for Research on Phytochemicals in Neurological Diseases

Srishti Raja, Chandini Sengupta, Kamarajan Rajagopalan, Inbathamizh L, Sudha S, and Jackson Durairaj Selvan Christyraj

12.1 INTRODUCTION

Phytochemicals are produced in the biosynthetic laboratory of plants. These naturally occurring substances have the potential to be bioactive. The Greek word for plant is the source of the prefix "Phyto." The term "secondary metabolites" is frequently used to describe these compounds. Terpenes, polysaccharides, glycosides, terpenoids, coumarins, gums, flavonoids, tannins, alkaloids, and phenols are some of the different types of substances that fall under this category. Phytochemicals in plants give host plants a natural defence mechanism in addition to giving them flavour, colour, and scent (Okwu, 2004). Contrary to minerals and vitamins, phytochemicals are essential for preventing age-related chronic diseases but are only minimally necessary for preserving cell viability.

The loss of the structure and functionality of the neural system results in the syndrome known as neurodegeneration. The risk factors for the disease, such as oxidative stress, hypertension, abnormal antioxidant enzymes, cytoskeletal abnormalities, advanced age, genetic defects, autoimmunity, mineral deficiencies, metabolic toxicity, and other vascular disorders, are revealed by numerous experimental and epidemiological studies. Numerous substances have been investigated for the treatment of neurodegenerative diseases (NDs), but because of their adverse effects, they only provide symptomatic relief. By altering the receptor activity of particular inhibitory neurotransmitters, phytochemicals play a critical part in preserving the chemical equilibrium of the brain (Yadav, 2021).

The majority of NDs progress over time. Because most brain illnesses result in a steady decline in health, early detection and therapy are essential to managing such anomalies. It is critical to have fresh and current knowledge of the cellular targets connected to neurological illnesses.

DOI: 10.1201/9781003389781-12

12.2 MOLECULAR MECHANISMS ASSOCIATED WITH BENEFICIAL EFFECTS OF PHYTOCHEMICALS

By interacting with the molecular and cellular structure of the brain involved in memory formation, phytochemicals may inhibit and even partially reverse the age-related loss in memory function. These interactions include the capacity to modulate growth factors in different brain regions, which in turn can up-regulate signalling pathways important for regulating and sustaining synaptic plasticity (Spencer et al, 2009). The molecular processes by which phytochemicals prevent neurodegenerative, neuropsychiatric, and neurotraumatic illnesses brought on by oxidative stress are correlated with their capacity to inhibit the harmful generation of both reactive oxygen species (ROS) and reactive nitrogen species (Calabrese et al, 2010a, 2010b; Son et al, 2008; Tosetti et al, 2009) (Figure 12.1).

Phytochemicals function at the molecular level by activating the protein kinase B (PKB/Akt) and extracellular signal-regulated kinase (ERK1/2) signalling pathways, which then activate the transcription factor cAMP response element-binding protein (CREB). This increases the expression of several neurotrophins crucial for mediating memory formation (Spencer, 2010). Resveratrol, among other phytochemicals,

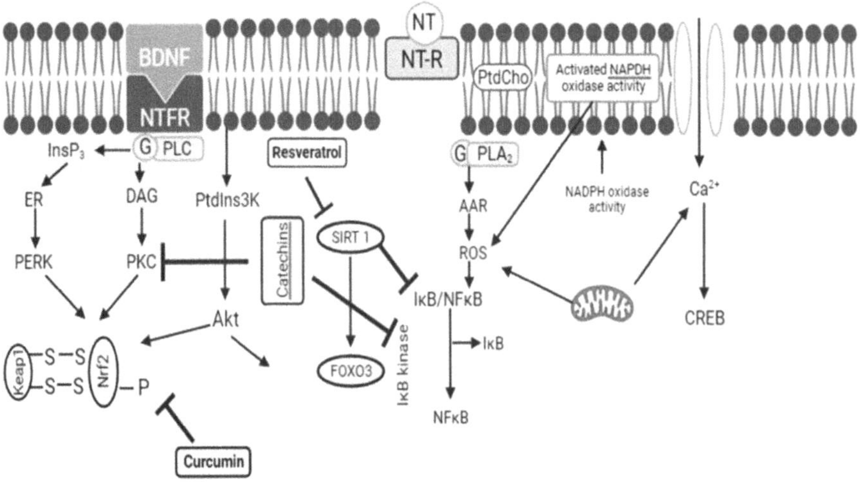

FIGURE 12.1 Neurohormetic responses linked to signal transduction pathways mediated by phytochemicals. Neurotransmitter receptor (NT-R); neurotrophic factor-activated specific receptor (NTFR); phospholipase C (PLC); diacyl-glycerol (DAG); protein kinase C (PKC); phosphatidylinositol 3-kinase (PtdIns3K); inositol 1,4,5-trisphosphate (InsP 3); phospholipase A 2 (PLA 2); nuclear factor-κB (nuclear factor kappa B (NF-κB)); inhibitory subunit of NF-κB (IκB); protein kinase B (Akt); PRK-like endoplasmic reticulum kinase (PERK); phosphatidylcholine (PtdCho); reactive oxygen species (ROS); nuclear factor (erythroid-derived 2)-like 2 (Nrf2); Kelch-like erythroid Cap 'n' Collar homologue-associated protein 1 (Keap1); silent information regulator two protein1 (SIRT1); forkhead box O (FOXO) transcription factor; brain-derived neurotrophic factor (BDNF); cAMP response element binding protein (CERB); and sirtuins (SIRT 1).

modifies the activity of a wide range of other proteins, including the Forkhead box O (FOXO) family. Members of the FOXO family have been demonstrated to regulate cellular function in processes like cell survival through the control of apoptotic cell death, cell cycle progression, and cell longevity (Kops et al, 2002). The phosphatidylinositol 3-kinase (PI3-K)/AKT signalling pathway targets FOXO proteins, which are phosphorylated by PKB (Tang et al, 1999).

12.3 PHYTOCHEMICALS IN THE REGULATION OF BIOCHEMICAL ACTIVITIES

12.3.1 Modulation of Enzyme Activities

At the protein level, inhibition of metabolic enzymes can be broadly categorized as competitive, non-competitive, or mechanism-based. Phytochemicals are susceptible to the body's detoxification processes, which include phase I-mediated oxidation, phase II-mediated conjugation, and phase III-mediated transport because they are viewed as xenobiotics or foreign chemicals. The phytochemicals may function throughout these processes as competitive or non-competitive inhibitors of enzyme interaction with co-ingested substrates, such as pharmaceuticals and carcinogens. The interaction between phytochemicals and metabolic enzymes may also lead to the creation of reactive intermediates, a few of which might cause the enzyme to become inactive. This kind of inhibition is known as "mechanism-based." Phenolic and sulphur-containing compounds are the most frequently mentioned examples of herbal components that hinder metabolic enzyme function (Mandlekar et al, 2006).

Numerous mechanisms of action that may prevent cancer have been linked to phytochemicals found in vegetables and fruits, including antioxidant activity or free radical scavenging, control of oncogene, tumour suppressor, or gene expression, induction of apoptosis, and modification of metabolic enzyme activity (Liu, 2004).

12.3.2 Stimulation of Adaptive Responses

Diets high in fruits and vegetables are associated with a lower risk of developing a number of severe diseases, including neurodegenerative disorders. It is becoming increasingly clear that many of the beneficial phytochemicals found in vegetables and fruits evolved as toxins that, at subtoxic doses, initiate adaptive cellular stress-response pathways in a wide range of cells, including neurons. This is true even though some beneficial phytochemicals may only act as antioxidants. The cell-survival signalling kinases, the transcription factors CREB and nuclear factor (erythroid-derived 2)-like 2 (Nrf2), and the sirtuin family of histone deacetylases are examples of such "preconditioning" or "neurohormesis" pathways. By promoting the production of neurotrophic factors, protein chaperones, antioxidant enzymes, and other proteins that aid cells in withstanding stress, neurohormetic phytochemicals including resveratrol, sulforaphanes, and curcumin help shield neurons against damage and disease (Mattson and Cheng, 2006).

Allium phytochemicals, which give garlic and onions their distinctive flavour and aroma, are found in large amounts in these meals. Animals exposed to the allium compounds display strong anti-cancer effects through the stimulation, meant to defend organisms against poisons, which are phase-2 enzymes (Tsai et al, 2005).

12.3.3 DNA Methylation

Recent research has emphasized the interaction between the epigenome and cancer metabolism. For epigenetic mechanisms such as histone and DNA methylation, histone acetylation, or histone phosphorylation, metabolites like acetyl-CoA, adenosine monophosphate (AMP), and S-adenosylmethionine (SAM), are necessary (Donohoe and Bultman, 2012). Therefore, the enzymes and metabolic pathways that provide these essential substances are essential for the upkeep and modification of the epigenome. In a methionine-deficient diet, there is a reduction in SAM levels, decreasing DNA and histone methylation and drastically affecting gene expression, all of which contribute to neurodegenerative disorders (Parasramka et al, 2012). Because betaine, methionine, choline, and folate metabolism are intertwined, deficiencies in any one of these nutrients can lead to abnormalities in metabolism and function. Global DNA methylation patterns can be quickly impacted by a diet lacking in methyl donor contributions (Niculescu and Zeisel, 2002).

12.3.4 Antioxidant and Anti-Inflammatory Effects

Similar to other antioxidant compounds, polyphenolic phytochemicals (PPs) work in vitro by neutralizing dangerous free radicals such as lipid peroxides and chelating divalent metal ions to lower their oxidative potential (Haslam, 1998). PPs have powerful antioxidant activity in the plant tissue from which they originate as well as in any food in which they are included, notably at relatively high concentrations along with the presence of other antioxidants such as tocopherols and ascorbic acid. They are considered to prevent oxidative deterioration of food's lipid component. However, according to recent research, the in vivo situation is presumably considerably different.

The complexity of the absorption, digestion, interactions, and metabolism of phytochemicals and foods undermines the comprehension and utilization of these anti-inflammatory phytochemicals to reduce chronic inflammation and hence prevent chronic diseases, even though several phytochemicals in each category exhibit anti-inflammatory effects (Borges et al, 2018) For instance, factors such as age, cultivation method, geographical distribution, species, harvesting season, part of the plant, and preservation method have a substantial impact on the phytochemical composition and quantity of plants (Lim et al, 2005).

Even at low dietary dosages in animals and humans, a variety of whole foods and phytochemicals appear to successfully prevent or reduce the indications of metabolic syndrome (Amiot et al, 2016; Babu et al, 2012; Si and Liu, 2008). For instance, theobromine, a phytochemical found in cocoa, synergistically improved the anti-hypertensive effect of (-)-epicatechin, one of the important phytochemicals in cocoa, by raising the level of (-)-epicatechin in humans' bloodstream. This study indicates that consuming whole cocoa (extract) is preferable to consuming pure (-)-epicatechin alone for lowering blood pressure.

12.3.5 Induction of Tumour Suppressor Genes

Bioactive phytochemicals, which are readily available and have fewer adverse effects, have been studied for their potential function in epigenetic gene regulation for the treatment and prevention of cancer. The manipulation of chromatin modification and DNA methylation by various bioactive phytochemicals in cancer may have affected the expression of important tumour suppressors, tumour promoters, and oncogene genes. These bioactive phytochemicals demonstrated promising results against various cancers whether used alone or in conjunction with other phytochemicals. The main focus of cancer prevention and treatment is the activation of tumour suppressor genes and the downregulation of oncogenes using dietary phytochemicals. All of the important epigenetic mechanisms, including histone changes, miRNAs and DNA methylation have shown potential for alteration by dietary phytochemicals. These epigenetic changes result in changes in the functioning of cellular regulatory and metabolic pathways, which cause altered cells to lose their carcinogenicity (Stefanska et al, 2012)

Due to their established anti-cancer activities in several studies, natural phytochemicals have generated significant attention both as chemotherapeutic adjuvants and chemo-preventive agents. Resveratrol (3,4,5 trihydroxystilbene, RSV) is a polyphenol prevalent in red grapes and peanuts and has been demonstrated to influence gene methylation patterns, as well as survival, apoptosis, and cell cycle (Lee et al, 2013, 2014; Stefanska et al, 2012; Venkatalakshmi et al, 2016), thus looking at a reduction of monogenic disorders in the future.

12.3.6 Stimulation of the Immune System

Innate and acquired immune system components work well together to produce immunological responses. The immune system has drawn a lot of attention over the past three decades as a possible target of toxicity after exposure to medications, chemicals, or environmental toxins. Phytochemicals in plants give host plants a natural defence mechanism in addition to giving them flavour, colour, and scent (Venkatalakshmi et al, 2016).

A study was conducted on polysaccharides obtained from *Helicteres angustifolia L* (HACP). When given to BALB/c mice carrying the 4T1 breast tumour, the polysaccharide made from HACP also showed strong immunomodulatory action. The weight of the tumour and the expression of tumor necrosis factor (TNF)-α and interleukin (IL)-1 were significantly reduced after 15 days of therapy with HACP at doses of 100, 200, and 300 mg/kg. In the HACP-treated group, there was also a reduction in lung metastasis. As a result, HACP significantly contributes to tumour suppression by moderating aberrant immune system activities (Vickers, 2017).

12.3.7 Antibacterial and Antiviral Effect

Numerous studies have demonstrated that multiple modes of action, including bacterial membrane damage, suppression of virulence factors, reduction of the activity of toxins and enzymes, and bacterial biofilm formation, are how phytochemicals exercise their antibacterial activity (Barbieri et al, 2017). It takes roughly 10 years to bring a new antibiotic to market, and discovering new antibiotics is time-consuming and expensive. As a result, finding new synthetic chemical compounds with bacteriophage and antibiotic capabilities has become increasingly important alongside the search for antibacterial agents derived from natural materials, such as phytochemicals (Mandal et al, 2014).

Numerous active phytochemicals have been shown to have therapeutic uses against various viruses with varying genetic makeups and biological functions. These substances' antiviral mechanisms can be explained in terms of their antioxidant properties, scavenging abilities, ability to prevent DNA and RNA synthesis, ability to prevent viral entry or reproduction, etc. By combining in vitro and in vivo studies in various biological tests, a large number of candidate substances, such as phytochemicals and their synthetic derivatives, have been discovered (Naithani et al, 2008) Aside from testing naturally occurring chemicals, various groups have created novel analogues with potential antiviral action. When compared to the parent molecule, the primary goals of the structural alteration are to increase bioavailability and decrease toxicity. In order to survive, viruses must have access to host cells. For each virus, a distinct invasion tactic might be used. It could be used to prevent both DNA and RNA replication of viruses because medicinal plants contain an infinite range of chemical components.

12.4 PHYTOCHEMICALS AND NEUROHORMETICS

Hormesis is the process through which a substance that is poisonous at greater concentrations has a positive effect on the organism or cell when it is exposed at low doses. The definition of hormesis, which is frequently used in the field of toxicology, is "an adaptive response characterized by biphasic dosage responses with usually identical quantitative properties with regard to amplitude and range of the stimulatory response that is either directly caused or the outcome of compensatory biological processes following an initial disruption in homeostasis" (Calabrese et al, 2011). According to the definition of neurohormesis, the adaptive process through which neurons (and consequently nervous systems and organisms) react to a moderate amount of stress by improving their resistance to more severe stress that may otherwise be fatal or cause dysfunction or disease (Mattson and Cheng, 2006).

12.4.1 Phytochemicals Protecting Neuroinflammation in Neurodegenerative Diseases

Numerous research has supported the biological benefits of *Cantella asiatica* (CA) and its phytochemicals, which have broad ethnopharmacological applications. These results indicate that CA imparts anti-inflammatory and pleiotropic neuroprotective advantages via its mitoprotective and antioxidative properties, which may be exploited for the treatment of ageing and NDs (Wong et al, 2021).

In placebo-controlled cross-over research with young college students, daily supplementation with 60 g of walnuts for 8 weeks significantly enhanced inferential verbal reasoning. The fact that there were no changes between the control and diet groups in terms of memory, mood, or nonverbal thinking may be due to the study being done on healthy, cognitively sound young adults. Walnuts might also aid in the prevention of Alzheimer's disease (AD) by regulating cognitive functions in the brain (Poulose et al, 2014).

12.4.2 Neuroprotective Mechanisms

The term "neuroprotection" refers to the methods and supporting systems that can protect the central nervous system (CNS) from the neuronal damage brought on by a variety of neuropsychiatric and neurodegenerative disorders, including Parkinson's disease (PD), cerebrovascular impairment, anxiety, seizures, AD, and others. Phytochemicals may be a useful tool in preventing neurodegenerative illnesses when it comes to neuroprotective techniques. There are numerous types of documented natural and manmade neuroprotective compounds. In recent years, both nationally and internationally, research institutions and enterprises have become much more conscious of herbal-based medicinal goods. Therefore, there has been a lot of attention paid to possible phytochemicals that can alter neuronal activity and prevent neurodegeneration. Herbal medicine, also known as phytotherapy, is an alternative and complementary therapy that makes use of plant parts for therapeutic purposes. It might be challenging to pinpoint which component of a herb is biologically active for a particular discussion because herbal products typically contain a range of bioactive phytochemicals. Researchers from all around the world are looking for bioactive phytochemicals from plants that are employed in traditional medical systems like the Indian Ayurvedic medical system, the Korean medical system, the Mediterranean medical system, the Chinese medicinal system, etc., as neuroprotective agents (Kumar et al, 2015).

Due to the lack of clinically relevant medications, a sizable portion of the elderly population at risk for NDs such as AD and PD requires nutritional intervention to improve health. To do this, numerous phytochemicals' anti-inflammatory mechanisms, including those of resveratrol, polyunsaturated fatty acids (PUFAs), propolis, ginsenosides, and curcumin have been thoroughly investigated (Wang et al, 2018).

12.4.3 Neurohormetic Phytochemicals Regulating Immune Function

Phytochemicals which are primarily found in plants, fruits, and vegetables can reduce the risk of cardiovascular illnesses, malignancies, inflammatory, immunological diseases, and neurological disorders. Plants can resist pests and pathogenic agents owing to their wide range of biological activities, which include antibacterial, antifungal, antioxidant, and anti-proliferative actions. But, when consumed by humans in relatively tiny concentrations, phytochemicals are not poisonous but do mildly arouse cellular stress responses (Calabrese et al, 2010a, 2010b). Sulforaphane, curcumin, resveratrol, catechins, allicin, and hypericin are some of the most prominent endocrine phytochemicals that can activate the adaptive stress response signalling pathways, boosting cellular resistance to injury and illness (Mattson and Cheng, 2006).

Resveratrol is one of the compounds that is most explored as an immunomodulator. Through extensive research, it has been concluded that resveratrol shows a hormetic biphasic dose-response relationship (Calabrese et al, 2012). Both low and high doses of resveratrol have been shown to drastically affect the cytotoxic activity of natural killer (NK) cells and cytotoxic T lymphocytes (CTL), protecting microglial cells and lowering the risk of neurodegenerative disorders. These results demonstrated the ability of resveratrol to either upregulate or inhibit immune response depending on concentration making it a potent and consumable neurohormetic phytochemical drug (Falchetti et al, 2001).

Another interesting phytochemical compound is curcumin which is a well-known primary ingredient in turmeric having immunomodulatory properties. It has the ability to control the activation of dendritic cells, NK cells, neutrophils, T lymphocytes, and B lymphocytes. Additionally, curcumin can lower the expression of pro-inflammatory cells and, at low dosages, boost antibody responses. It showed positive effects on a variety of diseases, including arthritis, asthma, AD, atherosclerosis, diabetes, and cancer because of its capacity to modify the immune system (Gao et al, 2004; Jagetia and Aggarwal, 2007) (Table 12.1).

12.4.4 Hormesis-induced Memory Performance

The ageing of the global population poses significant social and health challenges. Age-related neurodegenerative disorders are becoming more prevalent. None of the neurodegenerative disorders are currently curable due to their progressive nature, and the available treatments can only help with symptom relief and slow down the progression (Sahebnasagh et al, 2022). In regard to this, many people view herbal medicines and supplements as valuable health resources with a variety of uses (Eghbali et al, 2021; Hammad et al, 2019).

A plant-derived flavanol, epicatechin was studied for increasing memory performance in mammals. The results revealed that it boosted C57BL/6 female mice's cognitive response. Mice had enhanced memory, hippocampus angiogenesis, and neuronal spine concentration after 7 weeks of exercise and epicatechin (3 mg/kg) (Praag et al, 2007). Additionally, microarray observations showed that consuming epicatechin increased the expression of genes related to angiogenesis while decreasing the expression of genes linked to inflammatory response and apoptosis.

Epigallocatechin-3-gallate (EGCG) (10 mg/kg), an important polyphenol extracted from green tea, had positive effects in restoring the cognitive deficit in rats

TABLE 12.1
Neurohormetic Phytochemicals Inducing Memory-Performance.

Compound	Sources	Remarks	References
Epicatechin	Blackberries, apples, grapes, cocoa, tea leaves	Increases cognition Cytotoxicity to cancer cells	Shibani et al, 2019
Resveratrol	Peanuts, mulberries, grapes, blueberries	Improves cognitive deficit Anti-inflammatory action against cancer and diabetes	Gomcez et al, 2019
Oleuropein	Olive leaves	Improves spatial learning and memory impairments Anti-inflammatory response against diseases	Shibani et al, 2019

when administered a small dosage of 10 mg/kg by the modulation of S100B secretion, acetylcholinesterase, and antioxidant activity (Biasibetti et al, 2013).

Resveratrol, a natural substance found in large quantities in grapes, was injected to help 7–8-month-old mice establish long-term memory by reducing the expression of miR-124 and miR-134 and controlling the concentrations of CREB, resveratrol had this positive effect (Zhao et al, 2013).

12.4.5 Hormetic Studies

12.4.5.1 Parkinson's Disease (PD)

PD is a neurodegenerative condition marked by tremors, slow movement, and stiffness that impairs the ability of the muscles to move. It is defined as the death of dopaminergic nerve cells. Rats with PD have shown that resveratrol is a helpful chemical for easing their symptoms.

A study on PD-affected nerves overwhelmingly points out the possibility of resveratrol having a neuroprotective effect caused by 6-OHDA (6-hydroxydopamine) a neurotoxin serving as a PD animal model which may be associated with a diminished inflammatory response.

As a result, the current study offers a mechanistic justification for the use of resveratrol in PD treatment. Additionally, cyclooxygenase-2 and TNF-α mRNA and protein overexpression caused by 6-OHDA in the substantia nigra were inhibited by resveratrol. More research is needed to determine whether resveratrol can be used to cure PD (Jin et al, 2008).

Through the inactivation of BAX Bcl-2 associated X, MDM2 proto-oncogene (Mdm2), and reduction of B-cell lymphoma 2 and Bcl-w expression, epigallocatechin 3-gallate (0.1–10 M) displayed neuroprotective actions in a mouse model of PD (Levites et al, 2002).

Most plants contain a flavonoid named luteolin, which has hormetic and neuroprotective properties against rotenone-induced damage in BV2 cells and also protects microglia against rotenone. Low amounts of luteolin enhanced cell survival and decreased IL-1 and leucine-rich repeat kinase 2 mRNA levels (Elmazoglu et al, 2020) (Table 12.2).

TABLE 12.2
Some Neurohormetic Phytochemicals Reducing Symptoms of Parkinson's Disease.

Compound	Sources	Remarks	References
Allicin	Garlic Lesser quantities found in shallots and scallions	Protection mechanism against PD Prevents certain forms of cancer	Liu et al, 2015
Quercetin	Broccoli, apple, kale, and olive oil	Treats behavioural deficits Protection against cancer	Kanthasamy et al, 2017
Berberine	Oregon grape, bayberry, and tree turmeric	Improves deficits in behaviour movement	Zhang et al, 2017

TABLE 12.3
Some Neurohormetic Phytochemicals Reducing Symptoms of Schizophrenia.

Compounds	Sources	Remarks
Nicotine	Potato, tomato, and eggplants	Improves attention deficit
Stigmasterol	Legumes, vegetables, herbs, and seeds	antipsychotic effects
Kaempferol	Spinach, kale, and herbs such as dill and chives	Anti-inflammatory and neuroprotective effects against schizophrenia

12.4.5.2 Schizophrenia

Schizophrenia is often perceived to be a mental illness connected to psychosis. Schizophrenic patients frequently experience delusions and hallucinations that seem disconnected from reality along with difficulties focusing on regular activities. Treatment is typically lifelong and frequently entails a mix of phytochemical drugs, psychotherapy, and specialized medical services.

People with schizophrenia had higher levels of endocannabinoid anandamide in their blood and cerebrospinal fluids (Davies and Bhattacharyya, 2019). The terpenoid class of compounds known as cannabinoids is useful for treating schizophrenic patients. For example, at lower doses, tetrahydrocannabinol, a cannabinoid, reduced the symptoms of schizophrenia by acting on endocannabinoid receptors (Schwarcz et al, 2009).

A bioflavonoid called quercetin is used to reduce the signs and symptoms of schizophrenia because of its capacity to neutralize free radicals (Mert et al, 2019).

Studies on the anti-schizophrenic properties of xanthones, such as α-mangostin and magniferin, have also been conducted (Lum et al, 2021)

Antioxidant and anti-inflammatory characteristics are given by α-mangostin. In rat models of schizophrenia, it was proven to be effective and to inhibit 5-hydroxytryptamine 2A receptors and PDEs (phosphodiesterase).

Magniferin increased cognitive function by protecting mitochondrial processes, reducing dopamine, and acting as an anti-inflammatory agent (Table 12.3).

12.4.5.3 Huntington's Disease (HD)

The clinical symptoms of Huntington's disease (HD), a genetic neurological condition of the CNS, include escalating choreiform movements, personality abnormalities, deteriorating mental health, and early mortality.

HD has no known cure. However, some of the issues it causes, such as those brought on by depression, mood swings, and involuntary muscle contractions, can be lessened with the usage of some combinational drugs and support to assist in making daily duties easier.

Korean red ginseng was given at doses of 50, 100, and 250 mg/kg/day, and these dosages had therapeutic effects on the reduction of Huntington's symptoms. The NF-B and mitogen-activated protein kinases pathways were inhibited, whose mutations are responsible for exacerbating the disease's symptoms. Additionally, this phytochemical reduced TNF-α mRNA expression and microglial activation (Jang et al, 2013).

TABLE 12.4
Some Neurohormetic Phytochemicals Reducing Symptoms of Huntington's Disease.

Compounds	Sources	Remarks
Lycopene	Watermelon, pink guavas, apricots, and papayas	Prevents motor abnormalities and brain dysfunction
TQ-SLNs (solid lipid nanoparticles encapsulated thymoquinone) and thymoquinone	*Nigella sativa*	A low dose of TQ-SLNs is highly effective to reduce symptoms of Huntington's disease
Protopanaxtriol	*Panax pseudoginseng and Panax ginseng*	Enhances behaviour and body weight

Oral administration of fisetin and resveratrol (10 μM) was useful for the treatment of HD. They are readily available, small chemicals that can pass the blood-brain barrier and stimulate ERK signalling thus protecting brain activity (Maher et al, 2011).

Small doses of protopanaxtriol injection dramatically improved behaviour, decreased the generation of free radicals, enhanced Nrf2 translocation into the nucleus, and boosted NAD(P)H quinone oxidoreductase 1 and heme oxygenase-1 expression in the striatum (5, 10, and 20 mg/kg) (Gao et al, 2015) (Table 12.4).

12.4.5.4 Alzheimer's Disease

AD is a dreadful disease more prevalent in the United States. It is characterized by dementia and atrophy (shrinkage of the brain) followed by the death of brain cells. This progressive neurodegenerative disorder destroys mental functions and memory.

The excessive protein buildup in and around brain cells is what causes AD. Amyloid is one of the proteins involved. Amyloid-β peptide is a key component of the neuropathology of AD, and oxidative stress may be the cause of amyloid-β's neurotoxicity and deposits of it create plaques around brain cells (Flier et al, 2002).

Animal models of AD showed better learning and memory function when given ginsenoside extract from *Panax ginseng*. Gintonin when administered orally for three weeks in small doses reduced memory impairment and cholinergic dysfunctions brought on by scopolamine and amyloid through reductions in choline acetyltransferase, acetylcholine concentration, and initiation of acetylcholine esterase activity (Kim et al, 2015).

According to findings, oral administration of sulforaphane in small doses at levels between 10 and 50 mg/kg caused animal models to exhibit anti-AD-like action. The primary component of green vegetables, sulforaphane, reduced levels of tau, amyloid-β, neurodegeneration, inflammation, and oxidative stress in animal and cell models to ameliorate cognitive impairment in various doses but the reduction in symptoms is dose-dependent. Because higher doses of sulforaphane cause sedation, disturbance in motor coordination, and reduced muscle strength (Kim, 2021) (Table 12.5).

TABLE 12.5
Some Neurohormetic Phytochemicals Reducing Symptoms of Alzheimer's Disease.

Compound	Sources	Remarks	References
Curcumin	Turmeric, curry powder, and mango ginger	A small dose of curcumin was effective in lowering the symptoms of AD	Lim et al, 2001
Resveratrol	Peanuts, mulberries, grapes, blueberries	Reduction of Aβ aggregation	Han et al, 2004
Catechin	Black grapes, apricots, fresh tea leaves, and rock-rose leaves	Improves the conditions of cognitive deficits	Conte et al, 2003
Blueberry	Anthocyanins	Inhibits involuntary muscle contractions, reduces brain ageing and behavioural deficits by improving cognitive function	Joseph et al, 2016
Cannflavin A	*Cannabis sativa*	Lower doses of cannflavin produce anti-Alzheimer's disease effects	Eggers et al, 2019

12.4.5.5 Autism Spectrum Disorder (ASD)

Autism is a chronic developmental disorder that affects a person's capacity to interact and communicate. The neurological system is impacted by ASD, which also has an impact on the sufferer's general cognitive, social, emotional, and physical health. Symptoms can range significantly in scope and severity. The most common symptoms include difficulty in social interaction, communication problems, obsessive interests, and repetitive behaviours.

There is no known cause or effective therapy for autism to date. Therapy aims to enhance the child's functioning by minimizing the symptoms of ASD and fostering growth and learning.

When used as a neurotherapeutic, phytochemicals can help to reduce the oxidative stress that leads to mitochondrial dysfunction in ASD. Preclinical investigations on sulforaphane, hydroxytyrosol, and curcumin have produced promising outcomes.

It has been determined that autism causes a reduction in several functions. Sulforaphane and hydroxytyrosol are two substances that have been found to have metabolic effects on cellular stress responses which in turn activates Heat shock proteins (HSPs) and other mechanisms which are upregulated by sulforaphane-enhancing cortical connections (Calabresec et al, 2016). Also, by inducing the Nrf2 and HSF1-dependent genes, the substance also exhibits protective effects against a number of NDs (Singh et al, 2014; Zhang et al, 2011).

For example, a placebo-controlled, clinical demonstration revealed that Sulforaphane when administered orally every day for 18 weeks showed improved anti-autism symptoms. The aberrant Behaviour Checklist, The Social Responsiveness Scale, and the Clinical Global Impression Improvement Scale (CGI-I) were three reliable behavioural assessments that were used to conclude the results. The study revealed that verbal communication, aberrant behaviour, and social interaction were all improved, and these scores reverted to baseline levels after sulforaphane was stopped (Singh et al, 2014) (Table 12.6).

TABLE 12.6
Some Neurohormetic Phytochemicals Reducing Symptoms of Autism Spectrum Disorder (ASD).

Compound	Sources	Remarks	References
Sulforaphane	Cruciferous vegetables (kale, cabbage, etc.)	Improved autism symptoms in mice model	Nadeem et al, 2019
Curcumin	Turmeric, curry powder, and mango ginger	A neuro-psychopharmacotherapeutic drug Restores biochemical, behavioural, neurological, and molecular changes in autistic patients	Urdaneta et al, 2018

12.4.6 Drug Reaction to Neurological Medications

A neurohormetic biphasic dose-responses relationship is present in several medications. In other words, their positive benefits become more prominent at some low dose levels and become less significant at higher doses. Some of the medications with hermetic properties are anti-tumour drugs and anti-seizure drugs. In numerous animal models, these medications have additionally demonstrated a U-shape dose-response association.

Levodopa, a drug that is used to treat PD has been observed to exhibit a hormetic reaction by relieving motor symptoms (Quattrone et al, 1995). This is due to the fact that while levodopa has toxic effects at greater dosages, it has prolonged therapeutic efficacy several days after treatment at low cumulative doses (Zappia et al, 2000). This drug has the ability to affect many cognitive processes through mesocortical dopaminergic pathways (Miller and Cohen, 2001).

Another example is the U-shaped response association of the drug named Physostigmine, a natural component of Calabar bean used to treat AD. In AD, some of the nerve cells that use acetylcholine are lost so to prevent the normal hydrolysis of acetylcholine, Physostigmine is administered. Following this, a number of medications, including huperzine, tacrine, and arecoline were examined for this anti-acetylcholinesterase hypothesis. In numerous animal models, these medications have additionally demonstrated a U-shape dose-response association (Calabrese, 2008).

Furthermore, the most recent medication to be licensed called memantine which is an N-methyl-D-aspartate antagonist similarly works through an inverted U-shaped dose-response relationship that resembles hormetic (Calabrese, 2008).

12.5 ISSUES CONCERNING THE USE OF PHYTOCHEMICALS IN HUMAN HEALTH

Consuming plants for their goodness and health benefits, such as cancer prevention or treatment, or for therapeutic purposes has been a prevalent practice since the dawn of civilization. Fruits, vegetables, and many other plants contain phytochemicals, which are not necessary nutrients like vitamins or minerals but are frequently ingested or utilized as herbal medicines or dietary supplements due to their claimed therapeutic benefits.

In fact, some studies have indicated that phytochemicals like resveratrol and epigallocatechin gallate are useful for the treatment of cancer. Numerous dietary epidemiology data and animal studies for a wide range of dietary components point to potential protection against a number of malignancies, but these results have not yet been confirmed in randomized human clinical trials. In actuality, the majority of food intervention scientific trials have had relatively underwhelming results regarding phytochemicals consumed in the diet.

12.5.1 Phytochemicals and Biphasic Dose Responses

At lower doses, supplemental administration of phytochemicals helps to improve deficits in spatial learning and memory. While these compounds have hormetic effects in AD and PD at low levels, neurotoxicity has been linked to greater quantities. Both low and high phytochemical doses have been proven to have therapeutic effects in HD, though low doses are more efficient than higher levels.

Low dosages of plant extracts or metabolites derived from plants have a considerable impact on diseases associated with ageing through mitohormesis, whereas high amounts accelerate mitochondrial respiration in the brain.

With these variations in the data, we can conclude that more clinical research, particularly in In vivo models is required to fill the knowledge gap regarding phytochemicals' hormesis activity in NDs.

12.6 CONCLUSION

By interacting with the molecular and cellular structure of the brain involved in memory formation, phytochemicals may be able to reverse the decline in brain function caused by age-related neurodegeneration. Inhibition of metabolic enzymes can be broadly categorized as competitive, non-competitive, or mechanism-based. Phytochemicals are susceptible to the body's detoxification processes because they are viewed as foreign chemicals. Phenolic and sulphur-containing compounds are the most frequently mentioned examples of herbal components that hinder metabolic enzyme function. By interacting with the molecular and cellular structure of the brain involved in memory formation, phytochemicals may inhibit and even partially reverse the age-related loss in memory function. Numerous mechanisms of action that may prevent cancer have been linked to phytochemicals found in vegetables and fruits.

Numerous applications can be found with using herbal medicinal products and dietary supplements. It is advantageous to comprehend the underlying mechanism of action by using the hormesis phenomena in medicinal plants. Theoretically, hormetic models can account for a variety of effects in herbs, such as inhibition at greater dosages and stimulation at lower ones, which support their potential for regulating and treating disease. With recent advancements in medicine procuring drugs with a singular mechanism of action and/or inability to enter neurons, all pharmacological therapies for neurodegenerative illnesses appear to fail. Therefore, neurohormesis can be viewed as a potentially ground-breaking method of treating neurodegenerative disorders. Neurodegenerative and other neurological illnesses may benefit from neurohormesis, the adaptive component of the hormetic dose response in neurons (Sahebnasagh et al, 2022).

12.7 FUTURE DISCUSSIONS

There is still a disconnect between pharmacogenetics and pharmacological approaches to the treatment and cure of NDs, despite the common agreement in the field of pharmacology that it is preferable to delay or stop the progression of NDs. The systems, organs, and cells that are experiencing oxidative stress in ND states frequently experience imbalanced ionic gradients and are vulnerable to protein interference both inside and outside of the signalling pathways. The cells, organs, and systems are susceptible to damage in this environment. In vitro neuron cell models convincingly demonstrate the effectiveness of dietary phytochemicals with NGF potentiation effects, although numerous problems must be solved before clinical trials. To demonstrate the impact of dietary phytochemicals in preclinical in vivo models, additional research should be done.

Future studies should clarify whether the effects of phytochemical mixtures on the susceptibility of organs to ischemia are additive or synergistic. Future therapeutic approaches may also combine pharmaceuticals and phytochemicals to increase drug efficacy and/or lessen drug adverse effects (Chen et al, 2008). Even more so than before, socioeconomic factors influence the rate at which new medications are developed. This is related to growing interest in the repurposing and repositioning of outdated medications or other substances frequently used in conventional medicine. Therefore, it is anticipated that interest in medications based on phytochemicals will increase in the near future for disorders linked to oxidative stress and inflammation (Forni et al, 2019).

REFERENCES

Amiot MJ, C Riva, and A Vinet, "Effects of Dietary Polyphenols on Metabolic Syndrome Features in Humans: A Systematic Review," Obesity Reviews 17, no. 7 (July 1, 2016): 573–586, https://doi.org/10.1111/OBR.12409.

Babu PVA et al., "Genistein Prevents Hyperglycemia-Induced Monocyte Adhesion to Human Aortic Endothelial Cells Through Preservation of the CAMP Signalling Pathway and Ameliorates Vascular Inflammation in Obese Diabetic Mice," The Journal of Nutrition 142, no. 4 (April 1, 2012): 724–730, https://doi.org/10.3945/JN.111.152322.

Barbieri R et al., "Phytochemicals for Human Disease: An Update on Plant-Derived Compounds Antibacterial Activity," Microbiological Research 196 (March 1, 2017): 44–68, https://doi.org/10.1016/J.MICRES.2016.12.003.

Biasibetti R et al., "Green Tea (-)Epigallocatechin-3-Gallate Reverses Oxidative Stress and Reduces Acetylcholinesterase Activity in a Streptozotocin-Induced Model of Dementia," Behavioural Brain Research 236, no. 1 (2013): 186–193, https://doi.org/10.1016/j.bbr.2012.08.039.

Borges G et al., "Absorption, Metabolism, Distribution and Excretion of (–)-Epicatechin: A Review of Recent Findings," Molecular Aspects of Medicine 61 (June 1, 2018): 18–30, https://doi.org/10.1016/J.MAM.2017.11.002.

Calabrese EJ, "Hormesis and Medicine," no. Figure 1 (2008), https://doi.org/10.1111/j.1365-2125.2008.03243.x.

Calabrese V et al., "Novel Targets for Therapeutic Intervention in Neurodegenerative Disorders," British Journal of Clinical Pharmacology 66, no. 5 (2010a): 594–617.

Calabrese V et al., "Oxidative Stress, Redox Homeostasis and Cellular Stress Response in Ménière's Disease: Role of Vitagenes," Neurochemical Research 35, no. 12 (December 3, 2010b): 2208–17, https://doi.org/10.1007/s11064-010-0304-2.

Calabrese V et al., "Biochimica et Biophysica Acta Cellular Stress Responses, Hormetic Phytochemicals and Vitagenes in Aging and Longevity," BBA - Molecular Basis of Disease 1822, no. 5 (2012): 753–783, https://doi.org/10.1016/j.bbadis.2011.11.002.

Calabresec V et al., "Hormesis, Cellular Stress Response, and Redox Homeostasis in Autism Spectrum Disorders," Journal of Neuroscience Research 94, no. 12 (2016): 1–11, https://doi.org/10.1002/jnr.23893.

Calabrese EJ, EJ Stanek, and MA Nascarella, "Evidence for Hormesis in Mutagenicity Dose–Response Relationships," Mutation Research/Genetic Toxicology and Environmental Mutagenesis 726, no. 2 (December 24, 2011): 91–97, https://doi.org/10.1016/J.MRGENTOX.2011.04.006.

Chen CM et al., "Novel Regimen Through Combination of Memantine and Tea Polyphenol for Neuroprotection Against Brain Excitotoxicity," Journal of Neuroscience Research 86, no. 12 (September 1, 2008): 2696–2704, https://doi.org/10.1002/JNR.21706.

Conte A, S Pellegrini, and D Tagliazucchi, "Synergistic Protection of PC12 Cells from β-Amyloid Toxicity by Resveratrol and Catechin," Brain Research Bulletin 62, no. 1 (2003): 29–38, https://doi.org/10.1016/j.brainresbull.2003.08.001.

Davies C, and S Bhattacharyya, "Cannabidiol as a Potential Treatment for Psychosis," Therapeutic Advances in Psychopharmacology 9 (2019): 204512531988191, https://doi.org/10.1177/2045125319881916.

Donohoe DR, and SJ Bultman, "Metaboloepigenetics: Interrelationships between Energy Metabolism and Epigenetic Control of Gene Expression," Journal of Cellular Physiology 227, no. 9 (September 1, 2012): 3169–3177, https://doi.org/10.1002/JCP.24054.

Eggers C et al., "Novel Cannabis Flavonoid, Cannflavin A Displays Both a Hormetic and Neuroprotective Profile Against Amyloid β-Mediated Neurotoxicity in PC12 Cells: Comparison With Geranylated Flavonoids, Mimulone and Diplacone," Biochemical Pharmacology 169, no. August (2019): 113609, https://doi.org/10.1016/j.bcp.2019.08.011.

Eghbali S et al., "Therapeutic Effects of Punica Granatum (Pomegranate): An Updated Review of Clinical Trials," Journal of Nutrition and Metabolism 2021 (2021), https://doi.org/10.1155/2021/5297162.

Elmazoglu Z et al., "Luteolin Protects Microglia Against Rotenone-Induced Toxicity in a Hormetic Manner Through Targeting Oxidative Stress Response, Genes Associated With Parkinson's Disease and Inflammatory Pathways," Drug and Chemical Toxicology 43, no. 1 (2020): 96–103, https://doi.org/10.1080/01480545.2018.1504961

Falchetti R et al., "Effects of Resveratrol on Human Immune Cell Function," Life Sciences 70, no. 1 (2001): 81–96, https://doi.org/10.1016/S0024-3205(01)01367-4.

Flier VD et al., "Cognitive Decline in AD and Mild Cognitive Impairment Is Associated with Global Brain Damage." Neurology 59, no. 6 (2002): 874–879, https://doi.org/10.1212/WNL.59.6.874.

Forni C et al., "Beneficial Role of Phytochemicals on Oxidative Stress and Age-Related Diseases," BioMed Research International 2019, no. Figure 1 (2019), https://doi.org/10.1155/2019/8748253.

Gao X et al., "Immunomodulatory Activity of Curcumin: Suppression of Lymphocyte Proliferation, Development of Cell-Mediated Cytotoxicity, and Cytokine Production in Vitro," Biochemical Pharmacology 68, no. 1 (2004): 51–61, https://doi.org/10.1016/j.bcp.2004.03.015.

Gao Y et al., "Protopanaxtriol Protects Against 3-Nitropropionic Acid-Induced Oxidative Stress in a Rat Model of Huntington's Disease," Acta Pharmacologica Sinica 36, no. 3 (2015): 311–322, https://doi.org/10.1038/aps.2014.107.

Gocmez SS, TD Şahin, Y Yazir, G Duruksu, FC Eraldemir, S Polat, and T Utkan. "Resveratrol Prevents Cognitive Deficits by Attenuating Oxidative Damage and Inflammation in Rat Model of Streptozotocin Diabetes Induced Vascular Dementia," Physiology & Behavior 201, (2019): 198–207.

Hammad MA et al., "Prescribing Statins Among Patients With Type 2 Diabetes: The Clinical Gap between the Guidelines and Practice," Journal of Research in Medical Sciences 24, no. 1 (2019): 1–9, https://doi.org/10.4103%2Fjrms.JRMS_100_18.

Han YH et al., "Neuroprotective Effects of Resveratrol Against β-Amyloid-Induced Neurotoxicity in Rat Hippocampal Neurons: Involvement of Protein Kinase C," British Journal of Pharmacology 141, no. 6 (2004): 997–1005, https://doi.org/10.1038/sj.bjp.0705688.

Haslam, E, "Practical Polyphenolics: From Structure to Molecular Recognition and Physiological Action," Trends in Food Science & Technology 10, (1998): 339.

Jagetia GC, and BB Aggarwal, "'Spicing up' of the Immune System by Curcumin," Journal of Clinical Immunology 27, no. 1 (2007): 19–35, https://doi.org/10.1007/s10875-006-9066-7.

Jang M et al., "Korean Red Ginseng Extract Attenuates 3-Nitropropionic Acid-Induced Huntington's-Like Symptoms," Evidence-Based Complementary and Alternative Medicine 2013 (2013): 8–10, https://doi.org/10.1155/2013/237207.

Jin F et al., "Neuroprotective Effect of Resveratrol on 6-OHDA-Induced Parkinson's Disease in Rats," European Journal of Pharmacology 600, no. 1–3 (2008): 78–82, https://doi.org/10.1016/j.ejphar.2008.10.005.

Joseph JA, NA Denisova, G Arendash, M Gordon, D Diamond, B Shukitt-Hale, and D Morgan. "Blueberry Supplementation Enhances Signaling and Prevents Behavioral Deficits in an Alzheimer Disease Model," Nutritional Neuroscience 6, no. 3 (2003): 153–162, https://doi.org/10.1080/1028415031000111282. PMID: 12793519.

Kanthasamy AG, M Ay, J Luo, M Langley, H Jin, V Anantharam, and A Kanthasamy, Molecular Mechanisms Underlying Protective Effects of Quercetin Against Mitochondrial Dysfunction and Progressive Dopaminergic Neurodegeneration in Cell Culture and MitoPark Transgenic Mouse Models of Parkinson's Disease. Journal of Neurochemistry 141, no. 5 (2017): 766–782, https://doi.org/10.1111/jnc.14033. Epub 2017 May 9. PMID: 28376279; PMCID: PMC5643047.

Kim HJ et al., "Oral Administration of Gintonin Attenuates Cholinergic Impairments by Scopolamine, Amyloid-β Protein, and Mouse Model of Alzheimer's Disease," Molecules and Cells 38, no. 9 (2015): 796–805, https://doi.org/10.14348/molcells.2015.0116.

Kim J, "Pre-Clinical Neuroprotective Evidences and Plausible Mechanisms of Sulforaphane in Alzheimer's Disease," International Journal of Molecular Sciences 22, no. 6 (2021): 1–16, https://doi.org/10.3390/ijms22062929.

Kops GJ, RH Medema, J Glassford, MA Essers, PF Dijkers, PJ Coffer, EW Lam, and BM Burgering, "Control of Cell Cycle Exit and Entry by Protein Kinase B-Regulated Forkhead Transcription Factors," Molecular and Cellular Biology 22, no. 7 (April 2002): 2025–2036, https://doi.org/10.1128/MCB.22.7.2025-2036.2002. PMID: 11884591; PMCID: PMC133681.

Kumar GP, KR Anilakumar, and S Naveen, "Phytochemicals Having Neuroprotective Properties from Dietary Sources and Medicinal Herbs," Pharmacognosy Journal 7, no. 1 (2015): 1–17.

Lee YJ et al., "Synergistic Anti-Cancer Effects of Resveratrol and Chemotherapeutic Agent Clofarabine Against Human Malignant Mesothelioma MSTO-211H Cells," Food and Chemical Toxicology 52 (February 1, 2013): 61–68, https://doi.org/10.1016/J.FCT.2012.10.060.

Lee YJ et al., "Knockdown of Bcl-XL Enhances Growth-Inhibiting and Apoptosis-Inducing Effects of Resveratrol and Clofarabine in Malignant Mesothelioma H-2452 Cells," Journal of Korean Medical Science 29, no. 11 (November 4, 2014): 1464–1472, https://doi.org/10.3346/JKMS.2014.29.11.1464.

Levites Y et al., "Involvement of Protein Kinase C Activation and Cell Survival/Cell Cycle Genes in Green Tea Polyphenol (-)-Epigallocatechin 3-Gallate Neuroprotective Action," Journal of Biological Chemistry 277, no. 34 (2002): 30574–30580, https://doi.org/10.1074/jbc.M202832200.

Lim GP et al., "The Curry Spice Curcumin Reduces Oxidative Damage and Amyloid Pathology in an Alzheimer Transgenic Mouse," Journal of Neuroscience 21, no. 21 (2001): 8370–8377, https://doi.org/10.1523/jneurosci.21-21-08370.2001.

Lim W, KW Mudge, and F Vermeylen, "Effects of Population, Age, and Cultivation Methods on Ginsenoside Content of Wild American Ginseng (Panax Quinquefolium)," Journal of Agricultural and Food Chemistry 53, no. 22 (November 2, 2005): 8498–8505, https://doi.org/10.1021/JF051070Y.

Liu H et al., "Allicin Protects PC12 Cells Against 6-OHDA-Induced Oxidative Stress and Mitochondrial Dysfunction via Regulating Mitochondrial Dynamics" 710061, no. 277 (2015): 966–979, https://doi.org/10.1159/000430271.

Liu RH, "Potential Synergy of Phytochemicals in Cancer Prevention: Mechanism of Action," The Journal of Nutrition 134, no. 12 (December 1, 2004): 3479S–3485S, https://doi.org/10.1093/JN/134.12.3479S.

Lum PT et al., "Protective Effect of Mangiferin on Memory Impairment: A Systematic Review," Saudi Journal of Biological Sciences 28, no. 1 (2021): 917–927, https://doi.org/10.1016/j.sjbs.2020.11.037.

Maher P et al., "Erk Activation by the Polyphenols Fisetin and Resveratrol Provides Neuroprotection in Multiple Models of Huntington's Disease," Human Molecular Genetics 20, no. 2 (2011): 261–270, https://doi.org/10.1093/hmg/ddq460.

Mandal SM, A Roy, AK Ghosh, TK Hazra, A Basak, and OL Franco, "Challenges and Future Prospects of Antibiotic Therapy: From Peptides to Phages Utilization," Frontiers in Pharmacology 5 May (May 13, 2014): 105, https://doi.org/10.3389/fphar.2014.00105. PMID: 24860506; PMCID: PMC4027024.

Mandlekar S, J-L Hong, and A-N Tony Kong, "Modulation of Metabolic Enzymes by Dietary Phytochemicals: A Review of Mechanisms Underlying Beneficial Versus Unfavorable Effects," Current Drug Metabolism 7, no. 6 (2006): 661–675, https://doi.org/10.2174/138920006778017795.

Mattson MP, and A Cheng, "Neurohormetic Phytochemicals: Low-Dose Toxins That Induce Adaptive Neuronal Stress Responses," Trends in Neurosciences 29, no. 11 (2006): 632–639, https://doi.org/10.1016/j.tins.2006.09.001.

Mert DG et al., "The Influence of Quercetin on Recognition Memory and Brain Oxidative Damage in a Ketamine Model of Schizophrenia," Psychiatry and Clinical Psychopharmacology 29, no. 1 (2019): 1–7, https://doi.org/10.1080/24750573.2018.1442670.

Miller EK, and JD Cohen. "An Integrative Theory of Prefrontal Cortex Function." Annual Review of Neuroscience 24, (2001): 167–202, https://doi.org/10.1146/annurev.neuro.24.1.167. PMID: 11283309.

Nadeem A et al., "Nrf2 Activator, Sulforaphane Ameliorates Autism-Like Symptoms Through Suppression of Th17 Related Signalling and Recti Fi Cation of Oxidant-Antioxidant Imbalance in Periphery and Brain of BTBR T + Tf/J Mice," Behavioural Brain Research 364, no. November 2018 (2019): 213–224, https://doi.org/10.1016/j.bbr.2019.02.031.

Naithani R et al., "Antiviral Activity of Phytochemicals: A Comprehensive Review," Mini-Reviews in Medicinal Chemistry 8, no. 11 (2008): 1106–1133, https://doi.org/10.2174/138955708785909943.

Niculescu MD, and SH Zeisel, "Diet, Methyl Donors and DNA Methylation: Interactions between Dietary Folate, Methionine and Choline," The Journal of Nutrition 132, no. 8 (August 1, 2002): 2333S–2335S, https://doi.org/10.1093/JN/132.8.2333S.

Okwu DE, "Phytochemical and Vitamin Content of Indigenous Spices of South Eastern Nigeria," Journal of Sustainable Agriculture and Environment 6 (2004): 30–34, https://cir.nii.ac.jp/crid/1571135650688285824.

Parasramka MA et al., "MicroRNAs, Diet, and Cancer: New Mechanistic Insights on the Epigenetic Actions of Phytochemicals," Molecular Carcinogenesis 51, no. 3 (March 1, 2012): 213–230, https://doi.org/10.1002/MC.20822.

Poulose SM, MG Miller, and B Shukitt-Hale, "Role of Walnuts In Maintaining Brain Health With Age," Journal of Nutrition 144, no. 4 (2014): 561S–566S, https://doi.org/10.3945/jn.113.184838.

Praag HV et al., "Plant-Derived Flavanol (-)Epicatechin Enhances Angiogenesis and Retention of Spatial Memory in Mice," Journal of Neuroscience 27, no. 22 (2007): 5869–5878, https://doi.org/10.1523/JNEUROSCI.0914-07.2007.

Quattrone A et al., "The Subacute Kvodopa Test for Evaluating Long-Duration Response in Parclnson's Disease," Annals of Neurology 38, no. 3 (1995): 389–395. https://doi.org/10.1002/ana.410380308. PMID: 7668824.

Sahebnasagh A et al., "Neurohormetic Phytochemicals in the Pathogenesis of Neurodegenerative Diseases," Immunity and Ageing 19, no. 1 (2022): 1–16, https://doi.org/10.1186/s12979-022-00292-x.

Schwarcz G, B Karajgi, and R McCarthy, "Synthetic Δ-9-Tetrahydrocannabinol (Dronabinol) Can Improve the Symptoms of Schizophrenia," Journal of Clinical Psychopharmacology 29, no. 3 (2009): 255–258, https://doi.org/10.1097/JCP.0b013e3181a6bc3b.

Shibani F et al. "Effect of Oleuropein on Morphine-Induced Hippocampus Neurotoxicity and Memory Impairments in Rats," Naunyn Schmiedebergs Arch Pharmacol 392, no. 11 (2019): 1383–1391. https://doi.org/10.1007/s00210-019-01678-3. PMID: 31236657.

Si H, and D Liu, "Genistein, a Soy Phytoestrogen, Upregulates the Expression of Human Endothelial Nitric Oxide Synthase and Lowers Blood Pressure in Spontaneously Hypertensive Rats," The Journal of Nutrition 138, no. 2 (February 1, 2008): 297–304, https://doi.org/10.1093/JN/138.2.297.

Singh K et al., "Sulforaphane Treatment of Autism Spectrum Disorder (ASD)." Biological Sciences 111, no. 43 (2014): 15550–15555, https://doi.org/10.1073/pnas.1416940111.

Son TG, S Camandola, and MP Mattson, "Hormetic Dietary Phytochemicals," NeuroMolecular Medicine 10, no. 4 (December 10, 2008): 236–246, https://doi.org/10.1007/s12017-008-8037-y. PMCID: PMC2635914.

Spencer JPE, "Beyond Antioxidants: The Cellular and Molecular Interactions of Flavonoids and How These Underpin Their Actions on the Brain," Proceedings of the Nutrition Society 69, no. 2 (May 2010): 244–260, https://doi.org/10.1017/S0029665110000054.

Spencer JPE, D Vauzour, and C Rendeiro, "Flavonoids and Cognition: The Molecular Mechanisms Underlying Their Behavioural Effects," Archives of Biochemistry and Biophysics 492, no. 1–2 (December 1, 2009): 1–9, https://doi.org/10.1016/J.ABB.2009.10.003

Stefanska B et al., "Epigenetic Mechanisms in Anti-Cancer Actions of Bioactive Food Components – The Implications in Cancer Prevention," British Journal of Pharmacology 167, no. 2 (September 1, 2012): 279–297, https://doi.org/10.1111/J.1476-5381.2012.02002.X.

Stefanska B et al., "Comparative Effects of Retinoic Acid, Vitamin D and Resveratrol Alone and in Combination With Adenosine Analogues on Methylation and Expression of Phosphatase and Tensin Homologue Tumour Suppressor Gene in Breast Cancer Cells," British Journal of Nutrition 107, no. 6 (March 28, 2012): 781–790, https://doi.org/10.1017/S0007114511003631.

Tang ED et al., "Negative Regulation of the Forkhead Transcription Factor FKHR by Akt," Journal of Biological Chemistry 274, no. 24 (June 11, 1999): 16741–16746, https://doi.org/10.1074/jbc.274.24.16741.

Tosetti F, DM Noonan, and A Albini, "Metabolic Regulation and Redox Activity as Mechanisms for Angioprevention by Dietary Phytochemicals," International Journal of Cancer 125, no. 9 (2009): 1997–2003, https://doi.org/10.1002/ijc.24677.

Tsai CW et al., "Garlic Organosulfur Compounds Upregulate the Expression of the π Class of Glutathione S-Transferase in Rat Primary Hepatocytes," The Journal of Nutrition 135, no. 11 (November 1, 2005): 2560–2565, https://doi.org/10.1093/JN/135.11.2560.

Urdaneta KE, MA Castillo, N Montiel, N Semprún-Hernández, N Antonucci, and D Siniscalco. "Autism Spectrum Disorders: Potential Neuro-Psychopharmacotherapeutic Plant-Based Drugs." ASSAY and Drug Development Technologies 16, no. 8 (2018): 433–444.

Venkatalakshmi P, V Vadivel, and P Brindha, "Role of Phytochemicals as Immunomodulatory Agents: A Review," International Journal of Green Pharmacy 10, no. 1 (2016): 1–18.

Vickers NJ, "Animal Communication: When I'm Calling You, Will You Answer Too?," Current Biology 27, no. 14 (July 24, 2017): R713–R715, https://doi.org/10.1016/J.CUB.2017.05.064.

Wang J et al., "Connection between Systemic Inflammation and Neuroinflammation Underlies Neuroprotective Mechanism of Several Phytochemicals in Neurodegenerative Diseases," Oxidative Medicine and Cellular Longevity 2018 (2018), https://doi.org/10.1155/2018/1972714.

Wong JH, AM Barron, and JM Abdullah, "Mitoprotective Effects of Centella Asiatica (L.) Urb.: Anti-Inflammatory and Neuroprotective Opportunities in Neurodegenerative Disease," Frontiers in Pharmacology 12 (June 29, 2021): 687935, https://doi.org/10.3389/fphar.2021.687935. PMID: 34267660; PMCID: PMC8275827.

Yadav DK, "Potential Therapeutic Strategies of Phytochemicals in Neurodegenerative Disorders," Current Topics in Medicinal Chemistry 21, no. 31 (December 2, 2021): 2814–2838, https://doi.org/10.2174/1568026621666211201150217.

Zappia M, Oliveri RL, Bosco D, Nicoletti G, Branca D, Caracciolo M, Napoli ID, Gambardella A, Quattrone A. The long-duration response to L-dopa in the treatment of early PD. Neurology. 2000 May 23;54(10):1910-5. doi: 10.1212/wnl.54.10.1910. PMID: 10822428.

Zhang C et al., "Berberine Protects Against 6-OHDA-Induced Neurotoxicity in PC12 Cells and Zebrafish Through Hormetic Mechanisms Involving PI3K/AKT/Bcl-2 and Nrf2/HO-1 Pathways," Redox Biology 11, no. September 2016 (2017): 1–11, https://doi.org/10.1016/j.redox.2016.10.019.

Zhang Y et al., "Brief Communication by Sulfhydryl-Reactive Inducers of the KEAP1/NRF2/ARE Pathway," Chemistry & Biology 18, no. 11 (2011): 1355–1361, https://doi.org/10.1016/j.chembiol.2011.09.008.

Zhao YN et al., "Resveratrol Improves Learning and Memory in Normally Aged Mice Through MicroRNA-CREB Pathway," Biochemical and Biophysical Research Communications 435, no. 4 (2013): 597–602, https://doi.org/10.1016/j.bbrc.2013.05.025.

Index

Note: Locators in *italics* represent figures and **bold** indicate tables in the text.

Q

R

S

X

Y